UNDERSTANDING How Components Fail

THIRD EDITION

Donald J. Wulpi
Edited by Brett Miller

ASM International®
Materials Park, Ohio 44073-0002
www.asminternational.org

First printing, October 2013
Second printing, December 2018
Third printing, February 2020

In writing this book, the author has drawn upon both his personal experience and the available literature. As the text itself makes clear, there is often no single or simple remedy for component failure. The author makes no representation or warranty as to the appropriateness or success of any method or procedure herein suggested.

The terms "fail" and "failure" as used in this book are intended to mean cessation of function or usefulness, and carry no implication of negligence or malfeasance.

Comments, criticisms, and suggestions are invited, and should be forwarded to ASM International.

Prepared under the direction of the ASM International *Technical Books Committee (2012–2013)*, Bradley J. Diak, Chair; and the *Failure Analysis Committee*, Dustin A. Turnquist, Chair, and Larry D. Hanke, Past Chair.

ASM International staff who worked on this project include Scott Henry, Senior Manager, Content Development and Publishing; Karen Marken, Senior Managing Editor; Madrid Tramble, Manager of Production; and Diane Whitelaw, Production Coordinator.

Library of Congress Control Number: 2013933056
ISBN-13: 978-1-62708-014-9 (print)
ISBN-10: 1-62708-014-7 (print)
ISBN: 978-1-62708-015-6 (pdf)
ISBN: 978-1-62708-270-9 (electronic)

SAN: 204-7586

ASM International®
Materials Park, OH 44073-0002
www.asminternational.org

Printed in the United States of America

About the Editor

BRETT MILLER IS A metallurgical engineer with degrees from the Missouri University of Science and Technology and the University of Wisconsin-Milwaukee. He is a registered Professional Engineer in several states. For over 28 years, Brett has worked as a failure analyst, metallurgist, and expert witness, completing thousands of failure analysis investigations. He has authored numerous publications, including book chapters and journal articles. Brett has been active in ASM International since 1984 and is a former chairman of the ASM Failure Analysis Committee. He is also active in NACE International and the American Welding Society. He has experience in aerospace, oil field, and commercial laboratories. Brett is currently the technical director at IMR Metallurgical Services in Louisville, Kentucky.

About the Author

A METALLURGICAL CONSULTANT in Fort Wayne, Indiana, Donald J. Wulpi studied and analyzed failures of metal parts for more than 45 years. A graduate of Lehigh University, with a degree in metallurgical engineering, he spent most of his working life in several metallurgical laboratories of the International Harvester Company, now Navistar International Corporation. In retirement, Mr. Wulpi devoted himself to teaching the principles of failure analysis at the ASM International headquarters, in Materials Park, Ohio, and at various companies throughout the United States. He testified as an expert witness in product litigation cases for many years.

Contributors

Thomas N. Ackerson, IMR Metallurgical Services - Louisville
Jake Auliff, Sauer Danfoss
John A. Beavers, Det Norske Veritas (USA) Inc
Daniel J. Benac, Baker Engineering and Risk
Nicholas E. Cherolis, Rolls-Royce
Scott Chumbley, Iowa State University
Daniel P. Dennies, Exponent
Larry D. Hanke, Materials Evaluation and Engineering
John D. Landes, University of Tennessee - Knoxville
Brett A. Miller, IMR Metallurgical Services - Louisville
David M. Norfleet, Det Norske Veritas (USA) Inc
Ronald J. Parrington, IMR Test Labs
Wesley D. Pridemore, GE Aviation
Craig J. Schroeder, Element Materials Technology
Roch J. Shipley, Professional Analysis and Consulting
Ryan Spotts, Sauer Danfoss
Dustin A. Turnquist, Engineering Systems
Charles V. White, Kettering University

Contents

Understanding How Components Fail, Third Edition
Donald J. Wulpi
Edited by Brett Miller

Preface to Third Edition

THE FIRST EDITION of Don Wulpi's *Understanding How Components Fail* was handed to me at my first job, right after it was published. I was told to read it cover to cover because it had so much that I would need to know. It was easy to understand and provided a great deal of insight into failures. The book, and the subsequent second edition, became indispensable to my expanding technical bookshelves. Early in my career I often recommended it to colleagues. Now the book is one of the first reference books I hand to new engineers and technicians. Although it is written primarily for the novice, all technically interested people can learn a great deal from it.

Because I am a proponent of *Understanding How Components Fail*, preparing a new edition was a difficult task. However, as in most things, materials science and failure analysis are constantly changing. This new edition has attempted to add recent technical knowledge and analysis tools to bring the content up-to-date.

Engineering design, materials, and analysis techniques have all advanced significantly since the first edition. A substantial enhancement to this volume is the update to Chapter 10, "Fatigue Fracture." Fatigue striation counting, modeling, and crack rate prediction are now routinely performed for critical components, and this is reflected in this new edition. Corrosion failure analysis (Chapter 13) is updated to reflect the current greater understanding of the processes involved. DNA testing can now be performed to identify organisms responsible for microbiologically influenced corrosion (MIC), a technology that was not widely available in 1985. Chapter 14, "Elevated-Temperature Failures," also reflects our greater understanding of these failures, as engineering materials are used in progressively harsher conditions. Chapter 15, "Fracture Mechanics," was added to the second edition as the field became more prominent. In this edition, the chapter is updated to parallel modern thought and terminology. For younger readers this may sound odd, but most of us did not have computers on our desks in the mid-1980s. The advent of affordable, user-friendly (mostly) computers has been a boon to engineering designers and failure analysts. Complex calculations and computational modeling of huge volumes of data could not be approached before computers.

Don Wulpi obviously filled an unfilled niche with this book. Although many years have passed since the original publication, the book remains one of ASM International's best-selling technical books. I think that, beyond the comprehensive information the book provides, its informal manner and conversational tone are what keeps it in demand. Photographs and drawings illustrate concepts that are explained with copious real-world examples. I have spoken to numerous college professors, domestic and foreign, that use *Understanding How Components Fail* as an undergraduate course text. I think the practical, no-nonsense presentation of this book makes it a very good text for all materials and mechanical engineering students.

First, all thanks to Donald Wulpi for creating this book. It contains decades of practical failure analysis knowledge that has helped me throughout my career. In this third edition I believe we have remained faithful to his original intent for the book.

Many thanks to all of the chapter reviewers, who updated the material and tried to make the additions as seamless as possible. The reviewers consisted of engineering professors, industrial failure analysts, commercial failure analysts, and consultants. For many of us there is a compulsion to write in a more formal manner, so most were outside of their comfort zones. I think they were able to maintain the general tone of the book while making the necessary additions. It is a testament to the enduring legacy of this book that I was able to find so many willing and enthusiastic reviewers.

Thanks also to Karen Marken and the other professional staff at ASM International. Karen was able to keep everything moving forward and was a pleasure to work with.

Brett A. Miller, P.E.
Louisville, Kentucky

Preface to the Second Edition

SINCE THE FIRST publication of this work many years ago, I have been pleased with the reception that it has received. Care has been taken in the preparation of this book to make it as accurate and useful as possible, for analysis of failures is a critical process. Unfortunately, it is impossible to describe or predict every possible combination of type of part, assembly, material, metallurgical condition, service usage, and environment. For this reason, the book had to be written in general terms so that the principles of failure analysis can be applied to a very large number of specific failures. It is highly recommended that the study of any failure not be limited to this book but expanded to encompass many of the references and suggestions for further study. After more than 45 years of studying, writing, and teaching about failures, I am still learning. And so can you.

Again, I am indebted to all of those people who have suggested improvements, particularly J.L. Hess and J.L. Welker. Many suggestions are included in this revision.

Also, I thank E.J. Kubel, Jr., who prepared Chapter 15 on Fracture Mechanics, for his writing on this important subject. Ed is a metallurgical engineer with about 15 years experience in the foundry industry and nearly the same length of time as author and editor for several technical magazines in the areas of materials and manufacturing processes. He made this discussion as elementary as possible in keeping with the theme of this work on the principles of failure analysis and prevention.

Donald J. Wulpi
Fort Wayne, Indiana

Preface to the First Edition

THIS BOOK IS intended for use primarly by those who have little or no prior knowledge of the principles of metallurgical failures, such as in distortion, fracture, wear, and corrosion. In this sense, it can be considered to be a primer, for many complex technical concepts are explained in relatively simple terms—sometimes oversimplified for ease of understanding.

Failure analysis is a critical first step in identifying a problem that has occurred in a metal component of a mechanism or structure. Once the mode of failure has been identified, appropriate corrective measures may then be taken to try to prevent similar future failures.

Nowhere is the need for accurate failure analysis more dramatically obvious than after an aircraft accident. After a serious crash, trained teams of investigators examine all phases of the aircraft and its operation, seeking clues to the true cause of the accident. Lives literally hang on the accuracy of the failure analysis process.

Failure analysis frequently involves comparing an unexplained failure with examples of failures whose cause has been determined. Identification of fractures, for example, is facilitated by comparing them with photographs of various types of fractures such as those shown in this book. The field of failure analysis is so vast that it encompasses all of metal and metalworking technology. There are many books and articles in the technical literature that deal with all types of failures of virtually all metals. At the end of every chapter in this book are selected references for further study. The serious reader is urged to use these and other references to pursue specific problems beyond the scope of this volume.

There is an infinite variety of types of parts, metals, treatments, conditions, types of loading, applications, environments, and combinations of all of these. Since it is impossible for a single work to cover every conceivable possibility, basic principles are explained with examples using common metals, primarily ferrous. Also, common types of parts are used in the examples, for the principles involved in their analysis may also be applicable to other types of parts not considered.

This work is the distillation of a lifetime of interest in failures and the techniques used to prevent failures, in study of failures, and in teaching others how to identify failures and how to correct them. Failure analysis is a fascinating subject, frequently compared with autopsies in the medical field.

In preparing this book I have had help from many people. Reviewers who read the first draft of each chapter and gave me many valuable suggestions for improvements were F.J. Marcom, J.P. Sheehan, G.H. Walter, and J.L. Welker. E.J. Rusnak and R.W. Morris took many of the photographs that are used. Many other photographs were given to me with the understanding that the contributor not be identified. To all of these persons, and many others who have given assistance, my sincere thanks.

Finally, my wife, Jini, has put up with my erratic schedule and helped me with the proofreading for many months. To her, particularly, my sincere thanks. The effort will have been worth it if this book is able to help many readers understand how components fail.

Donald J. Wulpi
Fort Wayne, Indiana

Introduction—Why Perform Failure Analysis?*

Definition of Failure

Before we can understand how components fail, it is necessary to define failure. There is no universal definition for what constitutes a failure. One definition that has been developed by professional failure analysts is:

> *Failure. (1) A general term used to imply that a part in service (a) has become completely inoperable, (b) is still operable but is incapable of satisfactorily performing its intended function, or (c) has deteriorated seriously, to the point that it has become unreliable or unsafe for continued use. (2) Also commonly applied to manufacturing processes that produce components that do not meet specifications. (Ref. 1)*

Simply, failure can be described as the inability of a component to function properly. Failures can occur anywhere: during design, manufacturing, or with the end customer. It is a mistake to assume that the failure of any component is necessarily due to poor design or manufacturing flaws. Failures can be caused by a very large variety of factors, and many failures are the result of multiple related factors.

In a larger, more encompassing context, *failure* can be defined as "the inability of a component, machine, process, or culture/thinking to function properly" (Ref 2). A failure is not just a broken part but can sometimes

*Introduction written by Daniel P. Dennies and Brett A. Miller, IMR Metallurgical Services - Louisville, and Charles V. White, Kettering University

encompass all of the factors or attributes that are connected to the failure. In a rigorous and in-depth failure investigation, this may include the humans that are involved with the failure and the latent or overarching corporate and cultural environment that set the operational parameters of a company.

There is no way to eliminate failure entirely. Human ambition tends to outreach engineering, pushing the boundaries of safe, proven design. Designs and materials continue to become more complex, with new technologies developed to create them, and new instruments invented to analyze them. Engineers at all stages of the design and manufacturing process should appreciate the reasons why formal failure analysis is performed. A comprehensive understanding of how components fail is necessary for a proper failure analysis.

Why Do We Perform a Failure Analysis?

The goal of any failure analysis investigation is to discover the root cause of the failure. Discovering the root cause will identify corrective actions to prevent identical failures. It will also indicate if the failure was a unique situation, or if widespread failures can be anticipated. Proper root-cause investigation will help the judge or jury determine the party at fault in litigation-related failures.

This book focuses on the metallurgical or materials evaluation for failure mode identification. Engineers must realize that the materials analysis may be just a part of a larger root-cause investigation. Full root-cause analysis usually includes many other technical disciplines such as mechanical, civil, and structural engineering. Diverse nonengineering disciplines such as psychology, accounting, and criminal forensics may also be brought to bear.

The reason the failure analyst must strive to discover the root cause of the failure has many possible answers, as demonstrated in Fig. 1. Discovering the root cause will determine a definitive corrective action and prevent future occurrences of the same failure. Additionally, in failures that involve litigation, discovering the root cause will help decide the fault, or innocence, of the parties involved.

The benefits to a company or customer of a well-run and organized failure investigation are many. Failure investigation is an integral part of any company's Quality System, including such programs as Lean, Attention to Detail, Six Sigma, Total Quality Management, Continuous Improvement, and various other quality system concepts. A failure investigation can also assist in the redesign of a failed component that results in a better, more elegant, forgiving, and resilient engineering solution. It can solve a manufacturing problem that can save money and save time. In the end, a well-run and organized failure investigation can even save lives.

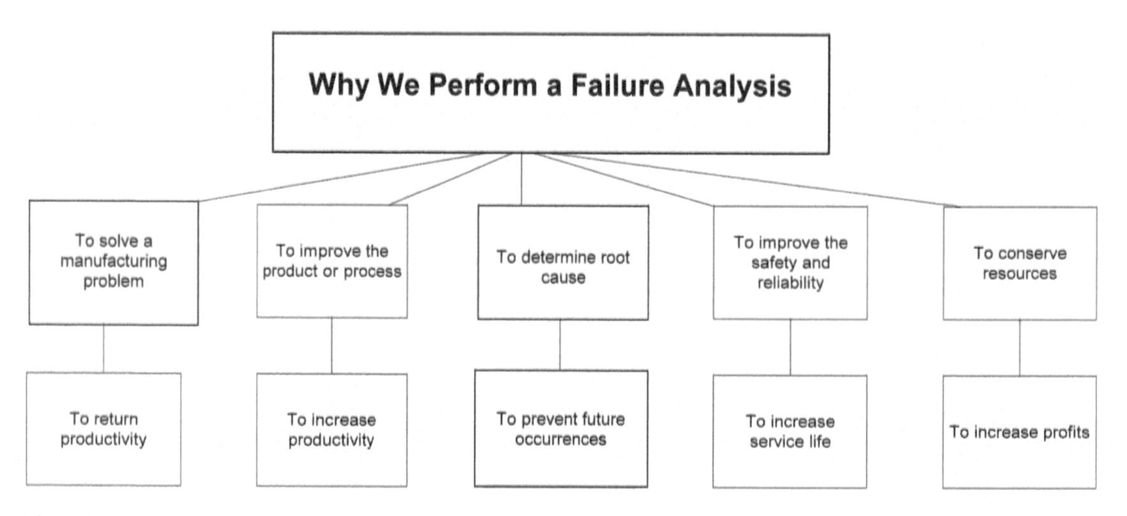

Fig. 1 Some of the reasons we perform failure analysis

The Responsibilities of the Failure Analyst

Failure analysts wield great power. Companies depend upon them to perform the task to the utmost of their capabilities, with due care, and with due diligence. Failure analysts must always remember the words of Voltaire, "With great power comes great responsibility."

Failure analysts must also always remember that failures can be associated with things that can be touched, like a broken component, or things that cannot be touched, like a process specification with conflicting or erroneous criteria. Pursuant to that goal, failure analysts must always strive to look at each and every failure as unique and different. They must not allow themselves to quickly dismiss a failure as "the same as all the others." A sound technical opinion must be developed to a reasonable degree of engineering certainty based on engineering principles, data, testing, and/or published literature for each failure. As failure analysts gain experience and expertise practicing this scientific art, they must use that intelligence and experience to enhance and sharpen the technical assessment and analysis, not abandon them. Lastly, failure analysts must consider that each failure is relative to the industry and requirements of a particular program or part. Many industries use the same material or processes but have radically different acceptance criteria or expectations. A good first question for any failure by the failure analyst is simply, "What is the failure?"

In the industrial world, failure analysts are often requested by their management or customer to "make the problem go away." Many times this type of directive may not appear to be asking the failure analyst to determine root cause. However, experience and time will teach the failure analyst that the best way to "make a problem go away" is to discover root cause.

This book may be the first of many books, papers, and articles you will read on the subject of failure analysis. However, there are many other books, papers, and articles that will assist you to hone your craft and expertise. Some are listed as references for your convenience.

REFERENCES

1. *Failure Analysis and Prevention,* Vol 11, *ASM Handbook,* ASM International, 10th ed., 2002, p 1066
2. D.P. Dennies, *How to Organize and Run a Failure Investigation,* ASM International, 2005

SELECTED REFERENCES

- C.R. Brooks and A. Choudhury, *Metallurgical Failure Analysis,* McGraw-Hill, 1993
- V.J. Colangelo and F.A. Heiser, *Analysis of Metallurgical Failures,* Wiley-Interscience, 1987
- D.P. Dennies, The Organization of a Failure Investigation, *Pract. Failure Anal.*, June 2001
- *Failure Analysis and Prevention,* Vol 11, *ASM Handbook,* 8th ed., ASM International, 1986, p 15–46
- *Failure Analysis and Prevention,* Vol 11, *ASM Handbook,* 9th ed., ASM International, 1995
- *Failure Analysis and Prevention,* Vol 11, *ASM Handbook,* 10th ed., ASM International, 2002
- C.H. Kepner and B. Tregoe, *The New Rational Manager,* Princeton Research Press, 1997
- R.J. Latino and K.C. Latino, *Root Cause Analysis: Improving Performance for Bottom Line Results,* CRC Press, 1999
- C.R. Nelms, *The Dynamics of Inculcating the Root Cause Mentality,* Failsafe Network, 1995
- C.R. Nelms, *What You Can Learn from Things That Go Wrong: A Guide Book to the Root Causes of Failure,* Failsafe Network, 1994
- N. Schlager and H. Petroski, *When Technology Fails,* Gale Research Inc., 1994
- Jim Scutti, Engineering Aspects of Failure and Prevention, *Failure Analysis and Prevention,* Vol 11, *ASM Handbook,* 10th ed., ASM International, 2002, p 1–78
- G.F. Vander Voort, Conducting the Failure Examination, *Pract. Failure Anal.*, April 2001
- C.E. Witherell, *Mechanical Failure Avoidance: Strategies and Techniques,* McGraw-Hill, 1994, p 31–65

Understanding How Components Fail, Third Edition
Donald J. Wulpi
Edited by Brett Miller

CHAPTER 1

Techniques of Failure Analysis*

What Is Failure Analysis?

Failure analysis is a systematic investigative procedure using the scientific method to identify the causes of a failure. "Forensic engineering" is often used as a synonym, but this term is more appropriate for litigation-based investigations. Analysis of failures is an integral part of design and manufacturing processes. In addition to providing incremental improvements for refinement of successful designs, proper failure analysis can help prevent catastrophic failures. Although failure analysis is essential, it is not typically emphasized as part of most engineering curricula. As a result, failure analysis methods and techniques are not always understood by many engineers.

Figure 1 shows how a failure investigation is the reverse of the design process. During the failure investigation, the failure analyst will revisit and reanalyze all of the service, manufacturing, and design aspects and decisions to identify the cause of failure.

In the study of any failure, the analyst must consider a broad spectrum of possibilities or reasons for the occurrence. Often a large number of factors, frequently interrelated, must be understood to determine the cause of the original, or primary, failure. The analyst is in the position of Sherlock Holmes attempting to solve a baffling case. Like the great detective, the analyst must carefully examine and evaluate all evidence available and then prepare a hypothesis—or possible chain of events—that could have caused the "crime." The analyst may also be compared to a coroner per-

**Understanding How Components Fail*, Second Edition, by Donald J. Wulpi, Chapter 1, Techniques of Failure Analysis, ASM International, 1999, reviewed and revised by Daniel P. Dennies, Exponent

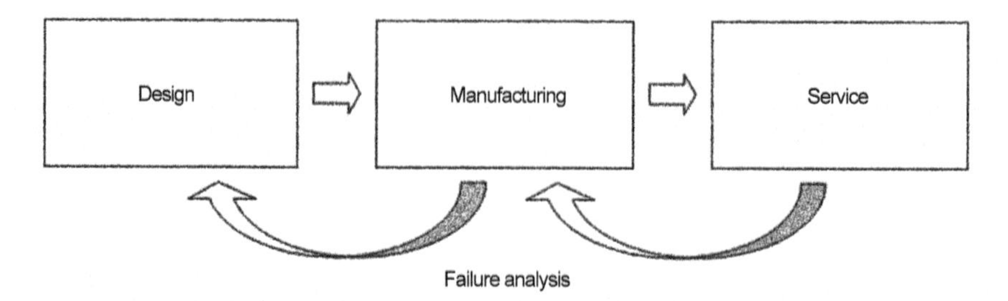

Fig. 1 Failure analysis is the reverse of the design process

forming an autopsy on a person who suffered an unnatural death, except that the failure analyst works on parts or assemblies that have had an unnatural or premature demise. If the failure can be duplicated under controlled simulated service conditions in the laboratory, much can be learned about how the failure actually occurred. If this is not possible, there may be factors about the service of the part or assembly that are not well understood.

Fractures, usually the most serious type of failure, are studied here in some detail. Usually undesired and unexpected by the user, fractures can have disastrous results when a load-bearing member suddenly loses its ability to carry its intended load. Distortion, wear, and corrosion failures also are important and sometimes lead to fractures. However, these types of failure can be reasonably well predicted and prevented.

Procedure for Failure Analysis

Reference 1 is a basic guide to follow in various stages of a failure analysis investigation. It must be emphasized that the most important initial step to perform in any failure analysis investigation is to do *nothing* but organize the failure investigation (Ref 2). In this first step, the failure analyst should collect data and information; study the evidence; think about the failed part or parts; ask detailed questions about the parts, the machine itself, and circumstances of the failure; and make accurate notes about the responses. When possible, it is highly desirable to use low-power magnification—up to about 25 or 50×—with carefully controlled lighting to study the failed part or parts.

A failure analyst must also realize that a failure investigation is evaluated using "real" or typical data derived from the failure investigation itself. Design values of mechanical properties taken from books and material certifications are less desirable substitutes for a tensile test and chemical composition analysis of the actual material involved in the failure. In addition, the streamlining and cost-reduction concepts of many industries over the last 20 years have resulted in a reduction of internal qualification testing for materials and processes certification. The testing has been delegated to the industrial supplier base, which results in the industrial database becoming pieces of paper that contain very little actual, or useful, technical information. The failure analyst must be prepared to generate the data required.

In addition, in some rare cases, design engineers who may not be well versed in failure investigation may have the unfortunate "*Titanic*" viewpoint that their design is too good to fail. It can sometimes be a real challenge for the failure analyst to convince these design engineers to rethink their design and "discover" a method to make their design fail.

For a complete evaluation, the sequence of stages in the investigation and analysis of failure, as detailed in Ref 1, is:

1. Collection of background data and selection of samples
2. Preliminary examination of the failed part (visual examination and record keeping)
3. Nondestructive testing
4. Mechanical testing (including hardness and toughness testing)
5. Selection, identification, preservation, and/or cleaning of specimens (and comparison with parts that have not failed)
6. Macroscopic examination and analysis and photographic documentation (fracture surfaces, secondary cracks, and other surface phenomena)
7. Microscopic examination and analysis using various light microscopy and electron microscopy techniques
8. Selection and preparation of metallographic sections
9. Examination and analysis of metallographic specimens
10. Determination of failure mechanism
11. Chemical analysis (bulk, local, surface corrosion products, deposits or coatings, and microprobe analysis)
12. Analysis of fracture mechanics (see Chapter 15, "Fracture Mechanics," in this book)
13. Testing under simulated service conditions (special tests)
14. Analysis of all the evidence, formulation of conclusions, and writing the report (including recommendations). Writing a report may not be necessary in many product litigation cases; it is best to follow the advice of the attorney or client with whom the analyst is working.

Each of these stages is considered in greater detail in Ref 1, and they are not discussed here. However, it must be emphasized that three principles be carefully followed:

- *Locate the origin(s) of the fracture.* No laboratory procedure should hinder the effort to find the location(s) where the fracture originated. Also, it is most desirable, if possible, to have both fracture surfaces in an undamaged condition.

- *Do not put the mating pieces of a fracture back together, except with considerable care and protection.* Even in the best circumstances, fracture surfaces are extremely delicate and fragile and are damaged easily, from a microscopic standpoint. Protection of the surfaces is particularly important if electron microscopic examination is to be part of the procedure. Many such examinations have been frustrated by careless repositioning of the parts, by careless packaging and shipping, and by inadequate protection from corrosion, including contact with fingers. If parts must be repositioned to determine deformation of the total part, fracture surfaces must be protected by paper or tape that will not contaminate the surface. Also, protect fracture surfaces and other critical surfaces from damage during shipping by using padding, such as an adhesive strip bandage for small parts.
- *Do no destructive testing without considerable thought.* Alterations such as cutting, drilling, and grinding can ruin an investigation if performed prematurely. Do nothing that cannot be undone. Once a part is cut, it cannot be uncut; once drilled, it cannot be undrilled; once ground, it cannot be unground. In general, destructive testing must be performed—if done at all—only after all possible information has been extracted from the part in the original condition and after all significant features have been carefully documented by photography. Caution is particularly necessary in product litigation cases because the details of destructive testing should be agreed upon by all parties in the lawsuit. Consult the attorney with whom the analyst is working.

If there are several fractures from one mechanism (a "basket case"), one should determine if any of the fractures is a fatigue fracture. If definite evidence of a fatigue fracture can be found, this is usually the source of the problem—the primary fracture. Fatigue fracture is the normal, or expected, type of fracture of a machine element after long service. However, there are many possible reasons for fatigue fracture and many different appearances of fatigue fractures, as described in this book. Fatigue fractures are quite common in mechanisms unless specific actions have been taken to prevent them during design, manufacture, and service.

Investigative Techniques

While not all failures require the same degree of effort needed to investigate a product litigation matter, it is imperative that the investigator follow a specific and organized plan during the analysis. The use of checklists and flow charts to keep the investigation on track is very effective to ensure that all elements of the analysis have been performed and properly documented.

The initial stages of the investigation are the most critical. This is the phase where information surrounding the failure is collected and documented. Without following an organized and well-developed plan, some vital piece of evidence may be overlooked (Ref 2). With the passage of time it may become difficult, if not impossible, to recall or obtain evidence that may prove to be the missing piece of the puzzle (Ref 2, 3).

Normal Location of Fracture

The analyst must be aware of the normal, or expected, location for fracture in any type of part, because any deviation from the normal, or expected, location must have been caused by certain abnormal factors or conditions that need to be discovered. In very simple relative terms, this type of failure might be considered a "good" failure because it occurred in the manner and time frame expected by the design.

An all-too-familiar type of fracture is that of the ordinary shoelace. A shoelace will inevitably fail at one of the two top eyelets, adjacent to the bowknot, as shown in Fig. 2. There are several logical engineering reasons that this is the normal location of fracture:

- When the knot is tied, the lace is pulled tightest at the upper eyelets; therefore, the service stress is highest at this location.
- Most of the sliding motion during tightening occurs at the lace as it goes through the upper eyelets. Therefore, the metal eyelets tend to wear, or abrade, the fibers of the lace.

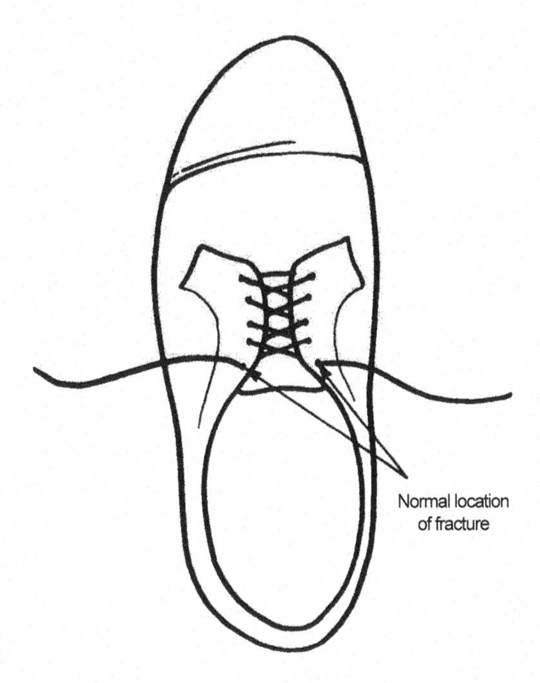

Fig. 2 Normal location of fracture for a shoelace

- Since the shoelace presumably has uniform mechanical properties along its length, it will eventually wear—and ultimately tear, or fracture—at the location where conditions are most severe, that is, at an upper eyelet.

If the shoelace were to fracture at any other location, such as at the lower eyelets or near the free ends, one would have to suspect that, for some reason, the shoelace had substandard mechanical properties at the location of failure. Or, alternatively, the lace could have been damaged, such as by burning from dropped cigarette ashes, thus causing it to be weakened and fractured at an abnormal location.

This familiar example of the normal location of fracture is easy to understand. The situation becomes considerably more complex in metal components that may have been manufactured—intentionally or unintentionally—with different mechanical properties at different locations in the part.

A metal part, however, can be expected to fail, or fracture, at any location where the stress first exceeds the strength, unlike the shoelace, which is expected to have uniform strength throughout its length and will fail at the location of greatest "wear and tear." Metal parts fracture in a much more complex version of the weakest-link principle: the weakest location in a part will not originate a fracture if the stress is below the strength at that location. On the other hand, a high-strength location in a part may suffer fracture if the stress is concentrated at that location. The point is that stress and strength are inseparably intertwined and must be considered together.

Other locations of normal fracture in metal parts are at geometric stress concentrations, such as the first engaged thread of a bolt within a nut or other tapped hole, or at a sharp-cornered fillet in a rotating shaft, or at the root fillet of a gear tooth. These represent some of the normal locations of fatigue fractures that may occur after long service. Fractures caused by abnormal events, such as accidents, may occur at locations other than those noted because of the unpredictable forces in an accident. The insidious problem with notches is that they concentrate the stress at specific locations; thus, the metal strength at a given location may not be able to survive the geometrical form that concentrates the stress at that location, no matter how high the strength level.

Note: The above discussion is considerably oversimplified. It ignores the different types of stress and strength; however, it is intended to point out, in general terms, the principles involved.

Questions to Ask about Fractures

The broad spectrum of considerations in any fracture investigation can be grouped into ten general areas of inquiry that may be answered with careful observation and study of a given fractured part. The sequence in which these interrelated areas are considered is unimportant; any one area may be the key in a particular situation. Some of the questions to ask and answer are:

1. *Surface of the Fracture*
   What is the fracture mode? The fracture surface can tell a story if enough careful attention is given to it in conjunction with other information to be learned. One must not make snap judgments about the fracture; all information must be evaluated before making a decision that is crucial. Examination of all regions of the part is necessary before this question can be answered.
   a. Is the origin (or origins) of the fracture visible? If so, is it (are they) located at the surface or below the surface? The location of the origin(s) depends on the relative stress and strength gradients, which are discussed in later chapters of this book.
   b. What is the relation of the fracture direction to the normal or expected fracture directions? The direction of a fracture usually has a specific relation to the direction of the stress that caused the fracture to occur.
   c. How many fracture origins are there? The answer concerns the relative magnitude of the actual stress to the actual strength of the part at the locations of failure.
   d. Is there evidence of corrosion, paint, plating, or some other foreign material on the fracture surface? This may indicate the presence of a preexisting crack, prior to fracture.
   e. Is the stress unidirectional, or is it reversed in direction? If the part is thought to be stressed in only one direction but the fracture indicates that it was stressed also in other directions, the assumed operation of the mechanism is not completely correct.

2. *Surface of the Part*
   a. What is the contact pattern on the surface of the part? This knowledge is extremely important because these "witness marks" of contact with the mating parts reveal how the part was loaded in service. These marks may be only slight polishing, or they may be severe wear or indentations from heavy contact with other parts of the assembly or from outside the assembly. The mating parts usually have corresponding indications of contact that should be matched. For example, rolling elements, such as balls, rollers, and needles in antifriction bearings, may leave indentations on the raceways that can aid in identifying the direction of the forces that caused the damage.
   b. Has the surface of the part been deformed by loading during service or by damage after fracture? The location and direction of deformation are very important in any examination of fractured parts.

The degree of deformation depends on the mechanical properties of the metal involved, as well as on the magnitude and type of force causing the deformation.

c. Is there evidence of damage on the surface of the part from manufacture, assembly, repair, or service? Tool marks, grinding damage, poor welding or plating, arc strikes, corrosion, wear, pitting fatigue, or fretting are possibilities. There are many ways in which the surface of the part can influence a fracture, because many fractures originate at the surface.

3. *Geometry and Design*
   a. Are there any stress concentrations related to the fracture? This refers to such common design features as fillets, oil holes or other holes, threads, keyways, splines, stamped identification marks, and any other intentional geometric notches.
   b. Is the part intended to be relatively rigid, or is it intended to be flexible, like a spring? The intent must be understood by the failure analyst.
   c. Does the part have a basically sound design? Occasionally, a part (or assembly) is found that seems to have been designed to fail: no amount of metallurgical help will be able to make it succeed. Parts of this type should have been recognized and corrected prior to serious problems, but occasionally this does not occur.
   d. How does the part—and its assembly—work? The function and operation must be thoroughly understood before analysis is undertaken.
   e. Is the part dimensionally correct? If possible, check the part against the drawing from which it was made, for it may be dimensionally inaccurate. If metal has been lost by wear or corrosion, however, dimensional checks may not be possible.

4. *Manufacturing and Processing*
   a. Are there internal discontinuities or stress concentrations that could cause a problem? All commercial metals contain microdiscontinuities that are unavoidable and are innocuous in normal service. However, a more serious problem that could interfere with normal service is a possibility.
   b. If it is a wrought metal, does it contain serious seams, inclusions, or forging problems, such as adiabatic overheating, end grain, laps, or other discontinuities, that could have had an effect on performance?
   c. If it is a casting, does it contain shrinkage cavities, cold shuts, gas porosity, or other discontinuities, particularly near the surface of the part? Frequently these are deep within the casting where the stress is often low, and they are harmless. However, machining may bring them near the final surface. Each case must be studied individually.
   d. If a weldment was involved, was the fracture through the weld itself or through the heat-affected zone in the parent metal adjacent

to the weld? If through the weld, were there problems such as gas porosity, undercutting, under-bead cracking, or lack of penetration? If through the heat-affected zone adjacent to the weld, how were the properties of the parent metal affected by the heat of welding?

e. If the part was heat treated, was the treatment properly performed? Many problems can be caused by inadequate heat treatment, including too shallow or too deep a case depth, excessive decarburization, very coarse grain size, overtempering, undertempering, and improper microstructure.

5. *Properties of the Material*
   a. Are the mechanical properties of the metal within the specified range and comparable to the typical properties of the metal, if this can be ascertained? If so, are the specifications proper for the application? The simplest mechanical property to measure is usually hardness; this test gives an approximation of tensile strength and is widely used for specification purposes. Measurement of other mechanical properties—tensile strength, yield strength, elongation, reduction of area, and modulus of elasticity—involves destructive testing and may not be possible in a fractured part.
   b. Are the physical properties of the metal proper for the application? These are considered to be physical constants, but they are critical in many applications. In some instances, such as close-fitting pistons and other precision parts, the coefficient of thermal expansion of both the piston and the cylinder is critical to the dimensions. Density, melting temperature, and thermal and electrical conductivity are other physical properties to be considered.

6. *Residual and Applied Stress Relationship*
   The residual stress system that was within the part prior to fracture can have a powerful effect—good or bad—on the performance of a part. Residual stresses cannot be determined by simple examination but may be deduced by an analyst familiar with residual stresses. See Chapter 4, "Residual Stresses," in this book for information on how to deduce their pattern. Applied stresses are more obvious than residual stresses. The magnitudes of both are algebraically additive.

7. *Adjacent Parts*
   a. What was the influence of adjacent parts on the failed part? One must always be aware that the fractured part may not be the primary, or original, failure. It may be damaged because of malfunction of some other part in the assembly.
   b. Were fasteners tight? A loose fastener can put an abnormal load on another part, causing the other part to fail. In this case, the loose fastener is the primary failure, while the other part is damaged, or a secondary failure.

8. *Assembly*
   a. Is there evidence of misalignment of the assembly that could have had an effect on the fractured part?
   b. Is there evidence of inaccurate machining, forming, or accumulation of tolerances? These could cause interference and abnormal stresses in the part.
   c. Did the assembly deflect excessively under stress? Long, thin shafts under torsional and bending forces, as in a transmission, may deflect excessively during operation, causing poor contact on the gear teeth and resultant failure by fracture or pitting of the teeth.

9. *Service Conditions*
   This is a difficult area to investigate because people are involved, and people may become defensive. But it is extremely important to question the operator of a mechanism and other witnesses to a fracture or accident to determine if there were any unusual occurrences, such as strange noises, smells, fumes, or other happenings that could help explain the problem. Also, these questions should be considered:
   a. Is there evidence that the mechanism was overspeeded or overloaded? Every type of mechanism has a design capacity or rated load; if that limit is exceeded, problems frequently arise.
   b. Is there evidence that the mechanism was abused during service or used under conditions for which it was not intended?
   c. Did the mechanism or structure receive normal maintenance with the recommended materials? This is particularly important when lubricants are involved in the failure, because use of improper lubricants can be extremely damaging to certain mechanisms, as well as to the seals and gaskets that are intended to keep them from leaking.
   d. What is the general condition of the mechanism? If it is a candidate for the scrap pile, it is more likely to have problems than if it is relatively new.

10. *Environmental Reactions*
    Every part in every assembly in every mechanism has been exposed to several environments during its history. The reaction of those environments with the part is an extremely important factor that may be overlooked in failure analyses. The problems relating to the environment can arise anywhere in the history of the part: manufacturing, shipping, storage, assembly, maintenance, and service. None of these stages should be overlooked in a thorough investigation.
    a. What chemical reactions could have taken place with the part during its history? These include the many varieties of corrosion and possible exposure to hydrogen (such as during acid pickling, electroplating, and certain types of service). Hydrogen exposure can, under certain conditions, result in fracture due to hydrogen em-

brittlement or formation of blisters. Another situation sometimes encountered is stress-corrosion cracking, in which exposure to a critical corrosive environment can cause cracking while the surface of the part carries a tensile stress—applied and/or residual.

b. To what thermal conditions has the part been subjected during its existence? This can involve abnormally high temperatures, which may cause melting and/or heat treatment of a very small area of the surface. Such accidental and uncontrolled heat treatment can have disastrous results. Problems frequently arise as a result of localized electrical arcing, grinding damage, adhesive wear, or other instances where frictional heat is encountered. Similarly, relatively low temperatures for the metal can result in brittle fracture of normally ductile metals with no change in microstructure. Also, low temperatures may initiate uncontrolled phase changes that may cause such problems as the temperature-induced transformation of retained austenite to untempered martensite in steels.

Following study of the fractured part or parts with consideration of these questions—along with others that inevitably arise—it is necessary to reach a conclusion about the reason for the observed fracture. This involves formulating a hypothesis of the sequence of events that culminated in fracture, along with recommendations to prevent the observed type of fracture in the future. Occasionally this process is quite simple; more often it is frustratingly difficult. In either instance, the facts of the situation must be set forth, either orally or in a carefully written report to the appropriate persons. All pertinent information must be documented thoroughly with carefully prepared photographs and other records that should be retained for a number of years. If a similar situation arises in the future, the previous work will serve as a guide.

Summary

Failure analysis is an extremely complex subject and involves areas of mechanics, physics, metallurgy, chemistry and electrochemistry, manufacturing processes, stress analysis, design analysis, and fracture mechanics, to name a few specialties. Because it is nearly impossible for any one person to be an expert in all these fields, it is extremely important to be organized and to know when to seek help. In any situation, it is very important not to leap to conclusions, for a misstep can be extremely hazardous for all concerned.

REFERENCES

1. *Failure Analysis and Prevention,* Vol 11, *ASM Handbook,* 8th ed., ASM International, 1986, p 15–46

2. D.P. Dennies, *How to Organize and Run a Failure Investigation,* ASM International, 2005
3. C.E. Witherell, *Mechanical Failure Avoidance: Strategies and Techniques,* McGraw Hill, 1994, p 31–65

SELECTED REFERENCES

- R.C. Anderson, *Visual Examination,* Vol 1, *Inspection of Metals,* American Society for Metals, 1983
- R.D. Barer and B.F. Peters, *Why Metals Fail,* Gordon and Breach, 1970
- B.E. Boardman, Failure Analysis—How to Choose the Right Tool, *Scanning Electron Microsc.,* Vol 1, SEM Inc., 1979
- C.R. Brooks and A. Choudhury, *Metallurgical Failure Analysis,* McGraw Hill, 1993
- *Case Histories in Failure Analysis,* American Society for Metals, 1979
- V.J. Colangelo and F.A. Heiser, *Analysis of Metallurgical Failures,* 2nd ed., John Wiley & Sons, 1987
- K.A. Esaklul, Ed., *Handbook of Case Histories in Failure Analysis,* Vol 1, ASM International, 1992
- K.A. Esaklul, Ed., *Handbook of Case Histories in Failure Analysis,* Vol 2, ASM International, 1993
- F.R. Hutchings and R.M. Unterweiser, Ed., *Failure Analysis: The British Engine Technical Reports,* American Society for Metals, 1981
- J.L. McCall and R.M. French, Ed., *Metallography in Failure Analysis,* Plenum Press, 1978
- "Fractography in Failure Analysis," STP 645, B.M. Strauss and W.H. Cullen, Jr., Ed., *Annual Book of ASTM Standards*, ASTM, 1978
- P.F. Timmons, *Solutions to Equipment Failure,* ASM International, 1999
- P.P. Tung, S.P. Agrawal, A. Kumar, and M. Katcher, Ed., *Fracture and Failure: Analyses, Mechanisms and Applications,* Materials/Metalworking Technology Series, American Society for Metals, 1981

Understanding How Components Fail, Third Edition
Donald J. Wulpi
Edited by Brett Miller

CHAPTER 2

Mechanical Properties*

WHILE A DETAILED STUDY of mechanical properties of metals is beyond the scope of this work, there are certain facts that must be understood if the task of failure analysis is to be successfully undertaken. Because both fracture and wear are closely related to mechanical properties, it is vital that the general relationships between mechanical properties under certain conditions be thoroughly understood.

Mechanical properties are defined as "the properties of a material that reveal its elastic and inelastic (plastic) behavior where force is applied, thereby indicating its suitability for mechanical applications; for example, modulus of elasticity, tensile strength, elongation, hardness, and fatigue limit" (Ref 1). Other mechanical properties, not mentioned specifically above, are yield strength, yield point, impact strength, and reduction of area, to mention a few of the more common terms. In general, any property relating to the strength characteristics of materials is considered to be a mechanical property. *Physical properties*—a term often improperly applied to mechanical properties—relate to the physics of a material, such as density and electrical, thermal, and magnetic properties. Chemical properties concern the reactions of a material with its environment, as well as the general chemical composition.

Elastic and Plastic Deformation

The terms *elastic deformation* and *plastic deformation* are used widely in failure analysis. Elastic deformation refers to the resilience of a metal, or its ability to return to its original size and shape after being loaded and unloaded. This condition is the state in which most metal parts are used

**Understanding How Components Fail, Second Edition*, by Donald J. Wulpi, Chapter 5, Mechanical Properties, ASM International, 1999, reviewed, revised and renumbered by Thomas N. Ackerson, IMR Metallurgical Services—Louisville.

during their period of service. As mentioned in Chapter 5, "Distortion Failures," in this book, most structural parts can be considered as springs, because they are intended to function in the elastic, or straight-line, portion of the stress-strain curve, as shown in Fig. 1. This means that they will return to their original shape and size after an applied load is released.

A very important feature of the stress-strain curve must be pointed out: the straight-line, or elastic, part of the stress-strain curve of a given metal has a constant slope. That is, it cannot be changed by changing the microstructure or by heat treatment. This slope, called the *modulus of elasticity*, measures the stiffness of the metal in the elastic range; changing the hardness or strength does not change the stiffness of the metal.

This point is expressed well in the following explanation, pertaining to steel, from Ref 2:

> The steel user should remember that the elastic deflection under load of a given part is a function of the section of the part rather than of the composition, heat treatment, or hardness of the particular steel that may be used. In other words, the modulus of elasticity of all the commercial steels, both plain carbon and alloy types, is the same so far as practical designing is concerned. Consequently, if a part deflects excessively within the elastic range, the remedy lies in the field of design, not in the field of metallurgy. A heavier section must be used, or the points of support must be increased, or some similar

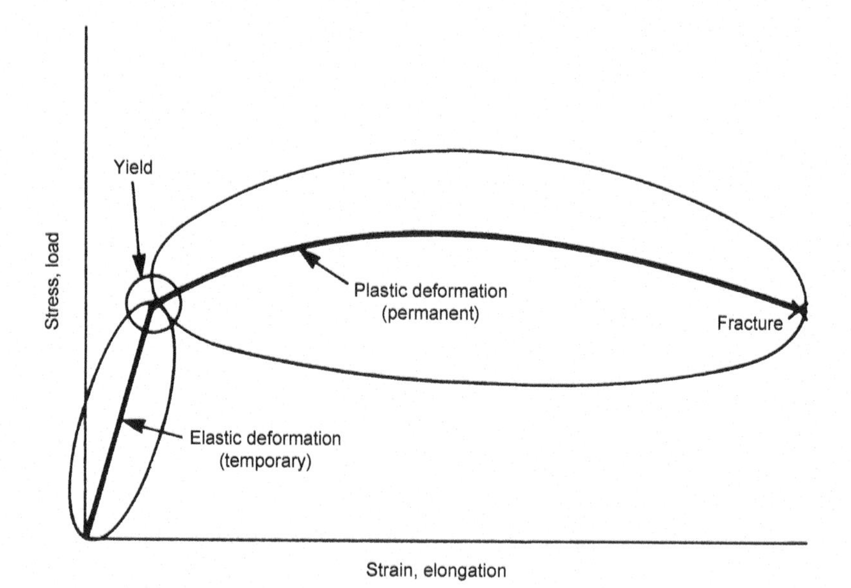

Fig. 1 General stress-strain curve showing elastic and plastic portions of a typical curve. Area marked "Yield" is the area of transition from elastic to plastic deformation. Yield strength, yield point, elastic limit, and proportional limit are all in this area. See Glossary for specific definitions of these terms.

change made, since under the same conditions of loading, all steels deflect the same amount within the elastic limit.

The same point may be made with a diagram, as in Fig. 2, which shows stress-strain curves for steels of different strength levels, which all branch from the same straight line (elastic portion). A very hard, brittle metal, A, such as very strong steel, goes straight up the elastic line with no deviation, and then fractures; a file behaves in this manner. Slightly less strong steel, B, has slight plastic deformation (ductility). Steel C is of intermediate strength, as is D. Steel E is of the relatively low-strength, high-ductility type desired for deep drawing and severe forming, such as deep-drawing-quality steel (DDQS). Note, however, that the straight-line (elastic) portion of the curve is identical for all.

If higher loads are applied to the part, however, the range of elasticity, or elastic deformation, is exceeded, and the metal is now permanently deformed; that is, it is in the plastic deformation range, as shown in Fig. 1. Examples of metals having very low elastic ranges but very great capacity for plastic deformation are aluminum foil, as used for household wrapping, and solder wire. Both are very readily deformed; indeed, aluminum foil must have tremendous ductility (plastic deformation) for it to be useful. If aluminum foil were too strong, it would not wrap properly and could be considered "defective." Figure 3 shows the type of stress-strain curve desired for this low-strength, easily deformable, highly ductile metal.

Indeed, the plastic deformation portion of the stress-strain curve is extremely important in manufacturing processes because it permits altera-

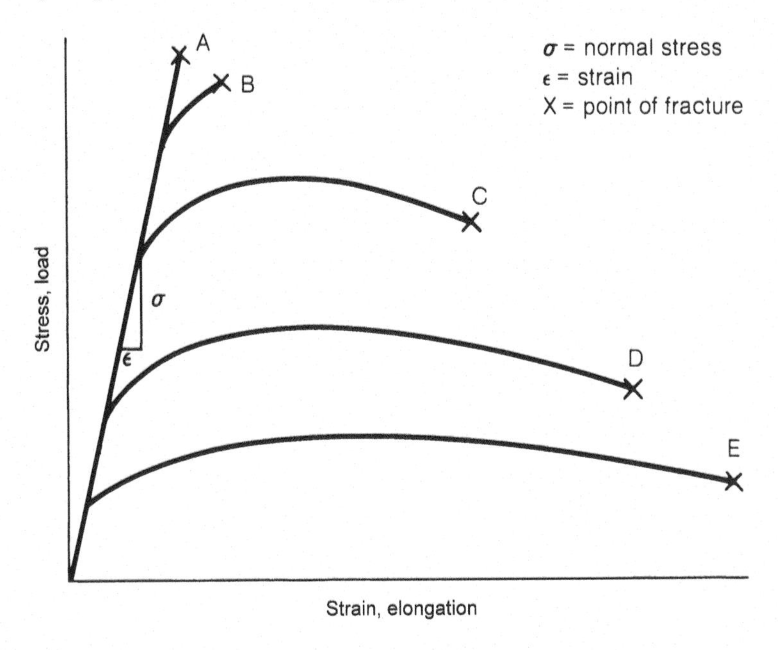

Fig. 2 Stress-strain curves for steels of different strength levels, ranging from A, a very hard, strong, brittle steel, to E, a relatively soft, ductile steel

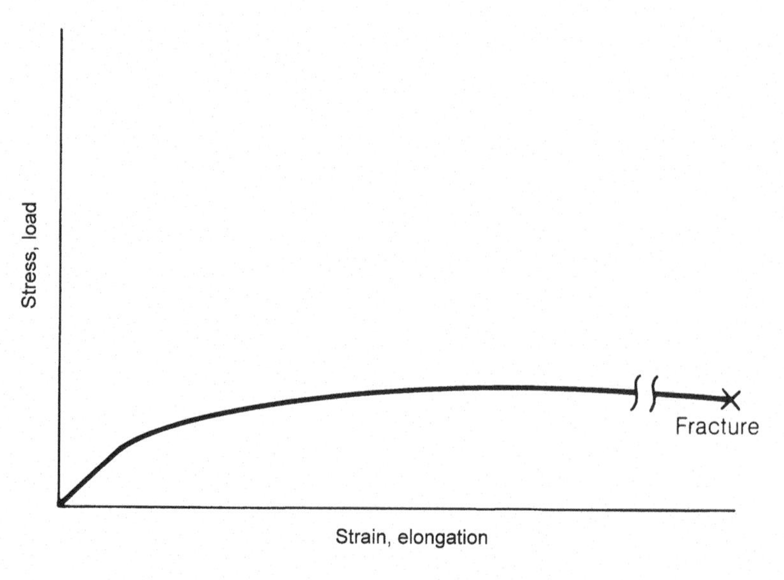

Fig. 3 Typical stress-strain curve for a low-strength, easily deformable metal, such as aluminum foil used for household wrapping

tion of the metal in order to change its shape, as in all cold-working processes. Also, for this reason, metals are heated to obtain increased ductility for shaping, as in forging, pressing, and upsetting operations. The reductions of the modulus of elasticity and yield strength with increasing temperature are discussed in the following section.

Effect of Temperature

There is only one condition that changes the stiffness of any given metal: temperature. The stiffness of any metal varies inversely with its temperature; that is, as temperature increases, stiffness decreases, and vice versa. This relationship is illustrated for four common alloy systems in Fig. 4. For example, steel alloys are usually considered to have a modulus of elasticity of 29 to 30 million psi, but this figure is valid only for room temperature. A spring will deflect more at an elevated temperature than at room temperature and must be designed accordingly. A spring used at low temperature will deflect less. Nominal room-temperature values are 15 to 19 million psi for copper alloys, 10 to 11 million psi for aluminum alloys, and 6 to 7 million psi for magnesium-base alloys, but these also decrease with increasing temperature.

Nonlinear Behavior

The above comments on the elastic portions of the stress-strain curves apply to nearly all metals. However, a few metals do not conform to Hooke's law, which states that stress and strain are linearly proportional within the elastic range. Figure 5 shows typical stress-strain curves for three classes of gray cast iron (Ref 3). This nonlinear behavior is caused by the graphite flakes embedded in the steel-like matrix that give gray cast iron its unique properties. The flakes act as internal notches, or stress concentrations, when the metal is loaded in tension. They tend to cause microscopic—and irreversible—yielding at the sides or ends of the flakes.

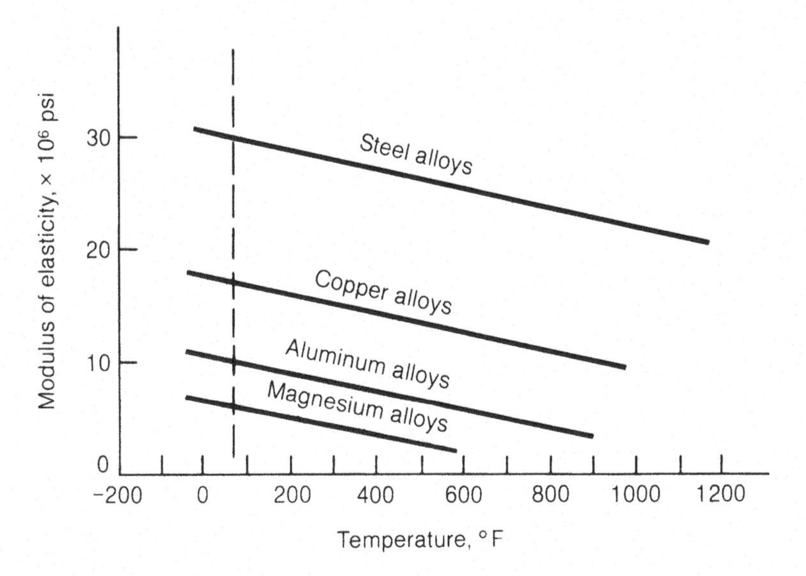

Fig. 4 Relationship of stiffness, or modulus of elasticity, to temperature for four common alloy systems

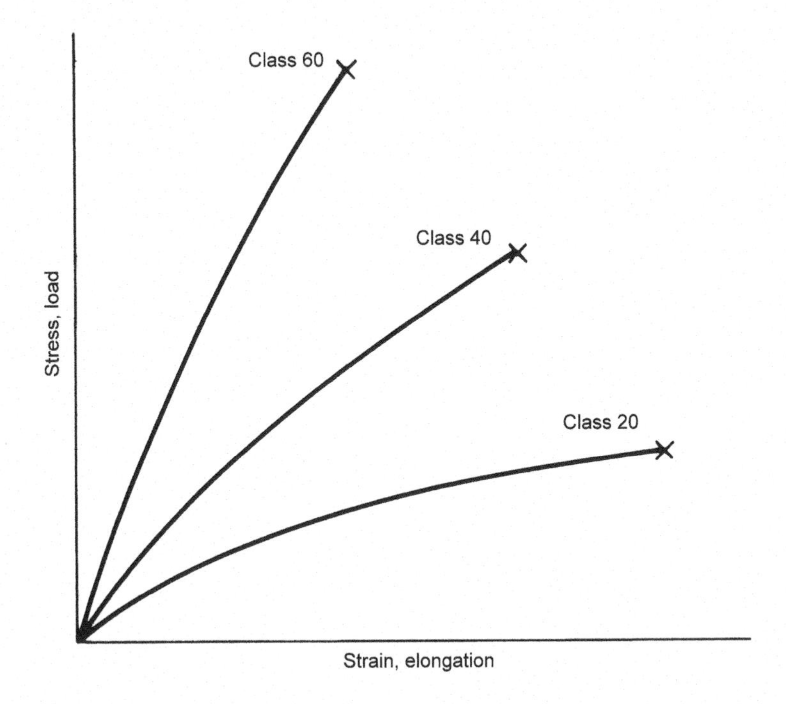

Fig. 5 Typical stress-strain curves for three classes of gray cast iron. This nonlinear behavior is caused by the graphite flakes, which act as internal stress concentrations, or notches, within the metal matrix. Source: Ref 3

Thus, the "elastic" properties of gray cast iron are determined, in part, by the size, shape, and distribution of the graphite flakes.

Sintered metals also have nonlinear stress-strain curves, and for the same reason: their internal pores, like the flakes in gray cast iron, act as internal notches. However, as the density of a sintered metal approaches the maximum theoretical density for that alloy, the curves tend to approach linearity (Ref 3).

Cold-drawn steel bars also have slight curves in the "elastic" regions due to the very high residual (internal) stresses caused by the cold-working process. However, heating at temperatures of about 370 to 480 °C (700 to 900 °F) relieves the stresses and restores linearity, or straightness, in the elastic region. It also simultaneously increases the yield strength. Consequently, cold-worked, stress-relieved bars are most commonly specified and are readily available (Ref 3).

Bidirectional Stresses

Another point that must be made about stress-strain curves is that they apply to bidirectional stresses. Normally, only the tensile part of the curve is shown, as in Fig. 1 to 3. However, the straight-line portion also extends into the compression region, as shown in Fig. 6. In metals that have yield strengths, the compressive yield strength is usually considered to be approximately equivalent to the tensile yield strength. With ductile metals in compression, there is no definite end point. Consequently, the end point must be an arbitrarily selected value depending on the degree of distortion that is regarded as indicating complete failure of the material (Ref 1). Certain metals fail in compression by a shattering type of fracture; these are normally the more brittle materials that do not deform plastically. Gray cast iron, which is relatively weak in tension because of the mass of internal graphite flakes, has a compressive strength that is several times its tensile strength (Ref 3).

Consolidating the information given in this chapter, it follows that the modulus of elasticity is reduced when a metal at elevated temperature is in compression compared with when it is tension at an elevated temperature. This is shown in Fig. 7. Temperature T represents an arbitrarily selected base temperature, such as room temperature, and T_1 and T_2 represent elevated temperatures. Note not only the decrease in the modulus of elasticity (the slope of the straight line portion) but also the decrease in yield and tensile (compressive) strengths with increasing temperature.

The preceding paragraph becomes exceedingly important in understanding residual (internal) stresses caused by thermal means, such as in welding. These stresses are the reason both for weldment distortion and for fracture resulting from tensile residual stresses caused by shrinkage during solidification and cooling of the weld. See Chapter 4, "Residual Stresses," in this book.

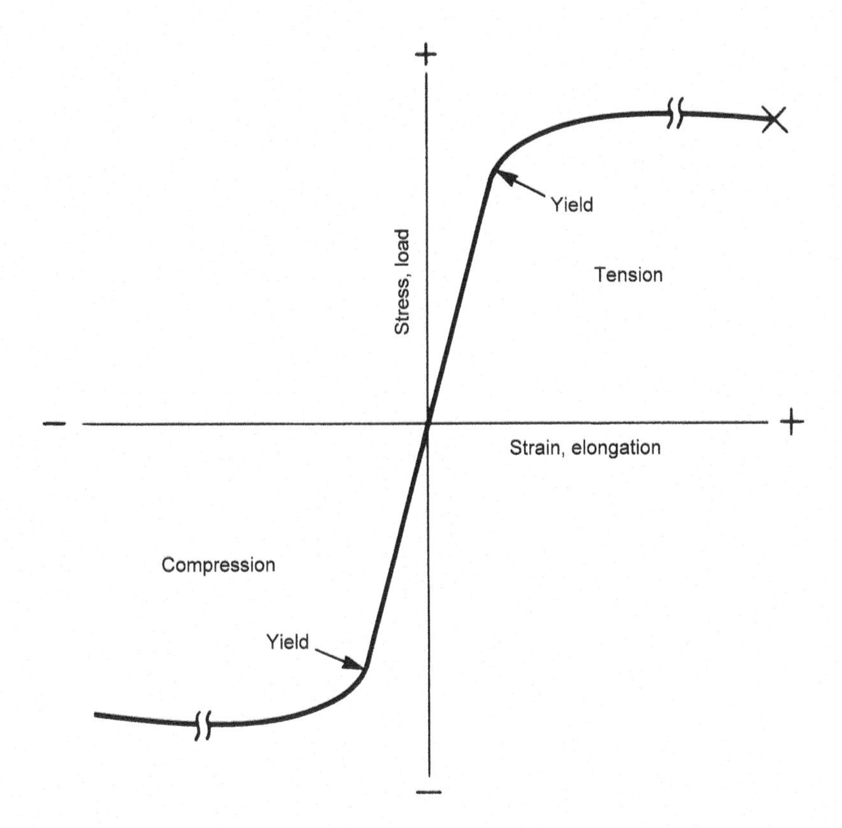

Fig. 6 Complete engineering stress-strain curve showing the normally considered tensile region (upper right) and the often neglected compression region (lower left)

Effect of Stress Concentrations

In general, the tensile strength of a metal changes in proportion to hardness, as shown in Fig. 8. However, this relationship does not always hold at high hardness levels or with brittle materials because these materials are more sensitive to stress concentrations, or notches, and may fracture prematurely when stressed in tension.

The harder and stronger the metal, the more sensitive it is to stress concentrations. Therefore, high-hardness, high-strength metals must be treated carefully; virtually everything becomes critical because such metals cannot easily tolerate stress concentrations. They cannot flow, or deform plastically, at the highly stressed regions of the stress concentrations as readily as more ductile metals of somewhat lower hardness. However, high-hardness, high-strength metals are extremely useful when carefully used, because of their high static and fatigue strength, as well as their high wear resistance.

Fatigue fractures are covered in Chapter 10, "Fatigue Fracture," in this book, but when discussing mechanical properties, it must be noted that

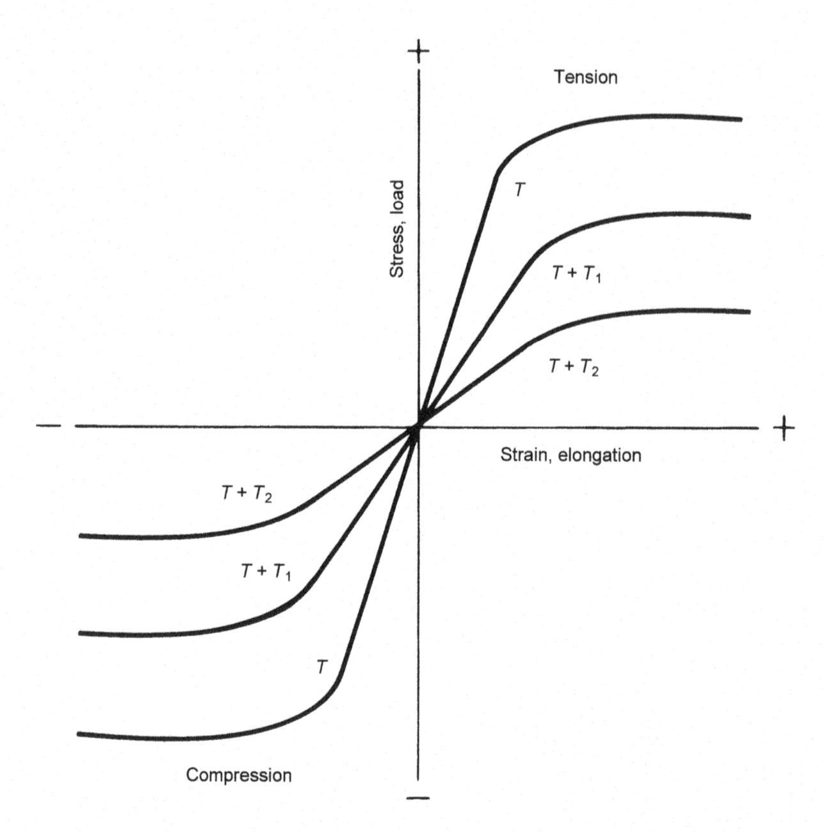

Fig. 7 Effect of elevated temperatures T_1 and T_2 on tensile and compressive properties of a typical metal

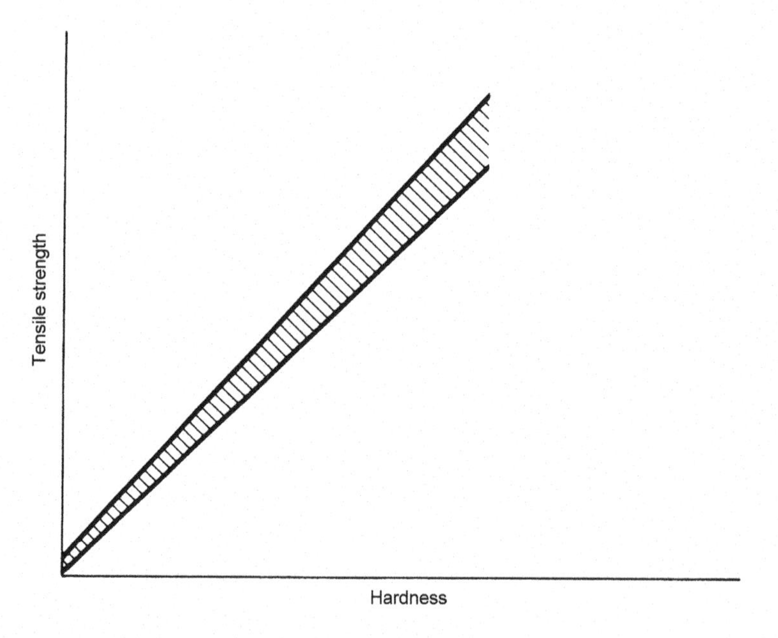

Fig. 8 Relationship between hardness and tensile strength of metals in the absence of stress concentrations

fatigue strength is greatly reduced by severe notches that have not been mechanically prestressed by compressive residual stresses. See Chapter 4, "Residual Stresses," in this book. Fig. 9 illustrates qualitatively the dramatic effect that severe notches can have on the "nominal" fatigue properties that are derived by testing polished bars without stress concentrations (Ref 4). In the real world, polished parts without stress concentrations are extremely rare; consequently, great care must be taken when "nominal" or book values are used for design. It is for this reason that parts and assemblies should be laboratory tested, as well as field tested, in order to have more confidence in the performance of the actual mechanism.

Summary

Because mechanical properties are closely related to fracture, the various relationships between mechanical properties must be well understood

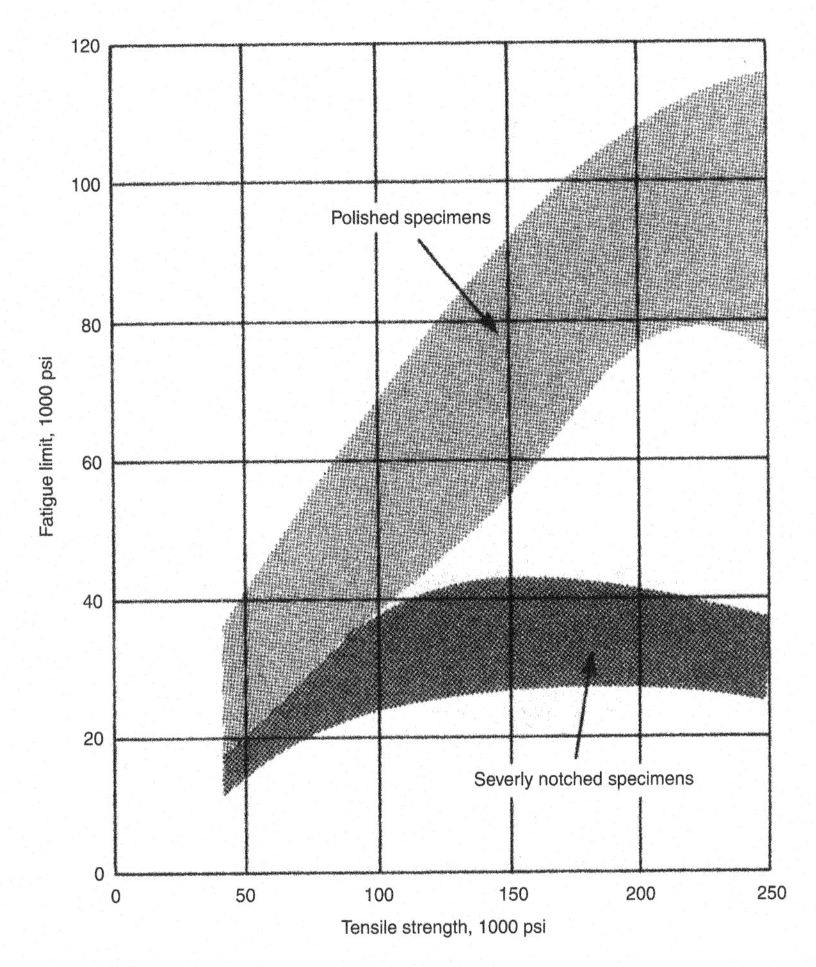

Fig. 9 Relationship between fatigue limit and tensile strength in polished and in severely notched steel specimens. Source: Ref 4

by the failure analyst. Effects of temperature variations on mechanical properties must also be considered, both in tension and in compression. Also, the general relationship between hardness and tensile strength and the effect of stress concentrations on high-strength metals and on parts subjected to fatigue stresses are critical.

REFERENCES

1. *Failure Analysis and Prevention,* Vol 11, *ASM Handbook,* ASM International, 1986, p 1–4
2. *Handbook,* Society of Automotive Engineers, J401, 1991, p 1.05–1.06
3. *Properties and Selection: Irons, Steels, and High-Performance Alloys,* Vol 1, *ASM Handbook,* ASM International, 1990, p 20, 255–257, 806
4. R.S. Archer, J.Z. Briggs, and CM. Loeb, Jr., *Molybdenum Steels-Irons-Alloys,* Hudson Press, 1948

SELECTED REFERENCES

- G.E. Dieter, *Mechanical Metallurgy,* 3rd ed., McGraw-Hill, 1986
- N.E. Dowling, *Mechanical Behavior of Materials: Engineering Methods for Deformation, Fracture, and Fatigue,* 2nd ed., Prentice Hall, 1999
- *Failure Analysis and Prevention,* Vol 11, *ASM Handbook,* ASM International, 1986
- P. Harvey, Ed., *Engineering Properties of Steel,* American Society for Metals, 1982
- J. Marin and J.A. Sauer, *Strength of Materials,* Macmillan, 1948
- *Mechanical Testing,* Vol 8, *ASM Handbook,* ASM International, 1985
- *Properties and Selection: Irons, Steels, and High-Performance Alloys,* Vol 1, *ASM Handbook,* ASM International, 1990
- *Properties and Selection: Nonferrous Alloys and Special-Purpose Materials,* Vol 2, *ASM Handbook,* ASM International, 1990
- S. Timoshenko and J.N. Goodier, *Theory of Elasticity,* McGraw-Hill, 1951

Understanding How Components Fail, Third Edition
Donald J. Wulpi
Edited by Brett Miller

CHAPTER 3

Stress versus Strength*

IN ANY FAILURE ANALYSIS involving metal fracture, the subject of both metal strength and the stresses within the metal part must be considered, because they cannot be divorced from each other. *Stress*, in the sense used in failure analysis, is defined as the force per unit area, often considered as force acting through a small area within a plane (Ref 1). *Strength*, in the metallurgical sense, is the property of a metal part that resists the stresses imposed upon the part, and when the applied stresses exceed the strength of material, the component will fail. From these simple definitions, the subject rapidly becomes much more complex because there are many different types of stress and strength.

In terms of general principles, however, it is good to emphasize a statement that will be elaborated upon in Chapter 7 in this book: "All fractures are caused by stresses, and a version of the 'weakest link' theory applies: fracture will originate wherever the local stress first exceeds the local strength." In this sense, *local* means any micro area within the metal where stress exceeds strength.

The failure analyst must be able to understand—actually visualize—the principal stress system that acted upon a fractured part. It is not enough to understand the stress system that is *supposed* to act on the part; one must understand the stress system that was *actually* imposed on the part to cause fracture. In order to accomplish this, it is necessary to be able to "read" the fracture surface to determine the origin(s) of fracture and whether or not this location is consistent with the intended stress system.

The elastic stress systems resulting from several common types of loading are discussed in this chapter. However, it is advisable to remember that elastic stress is only part of the stress system actually functioning in the

***Understanding How Components Fail, Second Edition*, by Donald J. Wulpi, Chapter 6, Stress versus Strength, ASM International, 1999, reviewed, revised and renumbered by Dustin A. Turnquist, Engineering Systems Inc.

part. The *applied* stresses that can cause fracture are shown schematically for several geometries and types of loading; however, it must be remembered that these are modified in actual parts by the *residual* (internal) stresses—such as from welding, heat treatment, assembly, or permanent deformation—that may also be present in the part. Thus, the true resultant stress in any orientation at any location is the algebraic sum of both the applied and the residual stresses, as discussed in Chapter 4, "Residual Stresses," in this book. It is for this reason that it also is vital to understand the residual stress patterns, for they are frequently ignored in failure analyses simply because they are not as obvious as the applied stresses.

The following discussion and illustrations cover the principal elastic stress distribution in members of various shapes under different types of pure loads. In actual service, combined stresses are usually present; therefore, modifications of these elastic stresses are necessary to fit individual cases. Also, component shapes need to be modified from the simple shapes shown. Obviously, if the elastic stress range is exceeded, plastic (permanent) deformation will occur at the locations of highest stress, and residual stresses will become more of a factor after the applied stress is released. Again, see Chapter 4, "Residual Stresses," in this book for more information on understanding the residual stress pattern.

For the purposes of failure analysis, it is usually satisfactory simply to keep in mind the general shapes of these elastic stress distributions. It is necessary to be aware of where the highest and lowest applied stresses are located, modified by any stress concentrations. However, in certain cases, it may be necessary to perform stress calculations to quantify the values in the part under the specific types of loading and combinations thereof. Quantitative analysis is beyond the scope of this work but may be pursued in basic strength of materials texts, some of which are shown with the references.

Elastic Stress Distributions for Simple Shapes

Tension. In pure tension, the tensile stress is uniform across the section if there are no local stress concentrations or discontinuities present, as shown in Fig. 1(a). A transverse, or annular, groove or notch causes a stress concentration that increases with the sharpness of the notch. At the root of the notch, the axial tensile stress increases, as shown in Fig. 1(b). A transverse hole also concentrates the tensile stress at the sides of the hole, as shown in Fig. 1(c).

Torsion. A cylindrical shaft under pure torsional loading (twisting) has stresses that are maximum at the surface and zero at the center or neutral axis of the shaft, as shown in Fig. 2(a). Pure torsion loading of a cylindrical shaft causes all stress components—tension, shear, and compression—to have equal magnitudes that are maximum at the surface and zero at the center. However, a stress concentration, such as a transverse hole, greatly

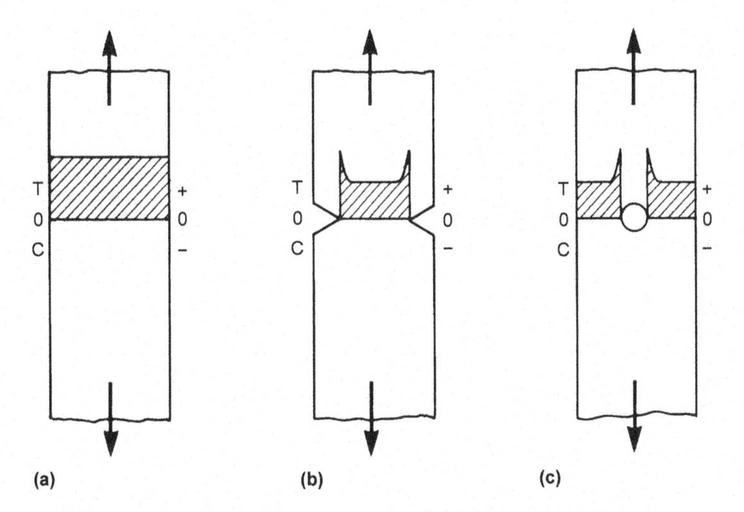

Fig. 1 Elastic stress distribution: pure tension. T, tension. C, compression. (a) No stress concentration. (b) Surface stress concentrations. (c) Transverse hole stress concentration

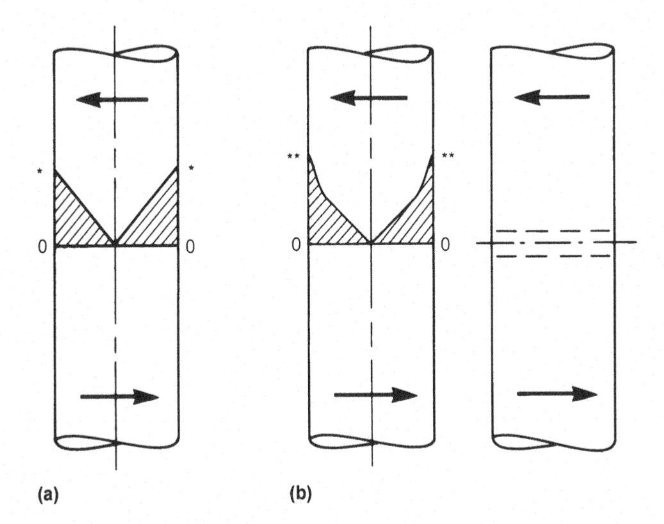

Fig. 2 Elastic stress distribution: pure torsion. (a) No stress concentration. *All stress components—tension, shear, and compression—have equal magnitude. (b) Transverse hole stress concentration. **Tension and compression stress components increase more than shear stress at a torsional stress concentration at the corner of the transverse hole.

increases the stress magnitude at the edges of the hole, particularly in the tension and compression planes, as shown in Fig. 2(b).

Longitudinal stress concentrations, such as grooves or splines that are in the same direction as the longitudinal shear stress component, may tend to direct any shear crack in that direction, as may a transverse or annular groove in the transverse shear stress direction. For explanation, see Fig. 1(b) in Chapter 7, "Stress Systems Related to Single-Load Fracture of Ductile and Brittle Metals," in this book.

Compression. In pure compression, the compressive stress is uniform across the section if no local stress concentrations or discontinuities are present, as shown in Fig. 3(a). A transverse, or annular, groove or notch acts as a stress concentration, with the magnitude increasing with the sharpness of the notch. At the root of the notch, the axial compressive stress increases greatly, as shown in Fig. 3(b). A transverse hole also concentrates the compressive stress at the sides of the hole, as shown in Fig. 3(c).

Bending. In bending a straight member, the convex side (outside) of the bend is stressed in tension at the surface, while the concave side (inside) of the bend is stressed in compression at the surface. In the absence of a stress concentration, the stress distribution is linear across the section, reaching zero at the center, or neutral axis, as shown in Fig. 4. Obviously, if there is a stress concentration perpendicular to the principal stresses, the distribution of stress is no longer linear but will be higher near the root of the notch. The neutral axis may or may not be at the geometrical center, although there must be zero stress at some location in the cross section.

Interference Fit (Press or Shrink). In an interference fit of a shaft or tube into a hole in a collar, hub, or other member, there is a tensile circumferential (hoop) stress around the hole tending to burst or split the outer member, as shown in Fig. 5(a). Similarly, the inner member has compressive circumferential stresses. In all cases, the magnitudes of the stresses are maximum at the interface between the members, as shown in Fig. 5(b).

Convex Surfaces in Contact. When convex surfaces, such as cylinders and balls, are under compressive forces in contact with other surfaces, they develop compressive stresses both parallel and perpendicular

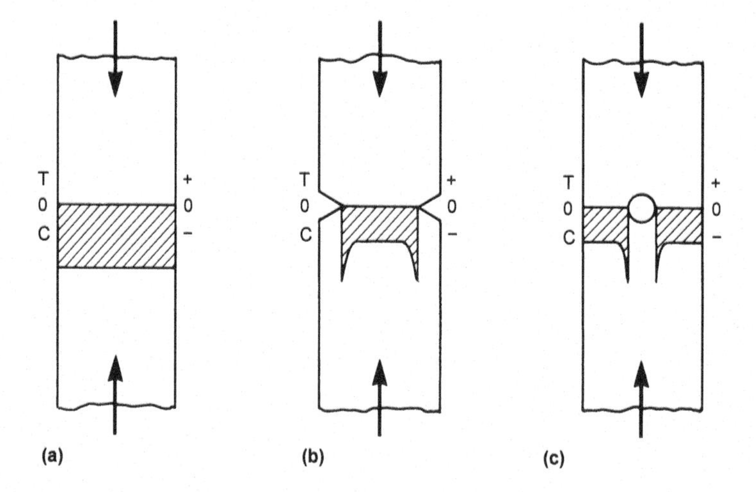

Fig. 3 Elastic stress distribution: pure compression. T, tension. C, compression. (a) No stress concentration. (b) Surface stress concentrations. (c) Transverse hole stress concentration

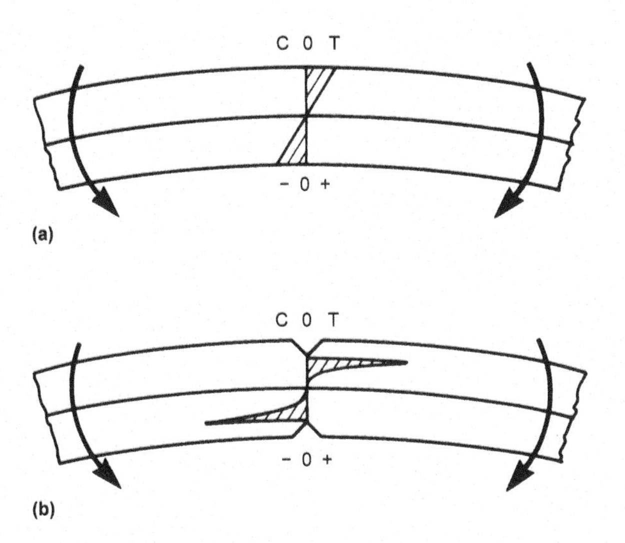

Fig. 4 Elastic stress distribution: pure bending. T, tension. C, compression. (a) No stress concentration. (b) Transverse surface stress concentrations

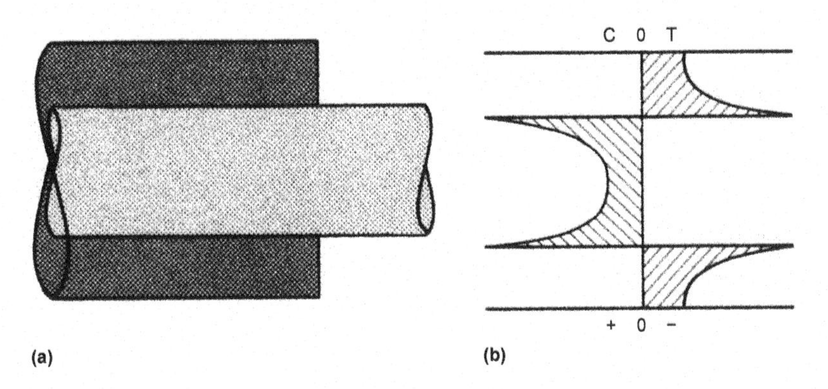

Fig. 5 Elastic stress distribution: interference fit (press or shrink). T, tension. C, compression.

to the line of contact in both members, as shown in Fig. 6(a). The maximum shear stress, the location of fatigue origin, is a short distance below the contact surfaces when the members are either stationary or rolling with respect to each other. Note in Fig. 6 that when under heavy pressure, the contact region distorts elastically and tends to flatten both convex members, bulging at the ends of contact. When rolling occurs, these bulges form stress waves that roll with the rolling elements.

When sliding forces are added to the rolling forces, the shear stress location and magnitude are modified. This happens at all locations on gear tooth active profiles (except at the pitch line) and in other nominally rolling elements. The maximum shear stress approaches the surface interfaces because of the friction between the rolling/sliding elements. This distorts

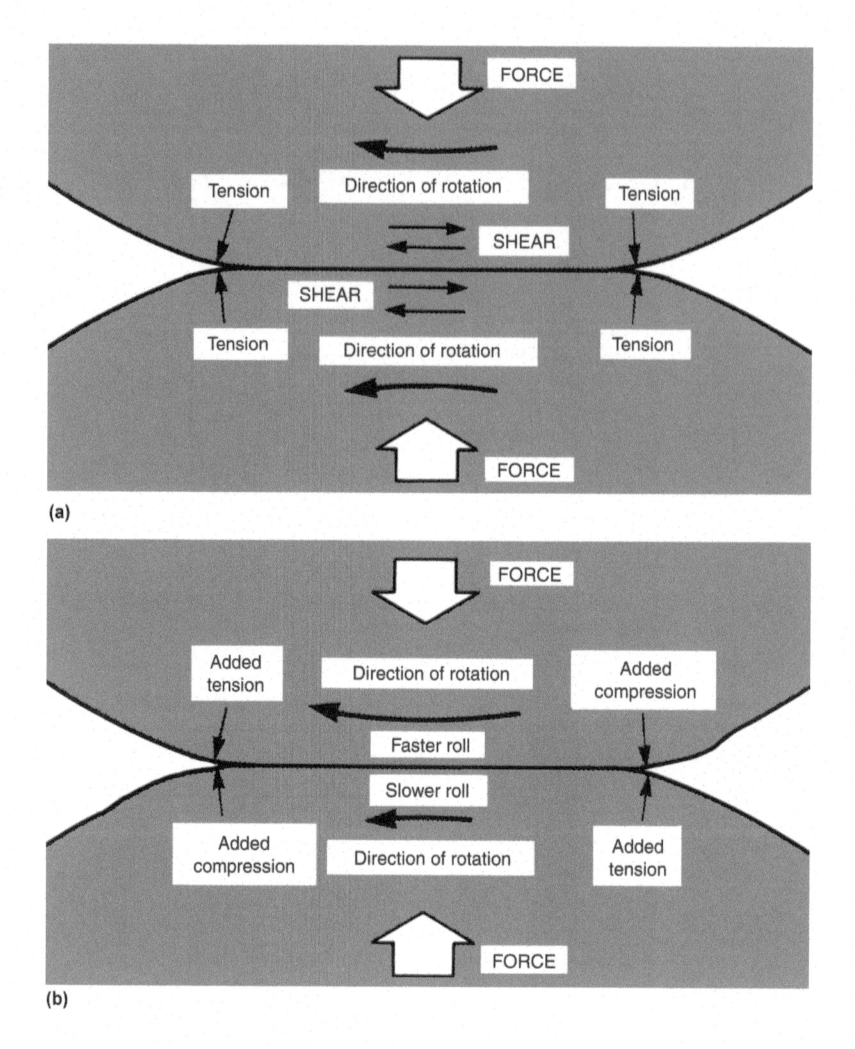

Fig. 6 Elastic stress distribution: convex surfaces in contact. (a) Rolls turning at same speed. (b) Rolls turning at different speeds

the bulges at the ends of each element, as shown in Fig. 6(b), and fatigue failures are likely to be of surface origin. See Chapter 12, "Wear Failures—Fatigue," in this book.

Direct (Transverse) Shear. There are two types of direct, or transverse, shear fractures: single shear and double shear. Both involve macroshearing fractures, with little or no bending distortion involved. In single-shear fractures, two closely fitting parallel surfaces with sharp edges, as shown in Fig. 7(a), can shear, or cut, a third material between them if they move toward each other. Shear or sliding fracture occurs on the dashed line at the interface as shown. Typical examples are scissors (shears), as well as punch presses and other machine tools that make cuts without actually removing metal chips. If the corners of the cutting edges are not sharp, or if the parts are not properly adjusted, sharp-edged burrs will be formed on the trailing edge of each of the cut surfaces.

Double-shear fracture occurs when two surfaces are stressed in shear simultaneously. This is characteristic of some bolted or riveted joints, as shown in Fig. 7(b), and in closely fitting clevis joints. True double-shear fracture occurs only when there is no bending of the member that becomes fractured; bending frequently occurs, however, if the joint is loose or the sides spread.

Thin-Wall Pressure Vessels. Compressed fluids within a thin-wall pressure vessel cause circumferential and longitudinal tensile stresses, as shown in Fig. 8. The maximum tensile stress—in the absence of a stress

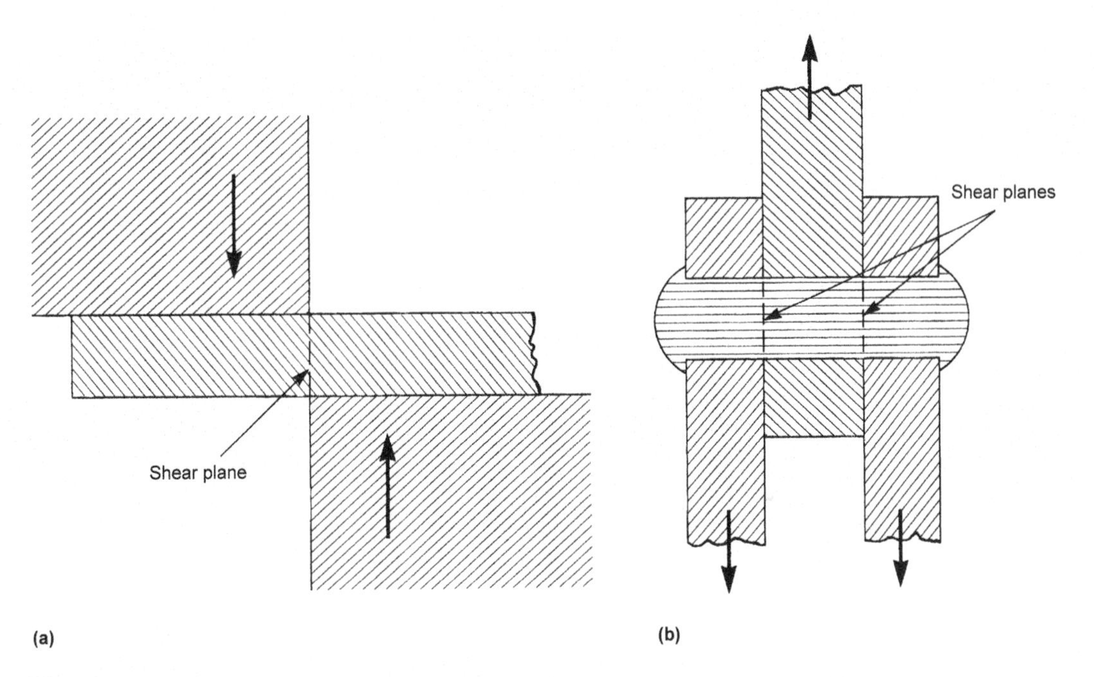

Fig. 7 Elastic stress distribution: direct (transverse) shear. (a) Single shear. (b) Double shear

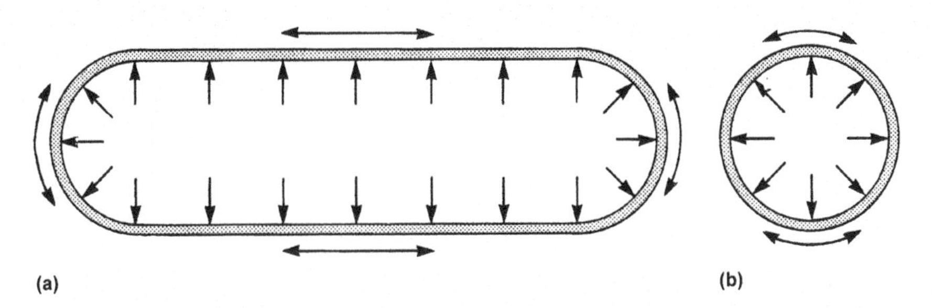

Fig. 8 Elastic stress distribution: thin-wall pressure vessel. (a) Longitudinal section. (b) Cross section

concentration—is normally near the center of the length and in the circumferential (hoop) direction. This tends to cause a longitudinal splitting rupture if the internal pressure causes the hoop stress to be too high for the strength of the material under the conditions involved.

In the real world, stress concentrations are invariably present, usually in the form of welded inlets, outlets, formed shapes, and the like. Each of these must be considered individually with respect to the type of fracture involved. The longitudinal splitting rupture noted above is sometimes associated with a longitudinal seam weld, either in the weld itself or in the heat-affected zone. Fatigue fractures are sometimes observed, for each pressurization represents one stress cycle; in this case there need be no permanent bulging or other plastic deformation, for the pressure vessel need not have been overpressurized.

Subsurface-Origin Fractures. Under certain conditions fractures can originate at significant depths below the surface. This is particularly true of fatigue fractures in parts that have no stress concentration on the surface but have a steep strength gradient, as frequently occurs with certain types of case hardening. Subsurface-origin fractures also occur in tension-stressed parts where there is an internal stress concentration, such as a discontinuity. Photographs of these types of fractures are shown in Chapter 10, "Fatigue Fracture," in this book.

Figure 9 illustrates the principle behind subsurface-origin fractures of parts such as shafts that are stressed in bending or torsion. These types of

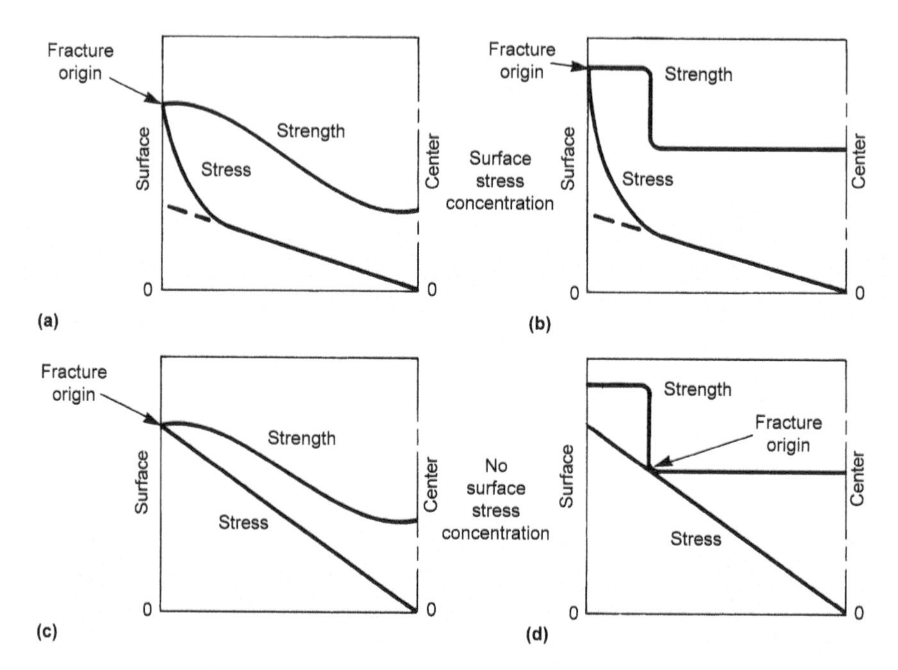

Fig. 9 Surface and subsurface fracture origins. Strength gradients: (a) and (c), gradual; (b) and (d), steep. Stress gradients: (a) and (b), surface stress concentrations; (c) and (d), no stress concentrations

loading cause the stress to drop to zero at the neutral axis, at or near the center of the part. Each of the four parts of Fig. 9 represents a different combination of stress gradients and strength gradients. All have the surface of the part at the left edge; the center, or neutral axis, is the dashed line at the right edge. Both stress and strength are shown on the vertical axis; the stress gradients pivot around the lower right corner (the neutral axis) as the stress increases (Ref 2).

Figures 9(a) and (b) show the same curving stress gradient because of a geometric stress concentration at the surface. The stress at the center (neutral axis) is zero and then rises sharply above the straight-line stress gradient (dashed line) near the surface where there is a stress concentration, such as a notch or groove. Figures 9(c) and (d) show the stress gradients that are present if there is no surface stress concentration; the stress is linear from zero at the center (lower right corner) to maximum at the surface (left edge).

Figures 9(a) and (c) both show the same gently dropping strength gradients, which are derived from hardness traverses. This type of strength gradient is characteristic of medium-carbon steels, plain carbon or alloy, with maximum hardness at the surface gradually dropping to moderate hardness at the center. On the other hand, Fig. 9(b) and (d) both show steep, precipitous strength gradients. This type of strength gradient is characteristic of some surface-hardened steels with very thin cases, steep drops, and/or a sudden change from a hard surface to a relatively soft core.

Keeping in mind the principle that fracture can originate wherever the local stress first reaches the local strength, it is easy to understand subsurface-origin fracture. Note that the shape of the stress and strength curves in Fig. 9(a), (b), and (c) is such that the stress gradient first touches the strength gradient at the surface. In these cases, surface-origin fracture would be expected. However, in Fig. 9(d), the steep strength gradient and the linear stress gradient intersect below the surface; therefore, a subsurface-origin fracture would be expected. An example of this type of behavior sometimes is seen in induction-hardened steel that has not been previously quenched and tempered.

The facts that can be learned from this exercise are:

- Surface-origin fractures can be expected at locations where surface stress concentrations exist, such as grooves, threads, slots, and transverse holes. Surface-origin fractures also can be expected where there are no surface stress concentrations but there is a relatively gentle strength (hardness) gradient sloping from maximum at or near the surface to minimum at the neutral axis.
- Subsurface-origin fractures can be expected if there are no surface stress concentrations perpendicular to the tensile stress but there is a very thin case, a steep hardness gradient, large nonmetallic inclusions, and/or a relatively soft core. This type of fracture can be prevented by

removing any or all of these conditions by increasing case depth, modifying heat treatment to eliminate the steep hardness gradient, using a cleaner steel, and/or increasing core hardness.

In the preceding discussion of subsurface fractures, which deals with the relationship between two different types of stress gradients and two different types of strength gradients, discussion of one vital factor was omitted for simplicity. As pointed out previously, residual stresses are just as important as the applied stresses discussed previously and can modify the location of fracture origin. Residual stresses can be considered as modifying the shape of the stress gradient; the resultant gradient must be considered in relation to the strength gradient according to the simple principles outlined.

As an example of modification of fracture origin, if a part with a moderate surface stress concentration fractures by fatigue, we expect the fatigue origin(s) to be at the surface, as shown in Fig. 9(a), (b), and (c). However, if the surfaces of similar parts are shot peened or rolled, any future fatigue fractures should originate below the surface, although after longer life or higher service stresses this may not necessarily be true. The reason is that the shot peening or surface-rolling operation induces compressive residual stresses in the surface that neutralize, or reduce, the tensile (actually shear) stresses that initiate fatigue fracture.

With respect to fatigue failures, one must also consider how the residual and applied stresses may have changed during the course of the progressive fatigue fracture mechanism. Additional information on residual stresses can be obtained from Chapter 4, "Residual Stresses," in this book.

Effect of Service Conditions on Applied Stresses

While a component may be designed to endure a certain load/stress criteria, the failure analyst must also consider how the surrounding environment may have played a role in a failure sequence. It is possible for loading conditions to change during the service life of a component. Factors such as wear or exposure to excessive temperatures can increase the internal stresses within a component or decrease the strength of the material, resulting in component failure. One example is a worn bushing or damaged bearing, which could result in a rotating shaft being exposed to bending stresses, in addition to the anticipated torsional stresses.

Another aspect that must be considered is the frequency and amplitude of loading cycles. While a certain criteria may have been used while designing a component, service conditions may differ from the design conditions. Intermittent spike loads can substantially alter life of a component. Spectrum loading (i.e., loading at variable frequency and amplitude) may be incorporated into the design and qualification of components.

Summary

The original premise of this discussion should be obvious by now. The failure analyst must have at least a basic understanding of both the stress and strength gradients of parts with and without stress concentrations and under different types of loading. It should also be clear that the relationship of stress and strength gradients must be considered simultaneously in analysis of a particular type of fracture. The complicating factor of residual stresses also must be understood, although quantitative analysis of residual stresses usually is more difficult to determine.

The analyst must also have an understanding of how a component may have been altered during service and whether any stress concentrations were unintentionally created in the component after manufacture.

REFERENCES

1. *Failure Analysis and Prevention,* Vol 11, *ASM Handbook,* 9th ed., ASM International, 1986, p 10
2. *Mechanical Testing and Evaluation*, Vol 8, *ASM Handbook*, ASM International, 2000, p 109–114

SELECTED REFERENCES

- H.J. Grover, *Fatigue of Aircraft Structures,* rev. ed., NAVAIR01-1A-13, U.S. Government Printing Office, 1960
- *Mechanical Testing,* Vol 8, *ASM Handbook,* ASM International, 1985
- W.D. Pilkey, R.E. Peterson, J.E. Peterson, and K.M. Clark, *Peterson's Stress Concentration Factors*, 2nd ed., John Wiley, 1997
- S. Timoshenko and J.N. Goodier, *Theory of Elasticity,* McGraw-Hill, 1951
- W.C. Young, *Roark's Formulas for Stress and Strain,* 6th ed., McGraw-Hill, 1989

Understanding How Components Fail, Third Edition
Donald J. Wulpi
Edited by Brett Miller

CHAPTER 4

Residual Stresses*

IT IS LITERALLY a shattering experience to witness a metal part suddenly split into two or more pieces after it has been resting on a floor, bench, or table under no external load. Yet this is not an uncommon situation and is within the experience of many persons who have worked with metals.

A spectacular example was that of a 40-ft-long I-beam that was lying on a shop floor after being torch cut at each end the previous day (Ref 1). It suddenly split longitudinally, as shown in Fig. 1. This fracture was caused by residual stress. Note the bowing of the beam as the transverse stresses relaxed following fracture. The extremely high forces involved can be imagined by trying to apply enough force to restore the beam to its original condition. This is the same force that was internal in the beam and caused it to fracture in the first place. The reasons for this fracture are related to the prior torch cutting, which was the trigger for releasing the energy stored in the beam.

Residual stresses are defined as those stresses that are internal or locked into a part or assembly, even though the part or assembly is free from external forces or thermal gradients (Ref 2). These internal stress systems, whether they are in an individual part or in an assembly of parts, are the result of mismatch, or misfit, between adjacent regions of the same part or assembly (Ref 3). This internal misfit, which can be caused by a variety of reasons, distorts the neighboring regions with elastic stresses.

Although residual stresses are hard to visualize, difficult to measure, and nearly impossible to calculate, they are just as important in the function of a part as are externally applied forces that are readily visualized, measured, and calculated. An individual grain or crystal of metal in a part

*_Understanding How Components Fail, Second Edition_, by Donald J. Wulpi, Chapter 7, Residual Stresses, ASM International, 1999, reviewed, revised and renumbered by Scott Chumbley, Iowa State University.

Fig. 1 Spontaneous residual stress fracture in a 40-ft-long I-beam under no external load. Source: Ref 1

reacts to a stress to which it is being subjected whether the stress is from an external or an internal source. A stress from either kind of source can result in serious problems of fracture and distortion. For this reason, residual stresses must be considered during failure analysis even though they are much more difficult to visualize and understand than applied stresses.

Residual stresses are internal forces not limited to metal parts or assemblies as covered in this work. Parts or assemblies made from any material are subject to uneven expansion and contraction due to variations in temperature, moisture, fastening, and the like. Water freezing in a pipe can cause the pipe to burst because water expands when it freezes. This is actually an applied stress but can be comparable to an internal stress. Even a large body such as the earth has cracks, which geologists call fault lines. When internal forces (residual stresses) cause uneven stick-slip movement along these fault lines, the adjacent plates slide against each other, causing earthquakes, which are the sudden release of enormous amounts of stored-up energy within the earth's crust.

It is a serious mistake to think that all residual stresses are harmful. Indeed, there are a number of manufacturing processes in which the sole purpose is to introduce beneficial residual stress patterns in critical parts. These processes include case hardening, shot peening, and surface rolling. On the other hand, there are manufacturing processes that must be carefully controlled in order to prevent unfavorable residual stresses; these processes include grinding, welding, and some machining operations.

In general, it is usually desirable to have high compressive residual stresses at the surface of parts that are subject to fatigue, stress corrosion, and fretting. However, one should be aware that there must be balancing tensile residual stresses within the part. In certain circumstances, these internal tensile residual stresses can be a problem.

There are several fundamental facts that must be understood about residual stresses:

- Residual stress systems are balanced. That is, if one part of the system is altered, the rest of the system will change or adjust to maintain the balance. This change or adjustment results in distortion, or dimensional change, of the part involved. This distortion can be used to estimate the magnitude and direction of the residual stresses. This static equilibrium is analogous to that of a boat floating in the water; if additional weight is added to the boat, it adjusts automatically by sinking slightly in the water until an equivalent weight of water is displaced.
- Residual stress systems are three-dimensional. For instance, in a shaft we can think of residual stresses in the longitudinal, circumferential (also called tangential or hoop), and radial directions. In a flat surface, such as in a sheet or plate, they are in the longitudinal, transverse, and thickness directions. In most cases, one of the directions is of little importance and can be ignored, such as the radial, or thickness, stresses noted above.
- Residual stress systems are described in terms of tensile and compressive stresses, although inevitably there must also be shear stress components. Since they are balanced systems of forces, each tensile residual stress (creating a tensile force) must be balanced by an equal and opposite compressive residual stress (creating a compressive force). This is a corollary to Newton's third law of motion: every force must have an equal and opposite reactive or balancing force.
- Residual stress systems may be described in three scales of magnitude: macro, micro, and lattice.

  a. *Macroscale:* This scale encompasses the entire cross section of a part. If the areas near the surface are in residual compression, for example, the areas near the center must be in residual tension to balance the system of forces. This is the scale usually considered for engineering purposes.
  b. *Microscale:* This scale is used in consideration of the stresses in an individual grain, or group of grains. The grains are affected by the

macrostress field in which they are located, but each grain is oriented randomly and has microdefects different from those of its neighboring grains. Residual stresses in clusters of grains are averaged when measured by the x-ray diffraction method of quantitative measurement.

c. *Lattice scale:* Since each grain is composed of a three-dimensional lattice structure of atoms, the distortions of the lattices in certain directions are actually measured by the x-ray method. The lattices are, in effect, submicroscopic strain gages. While lattice distortions can be measured on the nanometer scale using advanced transmission electron microscopy techniques, usually they are treated statistically and averaged on the microscale and macroscale.

- Residual stress systems are affected by foreign atoms that are introduced into the lattice structure. During heat treatment of steel, such as in carburizing, carbonitriding, nitrocarburizing, and nitriding, carbon and/or nitrogen atoms are deliberately diffused into the surface and near-surface regions at elevated temperatures. Because these atoms occupy space in the lattice, they tend to produce desirable compressive residual stresses parallel to the surface. Indeed, this is one of the major reasons for the use of these surface heat treatments, in addition to the desire for increased hardness, which provides enhanced strength and wear properties.
- For best resistance to fatigue fracture, the surface areas should have compressive residual stresses in directions that are perpendicular to the expected fatigue-crack direction, provided that the maximum applied tensile stresses are expected to be at the surface, as in the cases of bending and torsional loading. In other words, compressive residual stresses can be used to neutralize, or counteract, potentially damaging tensile applied stresses. Similarly, tensile residual stresses on or near the surface of most parts should be avoided because they will add to the tensile stresses from service and may cause premature failure.
- Residual stress systems may be formed, and altered, by many manufacturing processes and service conditions such as those that cause thermal, metallurgical or transformational, mechanical, or chemical changes in the metal. This is significant because nearly all processes and treatments in manufacturing—and many service conditions—have the distinct possibility of affecting the residual stress system for better or for worse.

Table 1 lists some of the many ways in which residual stresses can be altered in a given metal. Careful examination of this list reveals that nearly everything that can be done to a metal has the potential to affect the residual stress pattern.

Table 1 Factors that can affect residual stresses

Thermal action	Mechanical action	Chemical action
Heat treatment	Machining, grinding, and polishing	Etching
Stress relieving	Mechanical surface treatments	Corrosion
Annealing	Shot peening	Chemical machining
Hardening	Surface rolling	Surface coating and plating
Tempering	Hammer peening	
Diffusion treatment	Ballizing	
Carburizing	Cold forming	
Carbonitriding	Stretching	
Nitrocarburizing	Drawing	
Cyaniding	Upsetting	
Nitriding	Bending and straightening	
Decarburizing	Twisting	
Fabrication with heat	Autofrettage	
Welding	Interference fitting	
Flame cutting	Service overloads	
Hot forming	Explosive stressing	
Casting	Cyclic stressing	
Shrink fitting	Wear, chafing, bruising, gouging, and cracking	
Operation at elevated temperatures		
Electrical discharge machining		

Source: modified from Ref 4

The implications of these factors are explored in later sections. First, let us study the basic mechanisms of residual stress formation: thermal, transformational, mechanical, and chemical.

Thermal Residual Stresses

Thermal residual stresses are caused primarily by differential expansion when a material is heated and contraction when cooled. The changes that occur in the material as a result of misfit cause thermally induced plastic deformation, which produces residual stresses. Thermal residual stresses can especially be problematic when two materials having greatly different coefficients of expansion are joined together, which is often the case in composites. However, for purposes of the discussion below, only metallic single-phase materials are considered.

Formation of thermal residual stresses is the result of two factors: heat (including lack of heat, or refrigeration) and restraint. Both the thermal and restraint factors must be present to generate residual stresses or to affect the residual stress pattern.

A fundamental principle for understanding thermal residual stresses is as follows: *The area of the metal that cools last is in residual tension* (*if there is no hardening transformation*) (Ref 5). This principle should be memorized if an understanding of thermal residual stresses is desired.

Thermal residual stresses may be better understood by considering these examples:

- A metal bar expands in all directions when heated uniformly to a subcritical temperature, as shown in Fig. 2(a). That is, its volume increases

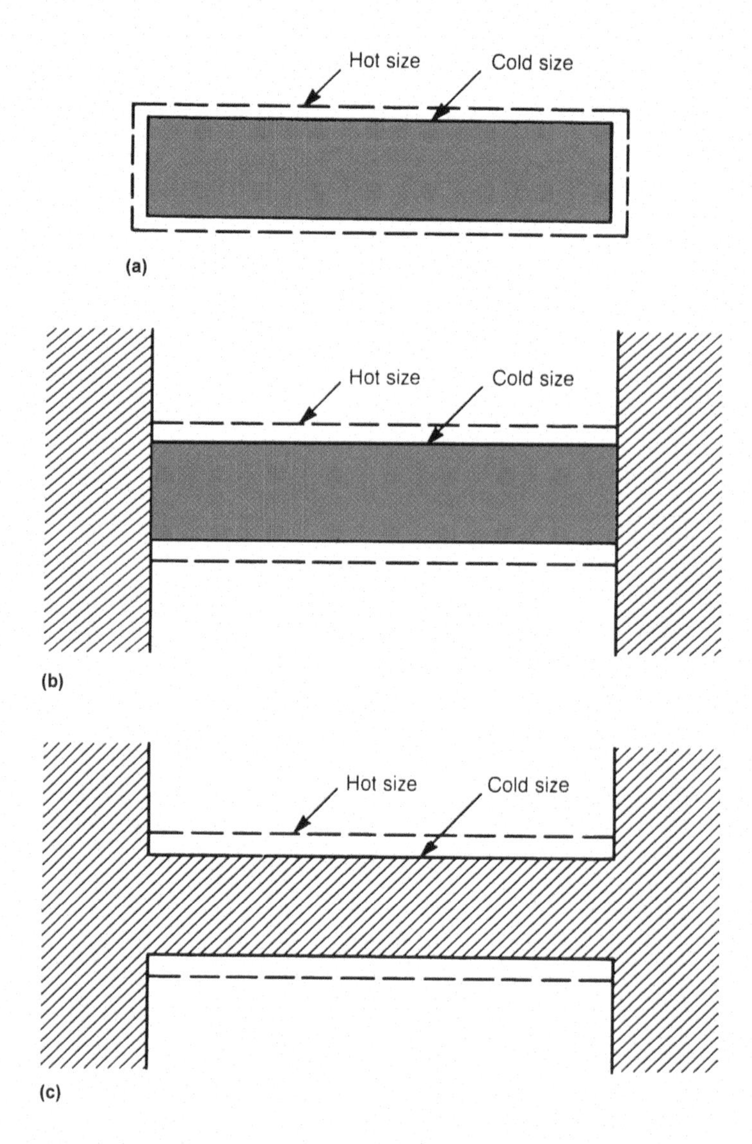

Fig. 2 Thermal residual stresses. (a) Unrestrained expansion and contraction. (b) Restrained expansion, unrestrained contraction. (c) Restrained expansion and contraction.

as a result of thermal expansion, as is shown by the dashed lines. Assuming that there are no hardening transformations or environmental effects, the bar will shrink back to the original shape and size when it cools uniformly to its original ambient temperature. Because heat—but not restraint—was present, no residual stresses were generated.

- If an identical metal bar is gently held between abutments (as in a vise) and similarly uniformly heated without heat transmittal to the abutments, the metal bar will again expand to the same volume that it had in the first example, as shown in Fig. 2(b). However, since it is restrained longitudinally, it cannot increase in length. Therefore, at the elevated temperature, it yields compressively (or is "upset") because the modulus of elasticity and the yield strength are lower at the elevated temperature. Because it must attain the same volume as the bar in Fig. 2(a), its transverse size when hot must exceed that in the first example. When cooled to its original temperature, it becomes shorter than it was originally and falls from between the abutments, or vise. As in the first example, no residual stresses are generated because the bar was restrained during heating but not during cooling.

- Now let us make the bar an integral part of the abutments, as shown in Fig. 2(c). It is now the same as in Fig. 2(b) except that there are no joints, or interfaces, at the ends of the bar. Assume that no heat is transmitted past the ends of the bar section. If the bar is heated to the same elevated temperature, it expands and yields compressively as it did previously, but during cooling it reacts differently. As it cools to the original temperature, it shrinks and tries to fall out of the abutments, as it did before. Now, however, it is axially restrained; therefore, a tensile residual stress is generated as the bar portion tries to contract away from the abutments. Since the bar is the only part that is heated, it is the last to cool, thus generating tensile residual stresses when it is restrained from dimensional change in both directions. If the tensile residual stress exceeds the tensile strength, the bar may fracture if made from a brittle metal.

To understand distortion resulting from thermal residual stresses, consider a stress-free plate, as shown in Fig. 3(a). Assume that the upper portion is uniformly heated, with no heat transfer to the lower part. The heated part expands in all directions—length, width, and thickness—but because it is restrained from free lateral expansion by the cold, strong lower portion, it causes bowing distortion, as in Fig. 3(b). The upper (convex) surface is compressively stressed at this time because of the restrained expansion. This compression of the hot, weakened metal causes it to exceed its compressive yield strength; again, in effect, it is upset, or compressed, parallel to the plate. The lower (concave) surface also is compressively stressed at this time simply because it has been forced into a concave shape. To maintain equilibrium, there must be a balancing tensile stress in the interior.

When the heated upper portion is allowed to cool to its original temperature and the thermal gradient with the lower portion disappears, the plate distorts as shown in Fig. 3(c). The reason for this reversal is that the lateral compressive plastic deformation of the upper layer when hot causes it to be shorter than it was originally at the ambient temperature. As it shrinks during cooling, it bows the lower portion as shown, forming a partial spherical shape. This shrinkage causes tensile residual stresses on the upper, formerly hot layer, because this is the last part to cool. The

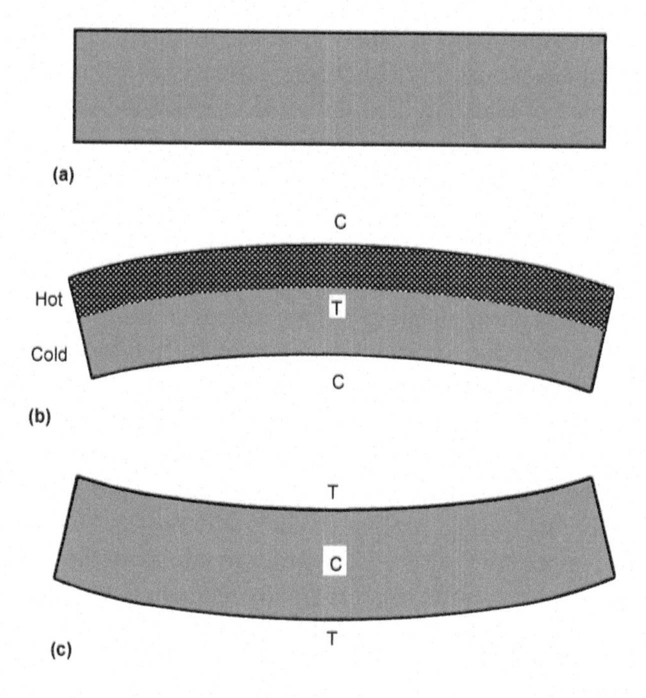

Fig. 3 Deformation caused by thermal residual stresses. (a) Flat, platelike metal at uniform temperature. (b) Lateral expansion of upper part on heating is restrained by cold, strong metal below, causing compressive stress (C) on upper (convex) and lower (concave) surfaces and tensile stress (T) in the interior. (c) Reversal of stresses on cooling.

lower (convex) surface also is stressed in tension simply because it was forced to be a convex surface, which forms tensile stresses, whether distorted by thermal or mechanical forces. To maintain equilibrium, there must be balancing compressive stress in the interior. One thing to note is that if the stresses induced by the thermal gradient stayed below the yield limit, there would be no permanent distortion. The plastic strains that occur when the material exceeds the yield limit are what are driving the distortion and residual stress.

In this example the stresses in the thickness direction are ignored for simplicity, because thickness expansion and contraction were unrestrained; therefore, they are insignificant and irrelevant in this discussion.

Now consider a thin, flat plate or sheet (Fig. 4a). When locally and rapidly heated, as in Fig. 4(b), the heated metal tries to expand laterally. However, it is restrained by the cold, strong metal surrounding the heated area; thus, as before, the weaker heated metal yields compressively. Shrinkage upon cooling (Fig. 4c) causes the formerly hot metal to generate tensile residual stresses parallel to the surface of the plate. It also contracts in the thickness direction, forming a slight, gentle cavity where it had been heated. This is the principle involved in the flame-straightening technique used for thin, flat metal (Ref 6).

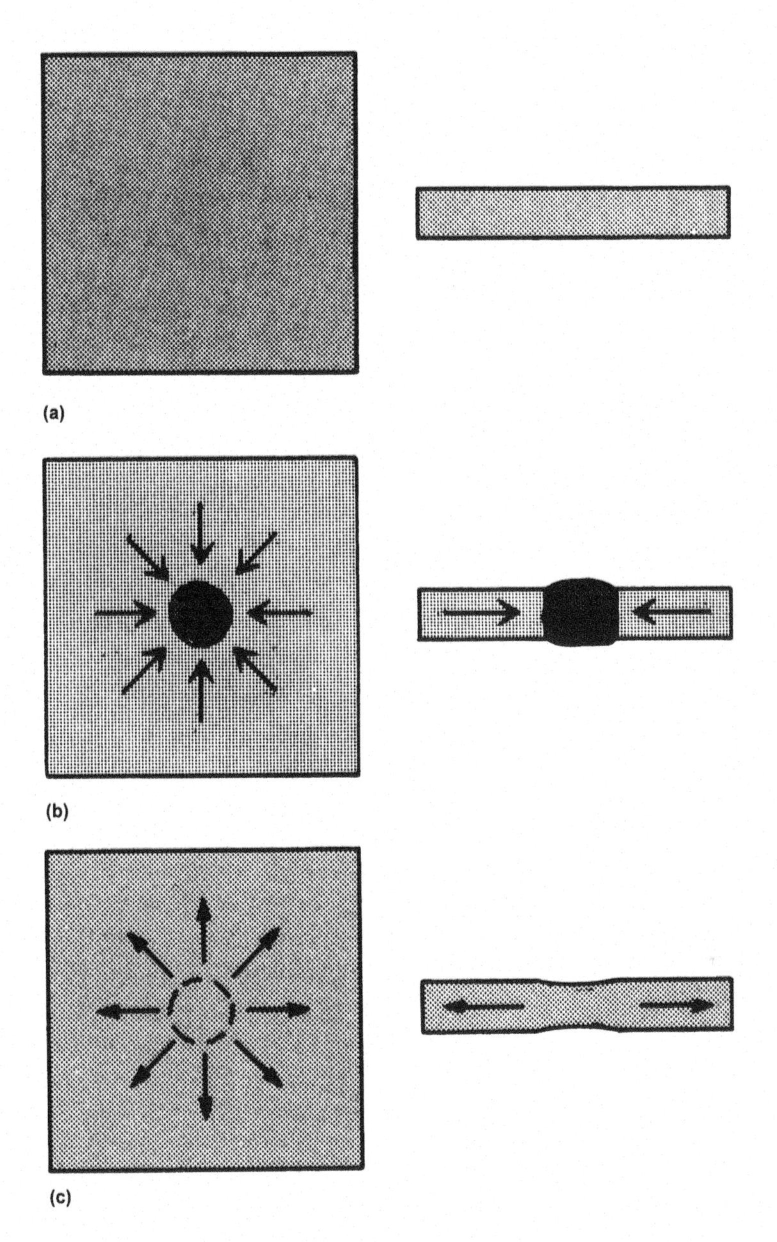

Fig. 4 Thermal residual stresses caused by spot heating. (a) Stress-free plate or sheet at uniform temperature. (b) When locally through-heated, plate expands laterally, generating compressive stresses; also bulges in thickness direction. (c) When cooled to original temperature, plate contracts laterally, generating tensile stresses; also contracts in thickness direction

Distortion and/or tensile residual stresses in weldments, particularly those involving large, boxlike, closed structures, can be a major source of problems. This was a contributing factor to brittle fractures of merchant ships during World War II, as discussed in Chapter 8, "Brittle Fracture," in this book. Careful attention must be paid to the sequence of welding, with the final welds placed at relatively harmless locations, if possible. When this is not possible, it may be necessary to thermally stress relieve the welds by localized heating, or by furnace treating the entire structure. For additional information on welding distortion, see Ref 6 and 7.

While the above discussion has been centered on metallic systems, thermally induced residual stresses also exist in ceramic systems. The classic example of thermally induced residual stress is found in tempered glass, which is processed in such a manner as to have compressive residual stresses on the surface. The compressive stress acts to close any small cracks that may develop on the surface, making the glass more resistant to impact. Introducing this favorable compressive stress is typically done by blowing cool air on the surface of the glass pane during the cooling process, causing the outside of the pane to thermally contract while the inside is still warm and has a high viscosity. Once the inside cools, this material desires to contract also but is constrained by the cooler (and now significantly harder and more viscous) outside shell. The result is that the interior of the glass pane is under tension while the outside is left with a significant compressive residual stress.

Transformational Residual Stresses

Induced residual stresses caused primarily by nonuniform expansion due to phase transformations occurring within a material are called *transformational residual stresses*. A classic example of this, which also is one of the most important industrial instances of residual stress in application, is the nonuniform expansion that occurs during the hardening process associated with the transformation of austenite to martensite in steel. The 3 to 4% volumetric expansion of martensite when it forms from austenite is responsible for these stresses.

The general principle that applies to steel is: *The area of metal that hardens (transforms) last is in residual compression.* Volumetric expansion causes compressive residual stresses when the area that expands is restrained, as illustrated by an analogy:

Imagine a small room with rigid walls into which are packed a large number of people of identical height. All are ordered to exhale so that more people may be packed in before the door is forced shut. The people stand, wedged together, all completely exhaled. Then all are ordered to inhale deeply; the expansion of all the chests creates a compressive stress, at chest level, that presses outward against the rigid walls. If the walls were nonrigid—that is, flexible—a smaller compressive stress would be generated, for the walls would bulge outward. In other words, there would be more distortion but less stress, because there is less restraint.

Because the hardening, or martensitic, transformation occurs during cooling, the general principles noted for thermal and transformational residual stresses seem to contradict each other. Both involve the regions that cool last, in one case (thermal) causing tensile residual stresses, but in the other (transformational) causing compressive residual stresses. It is difficult to imagine steel shrinking because of thermal contraction and simultaneously expanding because of expansion due to a change in crystal structure from the close packed face-centered-cubic structure of austenite to the less dense body-centered-tetragonal structure of martensite. In fact, these changes do occur at the same time in both directions during transformation to hard martensite and are responsible for much of the distortion that occurs during steel heat treating.

Keeping the general principles in mind, it is now easy to understand why deep-hardening steels, such as high-alloy and tool steels, can crack if used in relatively small parts or thin sections that are rapidly cooled, or quenched, during heat treatment. The last part to harden may be in the center of the section. If this occurs, the center will have compressive residual stresses, because this is the region that hardens last. However, the necessity for a balanced residual stress system ensures that surface areas will have tensile residual stresses, a dangerous situation that can lead to cracking of the surface if the tensile strength of the steel is exceeded by the tensile residual stress. Indeed, a part in this condition is similar to a pressure vessel (or a bomb) ready to fracture (explode) at any moment. Stress concentrations, such as seams, transverse holes, threads, and fillets, are likely sources of fracture origin.

It is for these reasons that the hardenability, or depth of hardening, of steel must be tailored carefully to the size of the part and the type of quench during heat treatment. Parts that consist of both large and small sections, or thin regions, must be treated very carefully, both in specifications of the steel and in its heat treating. Quench cracks, such as described here, are serious because they can lead to catastrophic fracture.

Residual stresses caused by the martensitic transformation and expansion discussed above are characteristic of many ferrous metals. Another type of strengthening and hardening mechanism is the precipitation-hardening reaction that occurs in certain aluminum, copper, and stainless steel alloys. This involves precipitation of a cloud of submicroscopic particles that restrict slippage of the shear planes in the lattice structure. From an engineering viewpoint, this type of strengthening tends to have little effect on residual stresses, although there may be a microeffect that can be measured with certain x-ray and electron diffraction techniques. For practical purposes, precipitation-hardening reactions seem to have little effect on residual stresses (Ref 8).

Mechanical Residual Stresses

To understand mechanical residual stresses, let us first examine the principle of the arch. As shown in Fig. 5, a stone arch is constructed in such a way that the critical joints between the inner arch stones are perpendicular to the inner contour of the arch. The greater the load of stone above the arch, the greater are the compressive forces squeezing the inner arch stones together and the more stable and secure is the structure.

The key element here is the compressive stress squeezing the inner stones together. They cannot slip because the joints are perpendicular to the compressive stress parallel to the inner contour of the arch. The same principle is used in many modern bridges and high dams. A sketch of Hoover (formerly Boulder) Dam, as in Fig. 6, shows that the curved dam,

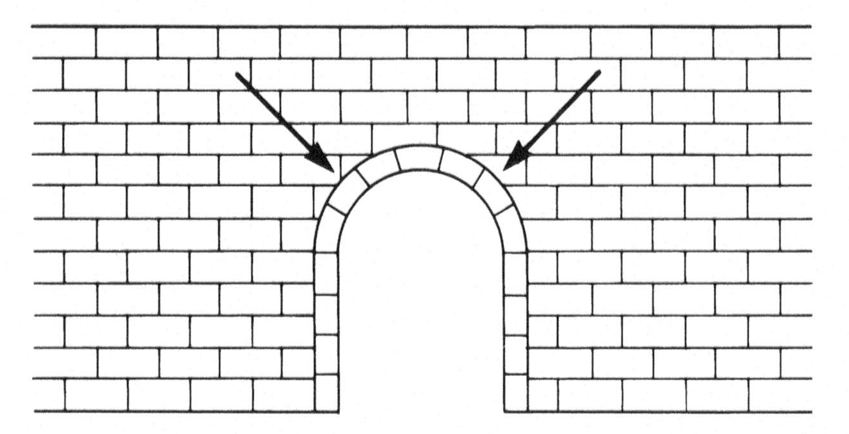

Fig. 5 The principle of the arch. Joints between stone blocks are radial, or perpendicular, to the inner surface of the arch. The greater the load above the arch, the more tightly the inner blocks are squeezed together by compressive forces, making a very stable structure.

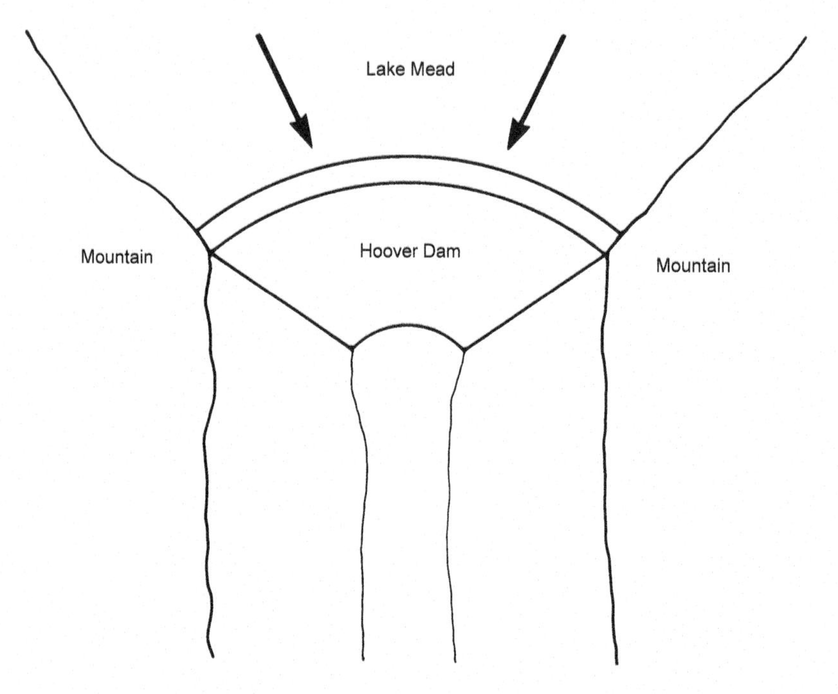

Fig. 6 Sketch of an aerial view of Hoover Dam showing arch-like construction. Again, high pressure is against the convex side.

buttressed by a mountain at each end, has the high pressure of the waters of Lake Mead trying to force the ends of the dam against the mountains, putting compressive stresses into the archlike dam. (Imagine the instability—and absurdity—of the structure if the high pressure was against the concave, rather than the convex, side of the dam!)

Now, transfer the thinking from blocks of stone or concrete to grains of metal; if the grains are squeezed together parallel to the surface, a crack cannot form to separate the grains. (Here the arch analogy breaks down slightly because the grain boundaries are not necessarily perpendicular to the surface; however, metal grains and grain boundaries have tensile and shear strength, while adjacent blocks of stone do not.)

The general principle that should be remembered about mechanically induced residual stresses is: *tensile* yielding under an applied load results in *compressive* residual stresses when the load is released, and vice versa. This simple principle is of profound significance, for it is the basis for all mechanical prestressing treatments intended to improve fatigue strength and resistance to cracking from stress corrosion and from fretting.

To illustrate this principle, Fig. 7(a) shows a very hard ball being pressed into a metal surface. At the moment of deepest penetration, the curved surface, which was originally flat, is stretched into a partially spherical shape. It has yielded in tension in all directions parallel to the surface while the load is applied. At the same time, the metal under the ball has yielded in compression in the radial directions—that is, perpendicular to the contour of the ball. These are the conditions at the moment of deepest penetration.

When the load on the ball is released, as shown in Fig. 7(b), elastic recovery takes place and the dent is forced outward into a slightly shallower depression than it had at deepest penetration. It is this elastic recovery that is of great value, for the surface of the indentation is forced into a compressive residual stress in all directions parallel to the surface of the indentation, while the internal (radial) stresses become tensile perpendicular to the surface. The most significant feature is that the compressive residual stresses force the grains of metal in the indentation to squeeze together, exactly as the stones of the arch are pressed together by the weight of the stone blocks above.

If the entire surface is mechanically dimpled with many small indentations, as shown in Fig. 7(c), it then has compressive residual stress. This is extremely useful in resisting certain types of cracking, such as from fatigue, fretting, and stress corrosion. The process just described is the common process of shot peening (Ref 9), in which a large number of small indentations are made in a surface to compressively prestress that surface. The same principles apply to other methods of mechanical prestressing (Ref 10), such as surface rolling of fillets, "ballizing" of holes, or pressing circular grooves around holes to prevent fatigue cracks from progressing from the hole. In all cases the principle is the same: *tensile* yield-

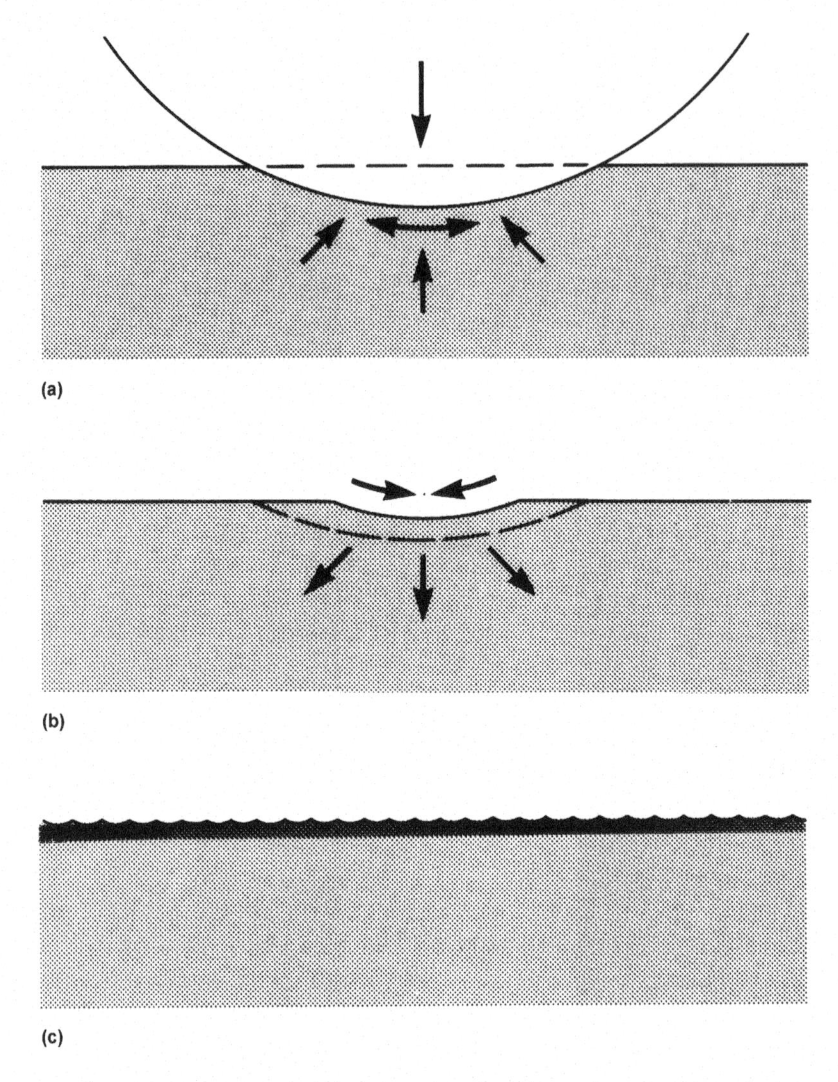

Fig. 7 Demonstration of the principle of mechanically induced residual stresses. (a) A hard ball pressed into a metal surface at point of greatest penetration. Note that the original surface (dashed line) is stretched (tension) into a spherical shape by the force on the ball. Radial reaction forces (below the ball) are compressive at this stage. (b) After the ball is removed, elastic recovery (or springback) causes a stress reversal: surface residual stresses in the cavity are now compressive, and radial reaction forces are tensile. (c) Creation of numerous small indentations in surface, as by shot peening, forms a compressive residual stress barrier that resists cracking.

ing under an applied load results in *compressive* residual stresses when the load is released, and vice versa. Mechanical prestressing also can be used as a manufacturing or salvage procedure to change the shape of metal parts, for example:

- Shot peening or laser peening can be used to form convex surfaces on relatively thin metal, such as aluminum sheet used for the skin of aircraft wings. This process is called peen forming. Long panels of aluminum are shot peened with the intensity varied to change the degree of curvature. Integral longitudinal stiffeners, or ribs, may be present on one side, which restricts the curvature to the transverse direction, rather than both longitudinal and transverse. The airfoil shape is desired primarily on the upper surface; if overcurvature occurs in a certain region, it may be corrected by peening the opposite side. As a side benefit, compressive residual stresses induced on both surfaces resist fatigue cracking during service (Ref 11).

- Surface rolling may be used in a similar manner to straighten bent or curved shafts. Selective roller pressure must be applied to the concave side of a curved shaft as it rotates on its axis. Increased compressive residual stress on the originally concave side tends to straighten the shaft. However, rolling pressure must be varied during each revolution of the shaft; pressure must be heaviest on the originally concave side and lightest on the originally convex side. This procedure is used to straighten very large shafts such as propeller shafts of ships, which commonly are uniformly surface rolled to prevent fatigue fracture. These shafts may be as large as 762 mm (30 in.) in diameter and 0.4 m (40 ft) or more in length. However, if they are curved for any reason, the selective roller pressure process may be used to straighten the shafts (Ref 12, 13).

Chemical Effects on Residual Stresses

Table 1 shows that certain chemical treatments are among the many factors that can affect residual stresses. Various methods of controlled corrosion, particularly etching and chemical machining, tend to change the residual stress pattern by removing metal from the surface. This permits the underlying metal to distort, if the balance of residual stresses is changed, or sometimes to crack, if high tensile residual stresses are present.

Surface coatings and platings also can affect the residual stress pattern, generally causing tensile residual stresses in the plated metals. These may cause cracking to occur in the plated metal, and possibly in the base metal being plated. Hard chromium plating, in particular, can be severely damaging to hardened steels. To address this problem, Federal Specification QQ-C-320B (Amendment 1), for example, requires steel parts harder than 40 HRC to be shot peened in accordance with MIL-S-13165 prior to electroplating and then baked for at least one hour at a temperature of 175 to 205 °C (350 to 400 °F), depending on section thickness, immediately after plating. Shot peening is intended to help prevent fatigue fracture; baking diffuses hydrogen out of the metal (Ref 14) and thus helps to prevent fracture due to hydrogen embrittlement.

As mentioned earlier, shot peening is also extremely effective in preventing stress-corrosion cracking, in which a tensile stress must be present with a metal in a susceptible environment. See Chapter 13, "Aqueous Corrosion Failures," in this book for further discussion of stress-corrosion cracking.

Helpful Hints

It is sometimes difficult to visualize residual stresses in materials because they are internal and are not obvious. However, the concept is greatly simplified if one visualizes tension and compression springs to represent tensile and compressive residual stresses, respectively. This spring analogy was first proposed by Heyn (Ref 15) and vividly illustrated by Baldwin (Ref 16). When presented with complex parts or difficult shapes, an attempt should be made to reduce the problem to the simplest, most basic shape that can be modeled using the spring analogy.

Illustrations of this spring analogy are shown in Fig. 8. Figure 8(a) shows a balanced, stable system consisting of two end blocks held together by a tension spring (or rubber band), while they are kept apart by two compression springs (or rubber pads). As shown in the sketch on the right, this represents a desirable system, with compressive residual stress on the outside and a balancing tensile residual stress in the interior.

Figure 8(c) is the reverse of the above, representing undesirable tensile residual stresses on the outside and a balancing compressive residual stress inside.

Each of these models may be rotated about the central axis to generate, in imagination, a bar or cylinder, or they may be moved laterally to generate, in imagination, a flat plate. One can then visualize that dimensional changes (distortion) will result when part of the structure is removed by machining or chemical removal. Indeed, these are the principles behind the dissection method of residual stress measurement.

For example, if part of the compressively residual-stressed surface of a bar is machined off, the resulting smaller-diameter bar tends to get shorter as the internal tensile residual stress is relieved. If, however, the center is bored out, then the resulting tube becomes longer than the original bar because of the relaxation of the compressive residual stresses. These examples refer to the stresses in the longitudinal direction; other dimensional changes also take place in the tangential, or circumferential, direction and in the radial direction as those residual stresses are relieved.

Similarly, if a plate with compressive residual stresses on the surfaces and balancing tensile residual stresses inside is fastened down to a flat table and the upper surface is machined off, the ends of the plate tend to bow upward when the plate is released from the table. The reverse is true if the plate originally had tensile residual stresses on the surface and compressive residual stresses in the interior.

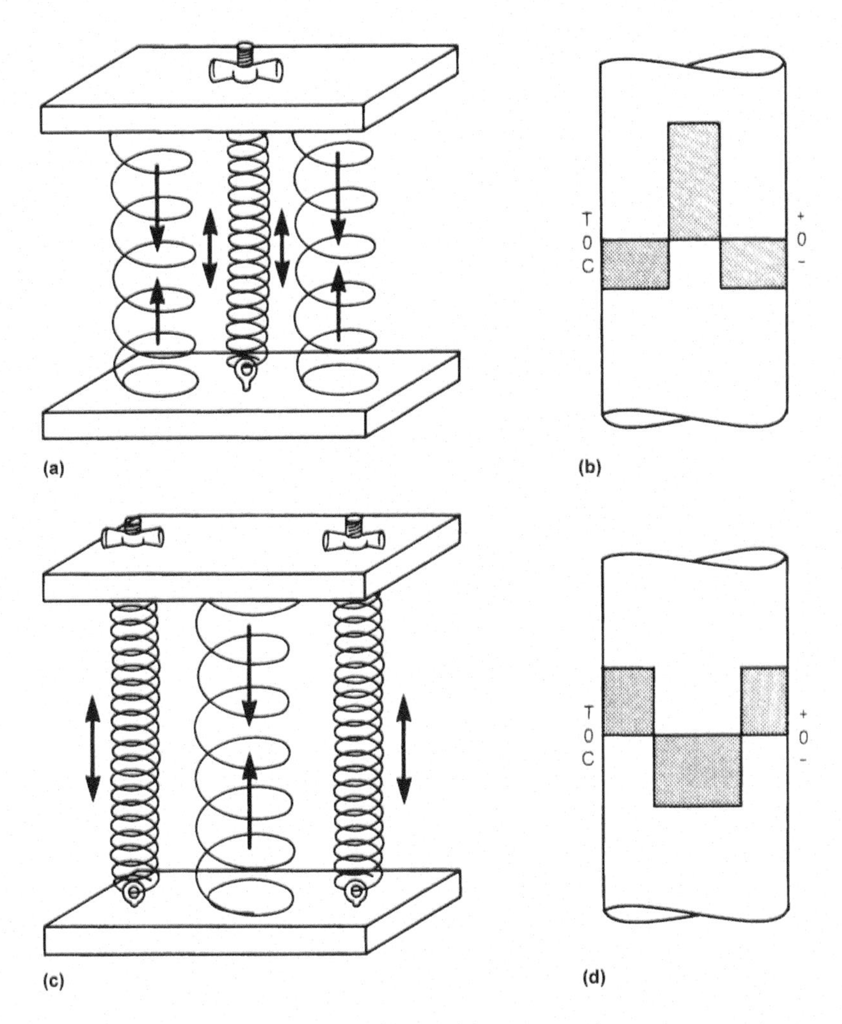

Fig. 8 Residual stress systems illustrated by spring analogy, (a) compressive stresses outside, tensile stress inside; (c) tensile stresses outside, compressive stress inside; (b) and (d) diagrams of corresponding residual stress systems. T, tension. C, compression.

Also, it helps to think of internal stresses in other closed systems to get a better feel for the elusive residual stresses. Two types of wheels may be used for examples:

- A bicycle wheel is a very light, strong structure because of the way in which the components are stressed. Referring to Fig. 9(a), if the wire spokes are assumed to be radial, tightening the spokes causes tensile radial stresses. These spokes pull the rim inward, stressing it in circumferential compression, while pulling the tubular hub outward, trying to pull it apart by circumferential tensile stresses.

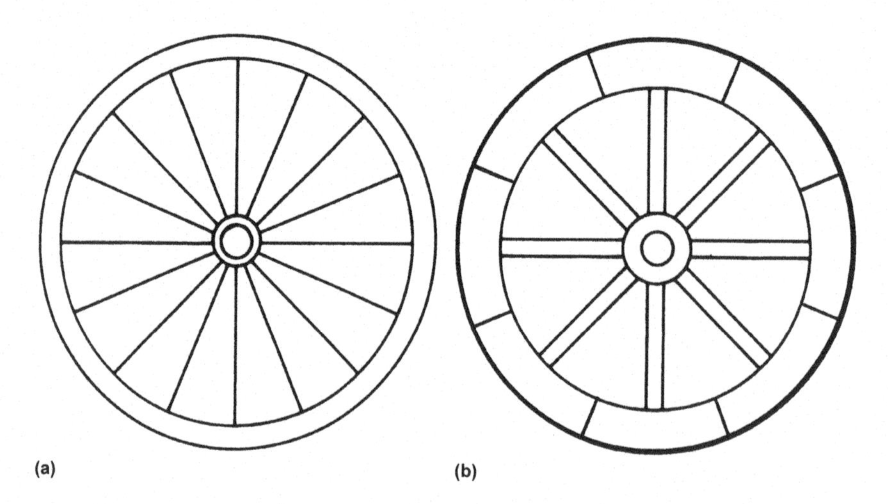

Fig. 9 Closed stress systems in two types of wheel. (a) Light, strong bicycle wheel (radial wire spokes) has tensile stressed spokes, compressively stressed rim (hoop stresses), and tensile hoop stresses in the hub. Longitudinal (axial) stresses are negligible. (b) Heavy wooden wagon wheel with a shrunk-on steel tire generates tensile hoop stresses in the tire, compressive radial and hoop stresses in the wooden rim, compressive radial stresses in the thick wooden spokes, and compressive hoop stresses in the hub. Longitudinal (axial) stresses are negligible.

- Conversely, an old wooden wagon wheel, Fig. 9(b), with a shrunk-on steel "tire" has a stress system just the opposite of that of the bicycle wheel. The shrinking of the steel "tire" creates tensile stresses in the steel, while it squeezes inward (radially) with compressive stresses in the wooden rim to press the segments together and in the large wooden spokes. The hub, in turn, also is squeezed compressively by the radial forces from the spokes. Note that all the wood is in compression.

In both examples, longitudinal residual stresses, axial to the shafts of the wheels, are negligible and can be ignored.

Summary

Residual, or "locked-in" internal, stresses are regions of misfit within a metal part or assembly that can cause distortion and fracture just as can the more obvious applied, or service, stresses. Residual stresses may be either extremely dangerous or highly beneficial, depending on their pattern and magnitude.

Residual stresses can be affected by virtually all manufacturing processes and by many service conditions. The principal factors that cause residual stresses are thermal, transformational, and mechanical imbalances within the metal.

Again, repeating the general principles:

- Thermal residual stresses are induced because the material that cools last is in residual tension (if there is no hardening transformation).
- Transformational residual stresses are induced in steels because the metal that hardens (i.e., transforms to martensite) last is in residual compression.
- Mechanical residual stresses are induced because *tensile* yielding under an applied load results in *compressive* residual stresses when the load is released, and vice versa.

Visualization and understanding of residual stresses can be aided if they are considered to be balanced systems made of tension or compression springs.

REFERENCES

1. F. Campus, Effects of Residual Stresses on the Behavior of Structures, *Residual Stresses in Metals and Metal Construction,* W.R. Osgood, Ed., Reinhold, 1954, p 1–21
2. *Failure Analysis and Prevention,* Vol 11, *ASM Handbook,* ASM International, 1986, p 8
3. D. Rosenthal, Influence of Residual Stress on Fatigue, *Metal Fatigue,* G. Sines and J.L. Waisman, Ed., McGraw-Hill, 1959, p 170–196
4. D.G. Richards, Relief and Redistribution of Residual Stresses in Metals, *Measurement of Residual Stresses,* American Society for Metals, 1952, p 129–204
5. H.C. Fuchs, Techniques of Surface Stressing to Avoid Fatigue, *Metal Fatigue,* G. Sines and J.L. Waisman, Ed., McGraw-Hill, 1959, p 197–231
6. "Distortion . . . How to Minimize It with Sound Design Practices and Controlled Welding Procedures Plus Proven Methods for Straightening Distorted Members," Bulletin G-261, Lincoln Electric, 1977
7. Control of Shrinkage and Distortion, *Design of Welded Structures,* James F. Lincoln Arc Welding Foundation, 1966, section 7.7
8. C. Barrett and T.B. Massalski, *Structure of Metals,* 3rd ed., McGraw-Hill, 1966
9. "Test Strip, Holder, and Gage for Shot Peening," Society of Automotive Engineers Standard J442, *Surface Rolling and Other Methods for Mechanical Prestressing,* Society of Automotive Engineers Handbook Supplement J811, 1962
10. *Forming and* Forging, Vol 14, *ASM Handbook,* ASM International, 1988, p 803
11. J.E. Ancarrow and R.L. Harrington, Main Propulsion Shafting Eccentricity Considerations, *Transactions,* Society of Naval Architects and Marine Engineers, Vol 81, 1973

12. L.L. Shook and C. Long, Surface Rolling of Marine Propeller Shafting, *Transactions,* Society of Naval Architects and Marine Engineers, Vol 66, 1948, p 682–702
13. *Surface Engineering,* Vol 5, *ASM Handbook,* ASM International, 1994, p 189–190
15. E. Heyn, Internal Strains in Cold Wrought Metals and Some Troubles Caused Thereby, *J. Inst. Met. (London),* Vol 12, 1914, p 1–37
16. W.M. Baldwin, Jr., Residual Stresses in Metals, Edgar Marburg Lecture, *Proceedings,* ASTM, Vol 49, 1949, p 539–583

SELECTED REFERENCES

- J.O. Almen and P.H. Black, *Residual Stresses and Fatigue in Metals,* McGraw-Hill, 1964
- *Failure Analysis and Prevention,* Vol 11, *ASM Handbook,* ASM International, 1986, p 97–98, 112, 125–126, 141
- *Fatigue and Fracture,* Vol 19, ASM International, 1996
- *Heat Treating,* Vol 4, *ASM Handbook,* ASM International, 1991, p 854–855
- "Injury in Ground Surfaces: Detection, Causes, Prevention," booklet, Grinding Wheel Division, Norton Co., Worcester, MA 01606
- *Internal Stresses and Fatigue of Metals,* Elsevier, 1960
- D.R Koistinen, The Distribution of Residual Stresses in Carburized Cases and Their Origin, *Transactions,* American Society for Metals, 1958
- *Materials Selection and Design,* Vol 20, *ASM Handbook,* ASM International, 1997, p 811–819
- *Mechanical Testing,* Vol 8, *ASM Handbook,* ASM International, 1985
- *Properties and Selection: Irons, Steels, and High-Performance Alloys,* Vol 1, *ASM Handbook,* ASM International, 1990, p 680–683
- *Residual Stresses for Designers and Metallurgists,* American Society for Metals, 1981
- *Residual Stress Measurement by X-Ray Diffraction,* Society of Automotive Engineers Handbook Supplement J784a, 1971
- "Shot Peening," a booklet reprinted from *Mechanical Engineering Handbook,* M. Kutz, Ed., John Wiley and Sons, 1986

Understanding How Components Fail, Third Edition
Donald J. Wulpi
Edited by Brett Miller

CHAPTER 5

Distortion Failures*

THE TERM *FAILURE*, as used in this work, means the inability of a part or assembly to perform its intended function for any reason. We usually think of failure in terms of fracture, wear, or corrosion. Even in the absence of any of these three factors, however, a part can also fail when distortion of size or shape prevents the performance of its intended function (Ref 1, 2).

Distortion failures are readily identified by the inherent change in size and/or shape; however, correction of a distortion failure may be far from simple. This is because distortion encompasses details of design and structural analysis, as well as materials technology. Another complication is that distortion may result from residual stresses within the metal, as well as from applied stresses. Study of Chapter 4, "Residual Stresses," in this book will help to clarify this extremely complex phenomenon.

Distortion failures are serious because they can lead to other types of failure or may even cause complete collapse of structures, such as bridges, ladders, beams, and columns. Distortion at elevated temperatures, or creep, depends upon the interrelationship between component design and the high-temperature properties of the metal.

Types of Distortion Failure

Distortion failures may be classified in different ways. One way is to consider them either as dimensional distortion (growth or shrinkage) or as shape distortion (such as bending, twisting, or buckling). They may also be classified as being either temporary or permanent in nature.

**Understanding How Components Fail, Second Edition*, by Donald J. Wulpi, Chapter 2, Distortion Failures, ASM International, 1999, reviewed, revised and renumbered by Roch J. Shipley, Professional Analysis and Consulting.

Temporary (Elastic) Distortion

Distortion is frequently fleeting and transient. Since metals are elastic, all metal parts deflect (or distort) even under relatively low stresses. In fact, most structural parts can be considered as springs, even though they are not called springs and do not look like typical springs. Most structural parts are intended to work in the elastic range and return to their original size and shape when unloaded, which is the definition of elastic, or temporary, deformation, usually referred to as deflection, as in springs.

In certain parts, this deflection may cause interference with another part. Such interference is particularly common with gear teeth, which are essentially carefully shaped cantilever beams. Evidence that the tips of the teeth from one gear have been digging into the flank, or lower portion, of the mating gear teeth is often seen. This is a major reason for modification of the tips of gear teeth, so that they will not interfere with the mating gear under load. Examination of gears after service should include careful inspection for evidence of such interference, which could lead to more severe damage.

Consider another example of temporary distortion: a blade on a high-speed rotor in a turbine at high temperature. The faster the shaft rotates, the higher will be the centrifugal stresses tending to elongate the blade, possibly causing either fracture or interference with the outer housing. Also, the modulus of elasticity, or stiffness, of the metal decreases with increasing temperature, as discussed in Chapter 2, "Mechanical Properties," in this book. Thus, increasing both rotational velocity and increasing temperature will tend to make the blade elongate and possibly make contact with the outer housing. Evidence of such contact will be evident on both the tip of the blade and the interior of the housing.

Permanent (Plastic) Distortion

Also serious is the permanent distortion that results from yielding during service, from creep, and from buckling (or compression instability).

Yielding during Service. If a part yields or distorts permanently during service after one or more load applications, the stress on the part has obviously exceeded the yield strength (actually, the elastic limit). If the part is a spring, we say that the spring has "taken a set," indicating that the spring is permanently distorted, as shown in Fig. 1, and can no longer perform its intended function. Identification of this type of failure is quite simple and obvious.

Less simple and obvious, however, may be the means for correcting the problem so that the same type of failure will not occur on other similar parts. In performing this function, the analyst may be strained to the limit of his or her ability—tracking down the specific cause of yielding may really become a challenge. It is vital to learn, for example, if this yielding is an isolated problem or is occurring on similar parts. If it is isolated, it is

Fig. 1 Distorted engine valve spring (left) compared with normal valve spring. Improper microstructure resulted in inadequate strength and hardness at the operating temperature. Source: Ref 1, 2

necessary to learn the details of what occurred to make this specific part yield. Obviously it is necessary to measure the distorted part and compare it with a new part of the same original dimensions. Photographic comparisons are often useful (as in Fig. 1), with the new and yielded parts in the same photograph, with a scale, if necessary, to show the distortion.

Detailed study of the part from a metallurgical standpoint is essential, with attention to such questions as: Is the microstructure the same as was originally specified, or has it been altered by exposure to an elevated temperature? Has the hardness been reduced by tempering during service? If so, what temperature would have caused that reduction in hardness level? Was the temperature abnormally high for the application? If so, why?

If investigation indicates that this is only one instance of a general and widespread problem, it may be necessary to redesign the assembly to reduce the stresses and temperatures, or to obtain a material with a higher yield strength at the operating temperature.

Creep. Somewhat similar to the problem of yielding during service is the problem of creep. However, creep differs from yielding in that creep is not a single, short-time phenomenon. Creep is defined as "time-dependent strain occurring under stress" and is discussed further in Chapter 14, "Elevated-Temperature Failures," in this book. Creep behavior is dependent on the material selected for a given application. Creep manifests as gradual distortion, generally occurring over a long period at relatively high temperatures. This imprecise description is necessary because of the many complications involved in this type of distortion. However, the reader should keep in mind that, for a given material, the amount of distortion is increased by higher stresses, higher temperatures, and longer times. For each of these contributing factors, the investigator should determine how the actual service conditions compared with the expected material performance and the expectation of the designer.

The following examples illustrate creep:

- Bolts holding certain engine parts together occasionally become loose and need to be periodically retightened. This is particularly true in hotter parts, such as the exhaust manifold and other parts that absorb some heat from it. Loosening of the bolts may be caused by gradual stretching of the bolts and similar gradual relaxation of compression in the structures joined by the bolts. When the bolts are tightened, they are satisfactory for another period of time but then gradually stretch until retightened again. This process may continue for a number of repetitions until the bolts can no longer be tightened because of thread deformation.
- In a diesel engine, a cup-shaped precombustion chamber must be clamped tightly to withstand combustion pressure. After a period of time, such as many hours of engine operation, the chamber may begin to leak because the tubular side wall has bulged outward as a result of the high temperature, internal pressure, and axial compressive force. Again, if the assembly is tightened to prevent leakage, the chamber will simply continue to gradually bulge outward and to shorten in length, resulting in additional leakage.

As noted previously, although the mode of distortion failure is readily identified, the method of correction is not necessarily simple. Possible corrective measures are to use an alloy with better resistance to creep at the operating temperature, to redesign the part to provide more resistance to deformation, and/or to reduce the temperatures and pressures encountered. However, some measures may be undesirable for performance reasons; others, for economic or availability reasons.

Buckling. This type of distortion failure is defined as collapse due to compressive instability. It is most common when long, slender columns are compressed in an axial direction, or when thin-wall tubes are compressed in either an axial direction or a diagonal direction as a result of torsional loading. This type of failure also can occur on the compressive (concave) side of a member under a bending load, such as a thin-wall tube or a flange of a channel or I-beam section.

A vital fact must be recognized in considering a part that has buckled, or in preventing buckling: the load at which a component buckles does not depend on the strength of the material, but on the dimensions of the part and the modulus of elasticity of the material at the operating temperature. These factors are shown in the column formulas given in references for strength of materials and engineering design (e.g., Ref 2). This means that the buckling load cannot be increased by heat treating the metal to increase the strength and hardness or by using stronger materials.

Buckling is a critical consideration in long, slender parts that must resist compressive axial forces. Typical examples are building and scaffolding columns, engine connecting rods and push rods, and tie rods in automotive steering linkages. Compression members such as these must be straight, because any bend between the points of load application greatly reduces the ability to resist buckling. It is also important to locate the compressively loaded material on the outer edges of an axially loaded member, as a simple demonstration shows:

An ordinary 8½ × 11 inch sheet of typing paper—on edge—can support a considerable weight if the material is properly located. As shown in Fig. 2, roll the paper to form a thin-wall tube approximately 1¼ to 1½ inches in diameter and 8½ inches long. Tape the full length of the outer edge to keep the paper from unwinding. Place the paper tube vertically on a horizontal surface and carefully balance an ordinary-size book on the top. Gently add more weight. If the experiment is performed carefully, a surprising load may be supported before the column buckles, or collapses.

As noted above, parts under bending load also are subject to buckling failure on the compressive (concave) side. Channel or I-beam sections, which commonly are used as extruded or rolled members in various frames and ladder sections, must be designed carefully to maximize the metal in the flanges, where it is most effective in resisting buckling. See Fig. 3 for an example.

Another simple demonstration illustrates the effectiveness of proper shape, or configuration, in preventing buckling failure in bending: The "tape" in a metal tape measure is usually made of thin spring steel that is slightly curved. If the tape is held horizontally, as in Fig. 4, and supported at one end only to make a cantilever beam, a fairly long length can be extended if the concave surface is the upper side. However, if the tape is reversed so that the convex surface is on top, only a short length can be

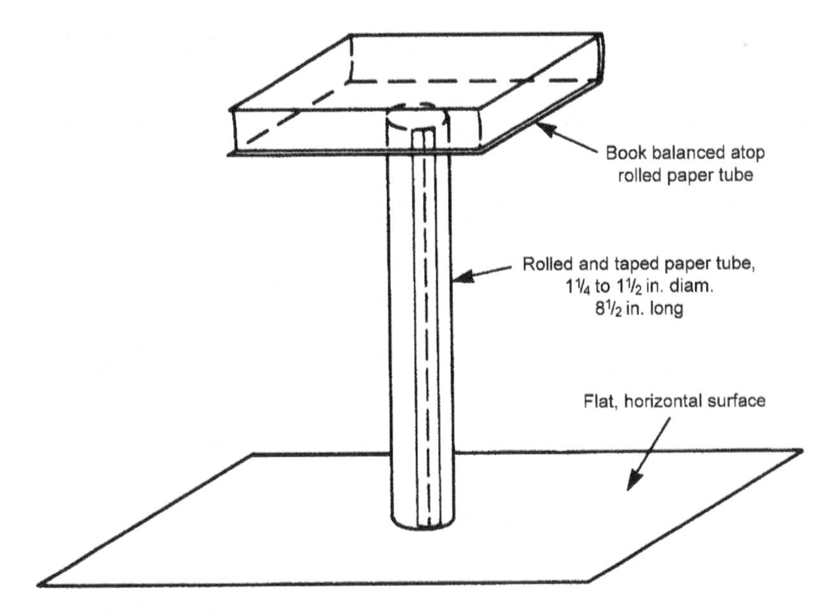

Fig. 2 Tube of ordinary typing paper supporting a balanced load. As additional weight is added to the column, the tube will eventually collapse, or buckle.

Fig. 3 Buckled flange (lower arrow) of an extruded aluminum channel section deliberately loaded with a lateral force (upper arrow). In service, the channel section is subjected primarily to axial compression, rather than the abnormal lateral force applied here.

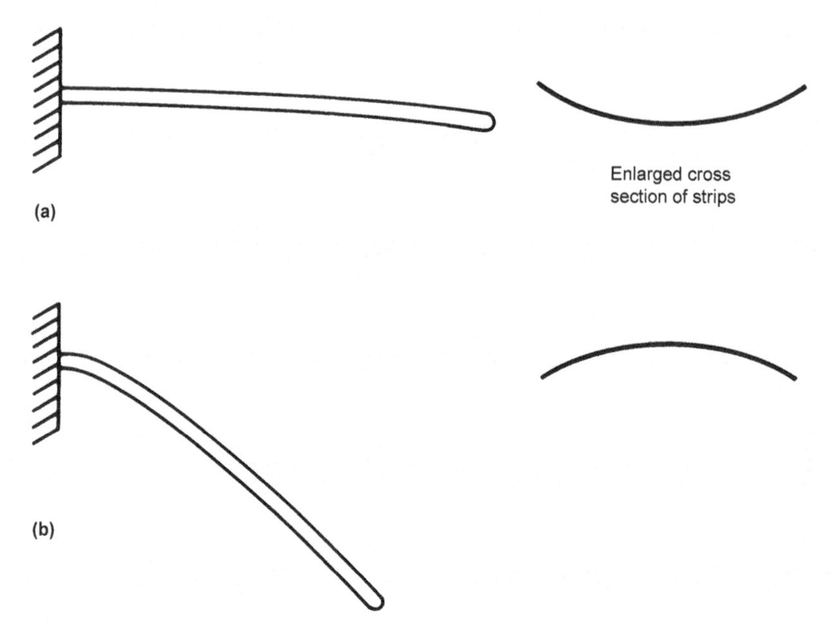

Fig. 4 Curved strip of thin spring steel, such as a tape measure or Venetian blind slat, supported horizontally as a cantilever beam (a) stable with concave side on top; (b) unstable with convex side on top, buckles readily.

extended before the tape buckles, or collapses. The same experiment can be performed with a metal Venetian blind slat, which is similarly curved. If the slat is supported at both ends as a simple beam, however, the reverse of the above is true: the slat will buckle quickly if the concave surface is on top. Because these parts are made from relatively high-strength steel, even though it is extremely thin, the distortion is only elastic, or temporary. There is no plastic, or permanent, deformation.

The purpose of these demonstrations is to emphasize the fact that buckling instability is really a geometrical problem, not a material problem. See strength-of-materials and/or design references for help with these design problems. The failure analyst must recognize that the only material property involved is the modulus of elasticity, which is essentially constant for a given metal (at a given temperature).

Summary

Many types of distortion failure are easy to identify. However, they may be difficult to correct because of limitations on materials, design, or economic factors.

Additional information relevant to the diagnosis and prevention of distortion failures can also be found in Chapters 2, 4, and 14 of this book, as well as in other design and materials references. A few of the many available references are listed in "Selected References," below.

REFERENCES

1. *Failure Analysis and Prevention,* Vol 11, *ASM Handbook,* 9th ed., ASM International, 1986, p 136–144
2. *Failure Analysis and Prevention,* Vol 11, *ASM Handbook,* 10th ed., ASM International, 2002, p 1045–1057

SELECTED REFERENCES

- F.P. Beer, E.R. Johnson, Jr., and J.T. DeWolf, *Mechanics of Materials,* 4th ed., McGraw-Hill, NY, 2006
- J.R. Davis, Ed., *ASM Specialty Handbook: Heat Resistant Materials,* ASM International, 1997
- N.E. Dowling, *Mechanical Behavior of Materials: Engineering Methods for Deformation, Fracture, and Fatigue,* 2nd ed., Prentice Hall, 1999
- R.W. Hertzberg, *Deformation and Fracture Mechanics of Engineering Materials,* 4th ed., Wiley, 1995
- R. Viswanathan, *Damage Mechanisms and Life Assessment of High Temperature Components,* ASM International, 1989

Understanding How Components Fail, Third Edition
Donald J. Wulpi
Edited by Brett Miller

CHAPTER 6

Basic Single-Load Fracture Modes*

FROM A FUNDAMENTAL STANDPOINT, there are only two modes, or ways, in which metals can fracture under single, or monotonic, loads: shear and cleavage. These two modes differ primarily in the way in which the basic metal crystal structure behaves under load.

Because most engineering metals at room temperature have either body-centered cubic crystals (primarily iron and its alloys, such as most steels) or face-centered cubic crystals (primarily aluminum and austenitic stainless steels), this discussion covers only these atomic crystal cell structures. Though metals with other crystal structures (hexagonal close-packed, tetragonal, etc.) fracture in somewhat similar ways, their fracture modes are more difficult to describe and are not covered here.

First, it must be recognized that almost all commercial solid metals are polycrystalline (i.e., having many crystals) in structure. Each individual crystal, or grain, is a structure composed of a very large number of atoms of the constituent elements. These atoms are arranged in cells within each crystal in a regular, repetitive three-dimensional pattern. For many years it was incorrectly thought that brittle fractures and fatigue fractures occurred because the metal had "crystallized." Understandably, this usage arose because certain types of fractures are bright and sparkling when they first occur and do appear to be crystalline. In fact, however, the metals were crystalline when they solidified from the liquid, or molten, state; the crystalline appearance reflects the type of fracture, not the cause of fracture.

Figure 1 illustrates the simplest, basic unit cell of a body-centered cubic structure, such as that of iron or steel. The cell is composed of eight atoms

---

**Understanding How Components Fail, Second Edition*, by Donald J. Wulpi, Chapter 3, Basic Single-Load Fracture Modes, ASM International, 1999, reviewed, revised and renumbered by Brett A. Miller, IMR Metallurgical Services-Louisville.

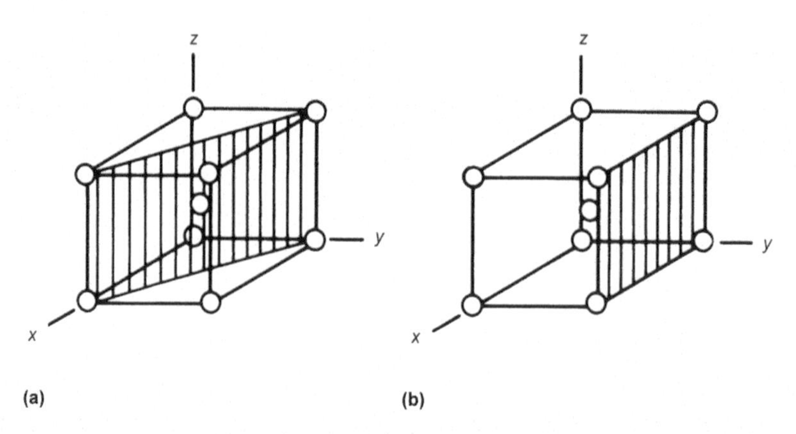

Fig. 1 Basic unit cell of a body-centered cubic structure, shaded to show (a) the diagonal plane along which shear deformation and eventual fracture can occur and (b) one face of the cell, the site of the cleavage mode of fracture

at the corners, plus another at the center of the cell. It must be recognized that this is only one of a very large number of similarly oriented cells in a given crystal. Adjacent cells share the corner atoms. The lines connecting the atoms represent the electrical forces that stabilize the position of the atoms by balanced forces of attraction and repulsion.

Shear Mode

The unit cell shown as Fig. 1(a) is shaded along a diagonal plane, which includes the center atom. This represents a shear plane by which the cell can be deformed by forces that move the two pairs of atoms at the end of the plane, or by forces that move the two pairs of atoms not in the diagonal plane toward the plane. Either of these forces can cause distortion of the cell and of all the other cells, which are similarly oriented, in an individual crystal or grain. This distortion tends to either lengthen or shorten the diagonal plane, changing the upper and lower cube faces from squares to parallelograms.

Shear deformation, then, actually represents a sliding action on planes of atoms in crystals. This sliding is analogous to that which occurs when a deck of cards is bent or otherwise moved laterally so that each card slides against its neighbors.

In a polycrystalline metal, slight deformation causes no permanent change in shape; it is called *elastic deformation.* That is, the metal returns to its original size and shape, like a spring, after being loaded. If a greater load is imposed, permanent (or plastic) deformation occurs because of irreversible slip between certain planes of atoms that make up the crystal structure.

If the force is continued until fracture occurs, the shear deformation causes tiny microvoids to form in the most highly stressed region; these tiny voids soon coalesce, or interconnect, and form fracture surfaces that have many half-voids, or dimples, on each side of the fracture face. This is the "dimpled rupture" fracture surface that is visible by electron microscope examination, as shown in Chapter 9, "Ductile Fracture," in this book.

Note that the shear deformation discussed herein refers to slip or sliding on a microscopic or submicroscopic scale within the metal crystals. It is slip or sliding on a scale of magnitude vastly different from the gross direct, or transverse, shear discussed in Chapter 3, "Stress versus Strength," in this book.

Cleavage Mode

Basically different is the cleavage mode of separation of the basic cell. Separation occurs very suddenly between one face of the cell (shaded in Fig. 1b) and the mating face of the adjacent cell. No deformation is present, at least not on a visible macroscale. The sudden separation is analogous to that of a rubber suction cup being pulled from a smooth surface: the suction cup remains fixed until the separating force exceeds the force of the atmospheric pressure holding the cup in place, and then it pops off abruptly.

In a polycrystalline metal, cleavage fracture usually occurs in relatively hard, strong metals, although under certain conditions—such as at lower temperatures—metals that normally fracture in the shear mode may fracture in the cleavage mode. However, metals with the face-centered cubic system, such as aluminum and austenitic stainless steels, do not fracture by cleavage. This is discussed in some detail in Chapter 8, "Brittle Fracture," in this book.

On a microfractographic scale, cleavage fractures occur along the faces of the cells but are seen as a splitting of the grains, with no relation to the grain boundaries. This is analogous to fracture through (not between) the bricks in a brick wall. In many cases the direction that the crack follows can be determined from study of the cleavage fracture faces under the electron microscope.

The differences between the shear and cleavage fracture modes are summarized in Table 1. When examined under an electron microscope, fracture surfaces are seldom entirely dimpled rupture or entirely cleavage. Depending upon the metal and its characteristics, areas of both fracture modes are often seen, even though either one may dominate. This is because each crystal—or grain—is an individual; accordingly, it may react differently to the separating force from its neighbors. Each grain is oriented differently, and each has its own weaknesses and microdefects, which are always pres-

Table 1 Differences between shear and cleavage fracture modes

Factor	Shear	Cleavage
Movement	Sliding	Snapping apart
Occurrence	Gradual	Sudden
Deformation	Yes	No
Behavior	Ductile	Brittle
Visual appearance of fracture	Dull and fibrous	Bright and sparkling
Microfractography	Dimpled rupture	Cleavage

ent. As with people, we are all basically similar, but we all have our own individual strengths and weaknesses that make us react to stress in ways that may be different from our neighbors' reactions.

Other Fracture Modes

The reader may be aware that there are fracture modes other than shear and cleavage. These include intergranular and quasi-cleavage fracture modes for single-load applications, and fatigue for multiple-load applications. However, they do not have their origins in basic cell structure, as shear and cleavage do.

For completeness, each of these other fracture modes are discussed briefly here and covered in more detail in other chapters.

Intergranular fractures can occur if the grain boundaries are weaker than the grains themselves. Fracture then occurs between the grains rather than through the grains, resulting in fracture surfaces that, under relatively high magnification, reveal the sides of the grains in many areas. This is analogous to fracture between the bricks (through the mortar) of a brick wall rather than through the bricks themselves. Intergranular fractures are usually caused by environmental factors such as hydrogen absorption, contact with liquid metals, high temperatures, and certain corrodents when the metal is under tensile stress, as well as any other mechanism that weakens the grain boundaries. Refer to Chapter 8, "Brittle Fracture," in this book for more information.

Quasi-cleavage fractures are seen frequently in electron microscopic examination of quenched and tempered steels. This mode is considered to be a combination of the shear and cleavage fracture modes because microdimples appear to be present on cleavage fracture planes. See Chapter 8, "Brittle Fracture," in this book.

Fatigue fracture is, of course, caused not by single-load applications, as are the others mentioned previously, but by the cumulative effect of a large number of load applications at stresses insufficient to cause fracture with one load application. The repetitive forces—which may reach into the thousands or millions of cycles (or even more) prior to fracture—exert a shearing or twisting action on the crystal structure that tends to cause the inevitable structural microdefects in the crystals to join together until a minute crack develops in certain vulnerable crystals. The repetitive stressing automatically locates the crystals with the weakest orientation in the regions of highest stress. Continued cycling causes the crack, or cracks, to enlarge, gradually becoming deeper at an increasing rate of growth. As the crack depth increases, the load-bearing capacity of the remaining metal ligament decreases. See Chapter 10, "Fatigue Fracture," in this book for more information.

Factors Affecting the Ductile Brittle Relationship

Several factors determine whether a metal will behave in a predominantly ductile manner, or whether it will be mostly brittle. These are trends only—generalities that are subject to many complications and qualifications for specific applications. However, they should be recognized as trends, always on an "other things being equal" basis. These factors and "most likely" trends are summarized in Table 2.

Temperature. With virtually all metals, ductile behavior is more likely at higher temperatures, unless there is some complicating environmental factor. Conversely, brittle behavior is more likely at lower temperatures, particularly with body-centered cubic metals, such as most ferrous metals.

Rate of Loading. Lower rates of loading tend to promote ductile behavior, allowing time for shear to occur on the crystallographic planes. Higher rates of loading tend to promote brittle behavior in many metals. This is because the metal appears to be less ductile because plastic deformation takes time; therefore, the shorter the time in which the load is applied, the less plastic deformation can occur.

Geometry. If there is no severe notch or stress concentration, shear deformation can occur, because the areas of highest stress can share the load and adjust gradually, promoting ductile behavior. A severe notch or stress concentration does not permit this adjustment of load; consequently, brittle fracture is more likely.

Size. For both metallurgical and geometrical reasons, smaller or thinner sections usually are more likely to have ductile behavior. Conversely, larger, thicker sections are more likely to have brittle behavior, partly because there is a higher probability that serious discontinuities—stress con-

Table 2 General trends of ductile brittle behavior

	Trend	
Factor	Ductile	Brittle
Temperature	Higher	Lower
Rate of loading	Lower	Higher
Geometry	No stress concentration	Stress concentration
Size	Smaller or thinner	Larger or thicker
Type of loading	Torsion	Tension or compression
Pressure (hydrostatic)	Higher	Lower
Strength of metal	Lower	Higher

centrations—will be present in the larger, thicker sections. Also, triaxial tensile stresses, which promote brittle fracture, are more likely in large sections.

Type of Loading. At a given strength level, a shaft loaded in torsion is more likely to have ductile behavior. The same shaft loaded in tension or compression is more likely to have brittle behavior. This effect is particularly noticeable at relatively high hardness and strength levels.

Pressure (Hydrostatic). A normally brittle material completely immersed in a pressure vessel under extremely high hydrostatic pressure is more likely to have ductile behavior than when the same material is at atmospheric pressure. Work by Bridgman many years ago (Ref 1) showed that normally brittle materials can be made to behave in a ductile manner if they are loaded while under extremely high pressure (up to several hundred thousand pounds per square inch). Practical use is made of this phenomenon in the commercial processing of extremely difficult-to-form metals and shapes (Ref 2).

Strength. In general, soft, relatively tough metals are more likely to have ductile behavior; harder, stronger metals tend to be more brittle. There are exceptions; for example, gray cast iron appears to be quite brittle even though it may be relatively soft. This behavior is caused by the large quantity of graphite flakes in the essentially steel-like metal matrix. The geometric effects of the sharp edged graphite flakes act as a very large number of stress concentrations, or notches, and override the ductility of the rupture, characteristic of ductile fracture.

Summary

Shear and cleavage are the basic fracture modes for metals under single-load applications. However, the type of metal behavior and ultimate fracture depend on several factors that can complicate the conditions. These factors should be taken into consideration in study of any single-load fracture.

REFERENCES

1. P.W. Bridgman, Fracture and Hydrostatic Pressure, in *Fracturing of Metals,* American Society for Metals, 1948, p 246–261
2. *Failure Analysis and Prevention,* Vol 11, *ASM Handbook,* 9th ed., ASM International, 1986, p 359–360

SELECTED REFERENCES

- *Failure Analysis and Prevention,.,* Vol 11, *ASM Handbook,* 10th ed., ASM International, 2002
- *Fatigue and Fracture,* Vol 19, *ASM Handbook,* ASM International, 1996

- M. Gensamer, *Strength of Metals under Combined Stresses,* American Society for Metals, 1941
- I. LeMay, *Principles of Mechanical Metallurgy,* Elsevier, 1981
- E.R. Parker, "Micro and Macro Mechanisms of Fracture," Technical Report No. GG 6-1.7, American Society for Metals, 1966

Understanding How Components Fail, Third Edition
Donald J. Wulpi
Edited by Brett Miller

CHAPTER 7

Stress Systems Related to Single-Load Fracture of Ductile and Brittle Metals*

IN ORDER TO UNDERSTAND how various types of single-load fractures are caused, one must understand the forces acting on the metals and also the characteristics of the metals themselves. All fractures are caused by stresses, and a version of the "weakest link" theory applies: fracture will originate wherever the local stress (load per unit of cross-sectional area) first exceeds the local strength. This location will vary depending on the strength gradients within the metal and the stress gradients imposed on the metal by applied and residual stresses. By understanding the ways in which single-load, or monotonic, fractures are caused, one can better understand fatigue fractures, which are the result of many thousands, millions, or billions of load applications at lower load levels.

When a force is applied to any member, components of other forces in other directions result, forming a stress system. To understand the forces, it is necessary to understand the stress systems acting on the part. A useful starting point is to study the stress systems acting on cylindrical members, such as rods or shafts. A variety of stresses can be applied to cylinders, and the same principles hold for noncylindrical parts. Shafts, and shaftlike parts, are very common and are widely used in construction of many assemblies and mechanisms.

**Understanding How Components Fail, Second Edition*, by Donald J. Wulpi, Chapter 4, Stress Systems Related to Single-Load Fracture of Ductile and Brittle Metals, ASM International, 1999, reviewed, revised and renumbered by Charles V. White, Kettering University.

Pure Loading Systems

Stress systems are best studied by examining free-body diagrams, which are simplified models of complex stress systems. Figure 1 shows the orientation of the normal (tensile and compressive) stresses and the shear (sliding) stresses, which are 45° to the normal stresses. Free-body diagrams of shafts in the pure types of loading—tension, torsion, and compression—are the simplest; they then can be related to more complex types of loading (Ref 1, 2).

Tension Loading

When a shaft or similar shape is pulled by tensile force, it becomes longer and narrower, just as a rubber band does when pulled. Similarly, the square in the free-body diagram in Fig. 1(a) is elongated in the direction of the tensile stress and contracted in the direction of the compressive stress. Note that the shear stresses are at 45° angles to the axial tensile stress and the transverse compressive stress. Note also that there are two sets of shear

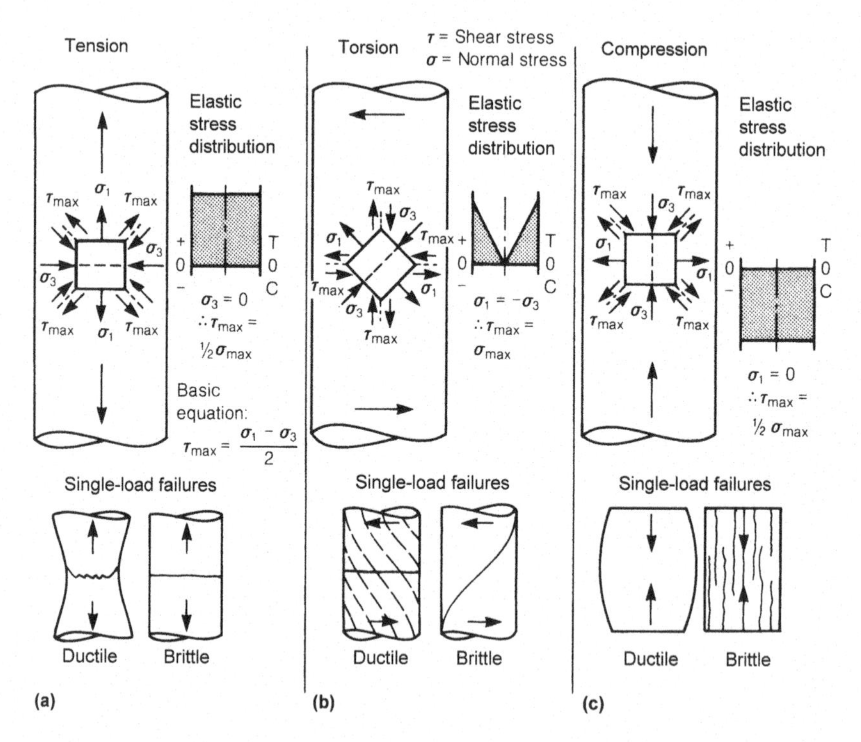

Fig. 1 Free-body diagrams showing orientation and elastic distribution of normal (tensile and compressive) and shear stress components in a shaft under pure (a) tension, (b) torsion, and (c) compression loading. Also shown is single-overload fracture behavior of ductile and brittle materials under these loading conditions (bottom diagrams). T, tension. C, compression. Adapted from Ref 1

stresses, each perpendicular to the other, diagonally between the normal stress directions. Since the magnitude of the stress is essentially uniform across the shaft, as shown in Fig. 1(a), fracture in pure tension can originate at any location in the cross section, in the absence of stress concentrations. (Figures 2 and 3 show examples of tensile fractures.)

Ductile metals are those that deform because the shear stress exceeds the shear strength before any other type of damage can occur. That is, the

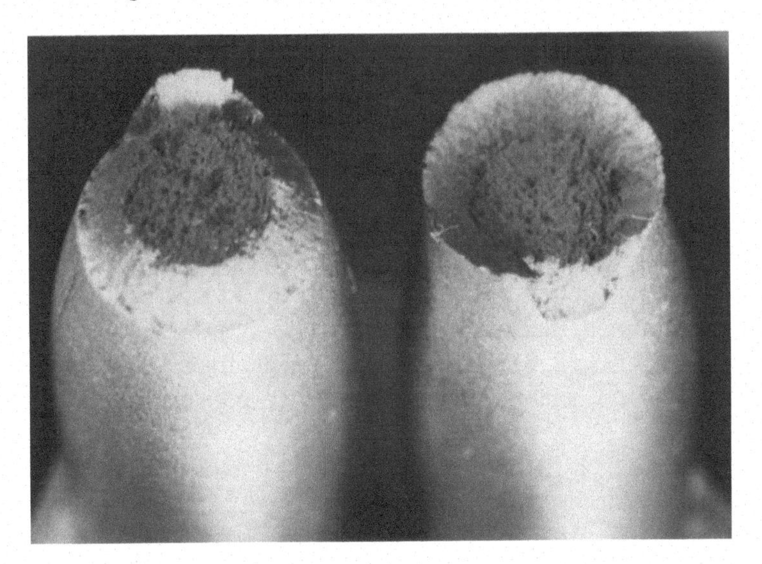

Fig. 2 Typical cup-and-cone fracture of a ductile annealed 1035 steel in a cylindrical tensile specimen. This type of fracture originates near the center of the section, with multiple cracks that join and spread outward until the 45° shear lip forms at the end of the fracture. Photo courtesy of Packer Engineering Associates, Inc.

Fig. 3 Two identical steel bolts that had been given different heat treatments and then pulled to fracture in a tensile test. The bolt at the left was annealed; when pulled, it had much deformation, as evidenced by the necking and thread separation. Fracture was of the cup-and-cone type with a large shear lip, similar to that in Fig. 2. The bolt at the right was austenitized and then brine quenched to a high hardness. The bright, transverse fracture and lack of deformation characterize a brittle fracture.

shear strength is the "weak link" in the system and is the controlling factor. Therefore, under a tensile force, the internal crystal structure of the metal deforms, or slips, permanently on the millions of microshear planes in the metal, resulting in lateral deformation— commonly called "necking"—prior to final fracture. This necking phenomenon occurs in the plastic region, which means that the deformation is irreversible. Of course, plastic—or permanent—deformation is characteristic of any single-load fracture of a ductile metal.

Fractures of ductile metals stressed in pure tension originate near the center of the shaft (provided there is no stress concentration). Toward the latter stages of the fracture process, many tiny internal voids, or cracks, develop and join to form a rough, jagged fracture surface. As these joining cracks progress outward, they eventually reach a region near the surface where the state of stress changes from tension to macroshear, forming a fracture at approximately 45° to the plane of the major fracture. This is the familiar "shear lip" (or "slant fracture") around the periphery. On a cylindrical specimen, this forms the dull and fibrous-appearing "cup-and-cone" fracture (Fig. 2, 3) that is typical of tensile fractures of ductile metals (Ref 1). (See Chapter 9, "Ductile Fracture," in this book for additional details.)

Brittle metals, by definition, are those that fracture because the tensile stress exceeds the tensile (or "cohesive") strength before any other type of damage can take place. Cohesive strength is now the "weak link" in the system and is the controlling factor. Brittle metals always have a fracture that is perpendicular to the tensile stress and have little or no deformation because fracture takes place before the metal can deform plastically as ductile metals do. Thus, a tensile fracture of brittle metal has a fracture plane that is essentially straight across. It also usually has a characteristic bright, sparkling appearance when freshly fractured.

Note that some brittle fractures may not be bright and sparkling, either because of the type of metal (such as gray cast iron and moderately hard steel) or because of the type of fracture (stress corrosion, some intergranular fractures, and fatigue). Also, atmospheric exposure usually rapidly dulls most initially bright fractures.

Torsional Loading

When a cylinder is twisted in pure torsion, the stress system characteristic of tension loading rotates 45° in one direction or the other, depending on which way the shaft is twisted. When twisted as shown in Fig. l(b), the entire stress system rotates 45° counterclockwise. Note that the normal stresses (tensile and compressive) are now at 45° to the shaft axis, while the shear stresses are longitudinal and transverse. Each pair of stress components remains mutually perpendicular to each other. Figure 4 demonstrates each of the stress components when they are concentrated in the "pure" stress directions.

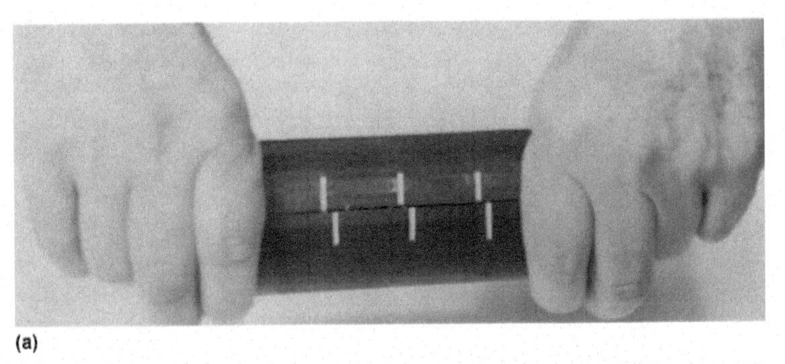

(a)

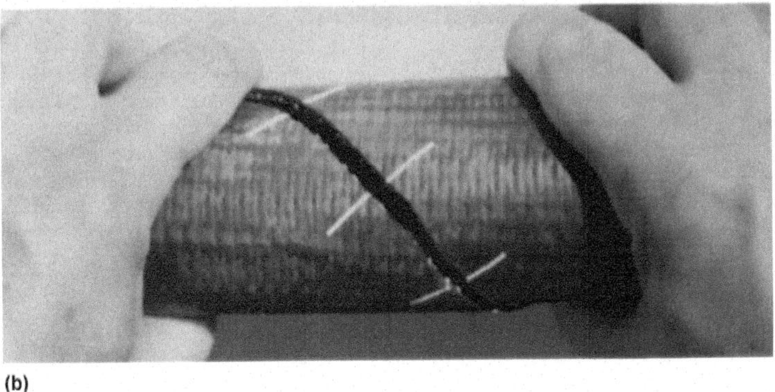

(b)

Fig. 4 Slit radiator hoses used to demonstrate concentrated "pure" shear stresses. (a) When hose is slit longitudinally, each side of the slit slides along the other side when the hose section is twisted back and forth. The sliding represents pure longitudinal shear stresses, because there is no opening (tension) or closing (compression). Transverse shear can be demonstrated similarly by rotating the smooth ends of two cylinders against each other. (b) When hose is slit at a 45° angle to its axis and then twisted as shown, opening represents tension, and closing represents compression forces. Imagine tiny rubber bands in line with the white marks; when twisted as shown, they are stretched in tension. When twisted in the opposite direction, compression forces are generated after the slit closes. Note that no sliding shear occurs when the 45° slit is opened and closed.

The elastic stress distribution in pure torsion is maximum at the surface and zero at the center of the shaft. This is true for all stress components—tension, shear, and compression—acting on the shaft in torsion. Figure 1(b) shows this with a V-shaped stress distribution pattern. For this reason, fracture originates at the surface where the stress is highest and then rapidly proceeds through the section as the shaft "unwinds" from its highly stressed condition. Also, the magnitudes of all elastic stress components—tension, compression, and shear—are identical in pure torsion, unlike other types of loading. Figures 5, 6, and 7 show examples of torsional fractures.

In a ductile metal, the shear strength is again the "weak link" when the shear stress exceeds the shear strength. Again, plastic or permanent deformation occurs, although in torsion the deformation may not be obvious unless there were longitudinal reference marks on the shaft prior to twisting (Fig. 5). Even in a cylindrical part without splines or other originally straight reference marks, twisting deformation can be made visible

Fig. 5 Single-overload torsional fracture on the transverse shear plane of a shaft of medium-carbon steel of moderate hardness. Note that the originally straight splines have been twisted in a counterclockwise direction. Final rupture was slightly off center due to a relatively slight bending force in addition to the torsional force. The fracture face is severely rubbed and distorted in a rotary direction by contact with the mating fracture surface at the moment of separation.

Fig. 6 Single-overload torsional fracture of a shaft of ductile steel similar to that in Fig. 5. The hole in the center is the lathe center from the original machining on the part

by revealing the longitudinal grain flow with chemical macroetching. An example of this type of etching to reveal torsional deformation is shown in Fig. 7.

The shape of a cylindrical part, such as a shaft, is not changed by torsional deformation. An example explains why: Imagine that the shaft consists of an infinite number of infinitely thin disks. When the pack of disks is twisted, each disk slips a very small amount with respect to its neighboring disks, but the diameter of each disk does not change with the slippage on the transverse shear planes. Eventually, fracture occurs on one of these transverse shear planes, which is essentially the interface between two adjacent disks.

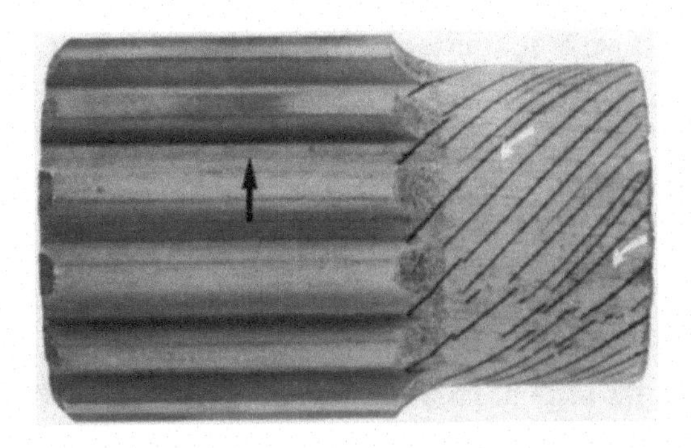

Fig. 7 Torsional fracture of a 1½-inch-diameter case-hardened steel shaft, illustrating cracking of the hard, brittle case and transverse shear fracture at the right end across the relatively soft, ductile core. Hot etched to reveal twisting and distortion of the originally straight grain flow in the cylindrical shaft (white arrows). Note that the black 45° brittle-fracture cracks (emphasized by the hot etching) are in the opposite direction to the twisting. This is because the cracks are always perpendicular to the tensile stress component that caused them, whereas the grain-flow distortion tends to proceed in the same direction as the tensile component. The black arrow on the spline indicates the direction this end was twisted.

Deformation also occurs on the longitudinal shear plane, but this normally does not cause single-load fracture unless the material is extremely weak in the transverse direction, as is wood.

When fracture does occur in pure torsion, the final rupture is at the center of the shaft; it is offset toward one side if a bending stress is also present (Fig. 5).

A brittle metal in pure torsion fractures perpendicular to the tensile stress component—as it does in tension—except that in torsion the tensile stress component is 45° to the axis of the shaft. This forms a spiral fracture of the type characteristic of torsional fracture of all brittle materials, including glass and chalk, if twisted carefully. The hard, relatively brittle case-hardened shaft may crack at the characteristic 45° angle, although the relatively soft, ductile core will fracture in the transverse shear plane. Figure 7 shows such a shaft, with many large, spiral cracks in the case, but the core—which formed the bulk of the cross section—fractured in the transverse shear plane opposite the splined end. Because this piece was etched, the twisting deformation of the originally straight grain flow of the shaft can be seen.

Compression Loading

When the cylinder is loaded in axial compression (assuming no instability such as buckling), the stress component system again rotates so that the compressive stress component is now axial, while the tensile stress component is transverse, as is shown in Fig. 1(c). Now the shear stress components are again 45° to the shaft axis, as they were during tension loading. Note that the elastic stress distribution in pure compression is the reverse of that in tension: uniform across the section (assuming no stress concentration), but in the compression—or negative—direction.

A ductile metal in compression does the opposite of what it did in tension: it becomes shorter and thicker because of slippage on the diagonal shear planes. In short, it bulges when squeezed by the compressive force. This is characteristic of metals being hot or cold headed and of "pancake" forgings under axial compression. However, there is no fracture; a truly ductile metal will simply bulge laterally as it becomes shorter.

A brittle metal in pure compression will, as always, fracture perpendicular to the maximum tensile stress component. Since this component is now transverse, the brittle fracture direction is now longitudinal, or parallel to the shaft. Brittle materials—such as very hard metals, glass, chalk, and rock—split or shatter longitudinally when loaded in compression. Indeed, this is the principle of rock crushing.

Figure 8 shows steel test specimens containing both brittle and ductile regions that illustrate axial cracking of the hard, brittle regions and lateral bulging of the soft, ductile region between the hard layers.

Bending Loading

When a straight part is loaded in pure bending, the convex surface has a tensile stress system similar to that shown in Fig. 1(a). Conversely, the concave surface is stressed in compression and has a stress system as shown for compression in Fig. 1(c). Approximately midway between the

Fig. 8 Compression test of two steel cubes deep case hardened only on the top and bottom surfaces. A compressive force perpendicular to the case-hardened surfaces caused cracking (arrows) in the very hard (66 HRC) cases on both surfaces. The soft, ductile cores simply bulged under the compressive force but did not fracture. As sketched in Fig. 1(c), these composite specimens illustrate behavior of both brittle and ductile metals and dramatically show the difference in properties of the case and core. (These 1-in.^{2} compression blocks were made from low-carbon alloy steels. They were water quenched and had $^3/_{16}$ in. case depth; both were compressed to 180,000 psi.) Photo courtesy of Xtek Corp., TSP Mill Products Division

two surfaces—depending upon the shape of the part—is the neutral axis, where all applied stresses are zero. Thus, fracture can be expected to originate on the convex (tensile) surface of the bend where the maximum tensile stress exists.

Fatigue

The preceding discussion is concerned with single-load deformation and fracture, which must be understood before the subject of fatigue can be understood. Fatigue is unique because the magnitude of the repetitive load applications need not be high enough to cause plastic deformation; that is, the stresses may be relatively low. However, after many relatively low-level load applications, microscopic changes take place in the structure that may lead to formation of fatigue cracks. (See Chapter 10, "Fatigue Fracture," in this book.)

The essential thing to remember is that the slow propagation, or growth, of a fatigue crack over a relatively long period of time is in exactly the same direction as the growth of a crack in a brittle material under the same type of loading. That is because fatigue cracks propagate in a direction that is perpendicular to the principal tensile stress—the brittle fracture direction.

Summary

By keeping in mind the principles outlined here, the analysis and understanding of single-load and fatigue fractures can be better accomplished. The principles are always the same; however, confusion can be caused by postfracture damage and also by misinterpretation and uncertainty as to the type of loading, particularly under combined stresses.

REFERENCES

1. *Fractography,* Vol 12, *ASM Handbook,* ASM International, 1987, p 12–71, 98–105
2. *Failure Analysis and Prevention,* Vol 11, *ASM Handbook,* ASM International, 1986, p 459–482

SELECTED REFERENCES

- M. Gensamer, *Strength of Metals under Combined Stresses,* American Society for Metals, 1941
- R.W. Hertzberg, *Deformation and Fracture Mechanics of Engineering Materials,* 4th ed., Wiley, 1995
- J.L. McCall and P.M. French, Ed., *Metallography in Failure Analysis,* Plenum, 1978

- T.R. Shives and W.A. Willard, Ed., *Mechanical Failure: Definition of the Problem,* NBS Special Publication 423, National Bureau of Standards, 1976
- C. Zener, The Micro-Mechanism of Fracture, *Fracturing of Metals,* American Society for Metals, 1948

CHAPTER **8**

Brittle Fracture*

THE TERMS *BRITTLE* AND *DUCTILE,* as used here and in the following chapter, refer to the extremes of metal behavior in single loading. *Brittle* means little or no permanent deformation prior to fracture, usually accompanied by high hardness and strength but with little tolerance for discontinuities. A brittle fracture occurs at stresses below the material's yield strength (i.e., in the elastic range of the stress-strain diagram). *Ductile* means considerable permanent deformation prior to fracture, usually accompanied by low hardness and strength, with considerable tolerance for discontinuities. A ductile fracture occurs at stresses greater than the material's yield strength (i.e., in the plastic range of the stress-strain diagram).

These terms, *brittle* and *ductile,* are opposites, like black and white. For many parts it is desirable to have metal properties that are predominantly hard, strong, and wear resistant, accepting the accompanying brittleness. Examples are gear teeth, antifriction bearings, and various types of cutting and forming tools. However, in many other types of material, it is desirable to have metal properties that are predominantly ductile, accepting the relatively low hardness, strength, and wear resistance. Examples are household aluminum foil; flat metal that must be shaped; and wire for cold forming into coat hangers, nails, or wire for wrapping, and so forth.

However, in a very large number of parts, the most desirable properties must be shades of gray, rather than the extremes of black or white. In these cases, neither extreme brittleness nor high ductility, with their accompanying properties, is entirely desirable; therefore, the properties must be balanced to a more desirable blend of behavior characteristics. Examples are springs, fasteners of various types, shafts, pressure vessels, blades, hand tools, and many other types of parts. These various conditions can be

***Understanding How Components Fail, Second Edition,* by Donald J. Wulpi, Chapter 8, Brittle Fracture, ASM International, 1999, reviewed and revised by Ronald J. Parrington, IMR Test Labs.

achieved by proper choice of the metal composition, as well as heat treatment or mechanical deformation, to achieve the optimum combination of properties desired for a particular part.

For the failure analyst, distinguishing between macroscale ductile and brittle fractures is an important first step in identifying the failure mechanism. Fractures accompanied by gross deformation (ductile fracture) are caused by one of three possible failure mechanisms: monotonic ductile overload, very low cycle fatigue, or stress rupture. In contrast, fracture in the absence of gross deformation (brittle fracture) is caused by either: (1) high cycle fatigue; or (2) monotonic overload of an inherently brittle material or of an embrittled, normally ductile material. Embrittlement may be due to conditions such as high constraint as caused by thick sections, sharp notches and cracks, or crack-like imperfections; low temperature; high strain rate or impact; improper heat treatment; and/or environmental embrittlement. Brittle fractures, rather than ductile fractures, are more often the subject of a failure investigation, because they are often sudden, unexpected, and sometimes catastrophic.

Similar to metals, many nonmetallic materials are subject to brittle fracture because they are inherently brittle or become embrittled under specific conditions. Examples are chalk, rock, brick, glass, ceramics, rigid plastics, elastomers, polymer matrix composites, and ceramic matrix composites. Discussion of brittle fracture in these materials is beyond the scope of this book. The remainder of this chapter concentrates on brittle fracture in metals and, more specifically, ferrous alloys.

Brittle metals are in daily use as normal engineering materials, and as long as they are properly handled, they are very satisfactory for many types of service. Hardened tool steels, gray cast iron, and many other metals are routinely used, within their limitations, with satisfactory results. However, if a tool such as a metal-cutting file is bent, it will suddenly snap with a brittle fracture; there will be little or no permanent deformation, and the pieces may be placed back together in perfect alignment.

In general, it is characteristic of very hard, strong, notch-sensitive metals to be brittle, although research work attempts to raise the useful strength of these metals without the danger of brittle fracture. Conversely, it is generally true that softer, weaker metals usually exhibit ductile behavior. Gray cast iron is an exception. This metal is brittle because it contains a very large number of internal graphite flakes, which act as internal stress concentrations and limit the ability of the metal to flow or deform, which is necessary for ductile behavior.

These and certain other metals are known and expected to be brittle, but they are normally used satisfactorily in applications that are suitable for brittle metals, where there is little or no danger of fracture. Certain other common metals—particularly low-carbon and medium-carbon steels, which are widely used in industry—are normally considered to have ductile properties and are normally used in applications where the ability to adjust by plastic deformation is desired. However, under certain combinations of circumstances, these normally ductile steels can fracture in a totally brittle manner. This completely unexpected behavior has been the cause of many disasters in the past and can still cause disasters if the lessons of the past are not heeded. Also, it is not necessary to have high applied stresses on the part or structure; brittle fracture can occur solely because of residual tensile stresses, with no applied load, or with any combination of applied and residual stresses. See Fig. 1 in Chapter 4, "Residual Stresses," in this book.

Brittle Fracture of Normally Ductile Steels

Brittle fracture of normally ductile steels has occurred primarily in large, continuous, boxlike structures, such as box beams, pressure vessels, tanks, pipes, ships, bridges, and other constrained structures. This was an extremely serious problem during World War II, when over 250 ships fractured or cracked; 19 of these broke completely in two. In some cases, fracture occurred in new ships that were still being outfitted and had never been to sea (Fig. 1). Reference 1 has many interesting discussions of ship fractures, as well as other catastrophic brittle fractures.

Fig. 1 New T-2 tanker, the *S.S. Schenectady,* which fractured in 1941 at its outfitting dock. This ship was one of 19 during World War II that had complete brittle fractures; over 200 other ships had partial brittle fractures of the hull. When tested, the fractured plates had normal ductility, as specified. Source: Ref 1

One of the most famous brittle fracture catastrophes was the Great Boston Molasses Tank Disaster in 1919. References 2 and 3 both discuss it and quote from the original accounts. The following gives a vivid picture of the tremendous damage that catastrophic brittle fracture can cause and why it must be guarded against:

> On January 15, 1919, something frightening happened on Commercial Street in Boston. A huge tank, 90 feet in diameter and about 50 feet high, fractured catastrophically, and over two million gallons of molasses cascaded into the streets.
>
> Without an instant's warning the top was blown into the air and the sides were burst apart. A city building nearby, where the employees were at lunch, collapsed burying a number of victims, and a firehouse was crushed in by a section of the tank, killing and injuring a number of firemen. On collapsing, a side of the tank was carried against one of the columns supporting the elevated structure [of the Boston Elevated Railway Co.]. This column was completely sheared off . . . and forced back under the structure . . . the track was pushed out of alignment and the superstructure dropped several feet. . . . Twelve persons lost their lives either by drowning in molasses, smothering, or by wreckage. Forty more were injured. Many horses belonging to the paving department were drowned, and others had to be shot (Ref 3).

This storage tank, which was of riveted construction, and the disaster itself were subsequently the subjects of a long investigation and trial in which the leading experts of the day tried to prove that: (a) the accident was the result of structural failure, in which case insurance would cover the monetary loss, or (b) the accident was the result of sabotage, such as by a bomb planted in or near the tank. The latter was a real possibility at the time, because of other sabotage in the area.

Following a lengthy investigation and trial, the court-appointed auditor handed down a decision that the disaster was the result of structural failure rather than an explosion (Ref 2, 3). His decision reflects the frustration felt by anyone trying to judge two real possibilities, particularly when each is strongly promoted by the technical authorities of the day. A fascinating part of his decision follows:

> Weeks and months were devoted to evidence of stress and strain, of the strength of materials, of the force of high explosives, of the bursting power of gas and of similar technical problems. . . . I have listened to a demonstration that piece "A" could have been carried into the playground only by the force of a high explosive. I have thereafter heard an equally forcible demonstration that the same result could be and in this case was produced by the pressure caused by the

> weight of the molasses alone. I have heard that the presence of Neumann bands in the steel herein considered along the line of fracture proved an explosion. I have heard that Neumann bands proved nothing. I have listened to men upon the faith of whose judgment any capitalist might well rely in the expenditure of millions in structural steel, swear that the secondary stresses in a structure of this kind were negligible and I have heard from equally authoritative sources that these same secondary stresses were undoubtedly the cause of the accident. Amid this swirl of polemical scientific waters it is not strange that the auditor has at times felt that the only rock to which he could safely cling was the obvious fact that at least one-half the scientists must be wrong. By degrees, however, what seem to be the material points in the case have emerged (Ref 3).

Although the Great Boston Molasses Tank Disaster was a landmark case for its era, similar disasters are not uncommon. A more recent brittle fracture disaster was the collapse of the Silver Bridge at Point Pleasant, West Virginia, in December 1967, resulting in the loss of 46 lives in cars that were plunged into the Ohio River. This accident was found by the National Bureau of Standards to be caused by stress-corrosion cracking resulting from long exposure to hydrogen sulfide (H_2S) in the air, because abnormal amounts of sulfur were found on the surfaces of the primary fracture (Ref 3–5).

The ship disasters during World War II were the impetus for a large body of research that has led to what we now refer to as fracture mechanics, as discussed in Chapter 15, "Fracture Mechanics," in this book. Study has made it apparent that the problem of brittle fracture of normally ductile metals results from a combination of circumstances. The absence of any one of the factors would prevent this type of fracture from occurring.

The factors that must all be present simultaneously in order to cause brittle fracture in a normally ductile steel are:

- *Stress concentration.* This may be a weld defect, a fatigue crack, a stress-corrosion crack, or a designed notch, such as a sharp corner, thread, hole, or the like. The stress concentration must be large enough and sharp enough to be a "critical flaw" in terms of fracture mechanics.
- *Tensile stress.* The tensile stress must be of a magnitude high enough to provide microscopic plastic deformation at the tip of the stress concentration. One of the major complexities is that the tensile stress need not be an applied stress on the structure but may be a residual stress that is completely within the structure. In this case, the stress is not obvious or easily measured, as is the applied stress. The part or structure can be completely free of an external or applied load—just lying on a bench or floor, for example—and still experience instantaneous, sudden, catastrophic brittle fracture. This type of occurrence is within

the experience of many persons who have worked with metals, particularly welded, torch-cut, or heat-treated steels. Chapter 4, "Residual Stresses," in this book explains how tensile residual stresses can form without obvious external forces.

- *Relatively low temperature for the steel concerned.* The definition of metal/temperature interrelationships is inexact, very much subject to the type of test used to try to understand whether or not a particular steel is actually subject to brittle fracture under certain conditions (Ref 6). However, regardless of the type of test used to try to establish the ductile/brittle transition temperature, the general results are the same: the lower the temperature for a given steel, the greater the possibility that brittle fracture will occur. For some steels, for example, the ductile/brittle transition temperature under certain conditions may be above room temperature.

As noted, the absence of any one of the factors—stress concentration, high tensile stress, relatively low temperature for the steel concerned—that contribute to brittle fracture of a ductile steel will prevent this problem from occurring. For the following reasons, however, the choices are sometimes limited on what can be done:

- Stress concentrations often are present by design, such as in necessarily sharp corners, threads, holes, and grooves, or in unintentional stress concentrations, such as fatigue cracks, stress-corrosion cracks, weld defects, and arc strikes. Great care can be taken to minimize these stress concentrations, but they may occur despite the best safeguards.
- Tensile stresses are usually inevitable during service loading, depending on the type of part or structure. Applied loads must be considered by designers when selecting materials and sizing components. Furthermore, care should be taken to ensure that damaging tensile *residual* stresses are eliminated or minimized. This is particularly true when shrinkage stresses from welding are involved.
- Temperature can be controlled in certain applications but not in others. For example, certain processing equipment may operate continually at elevated temperatures. In this case brittle fracture may not be a consideration unless there is a damaging environmental factor, such as absorption of hydrogen or hydrogen sulfide leading to hydrogen damage.

In many other applications, exposure to relatively low temperature for the steel involved may be a real possibility and thus a real problem if the other contributing factors are likely to exist. Therefore, the steel itself may be the only factor that can be controlled in order to prevent brittle fracture of normally ductile steel. The metallurgical trends that tend to decrease the likelihood of brittle fracture of steel are: low carbon content, moderate manganese content, high manganese-to-carbon ratio, inclusion of certain alloying elements, fine grain size, deoxidation, and heat treatment to produce tempered martensitic or lower bainitic microstructures. This subject is covered in more detail in Ref 6.

Macroscale Characteristics of Brittle Fracture

Brittle fractures have certain characteristics that permit them to be properly identified:

- There is no gross permanent or plastic deformation of the metal in the region of the brittle fracture, although there may be permanent deformation in other locations where relatively ductile fracture has occurred.
- The surface of a brittle fracture is perpendicular to the principal tensile stress. Thus, the direction of the tensile stress that caused the fracture to occur can be readily identified.
- Characteristic markings on the fracture surface frequently, but not always, point back to the location from which the fracture originated. In the case of flat steel, such as sheet, plate, or flat bars, and also case-hardened regions, there are characteristic V-shaped "chevron" or "herringbone" marks that point toward the origin of the fracture. In many instances, these marks are extremely fine and very difficult to recognize unless a strong light is positioned so that it just grazes the projections of the surface texture. (See Fig. 2 to 7 for illustrations of these marks.) Brittle fractures of some parts may have a pattern of radial lines, or ridges, emanating from the origin in a fanlike pattern. Again, it may be difficult to perceive the texture of the fracture surface unless the lighting is carefully controlled (see Fig. 8 to 11). Brittle fractures of extremely hard, fine-grain metals usually have little or no visible fracture pattern. In these cases, it may be very difficult to locate the origin with certainty.

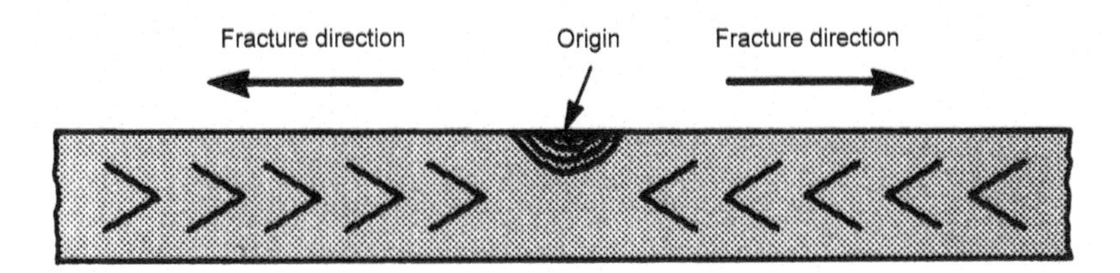

Fig. 2 Sketch of pattern of brittle fracture of a normally ductile steel plate, sheet, or flat bar. Note the classic chevron or herringbone marks that point toward the origin of the fracture, where there usually is some type of stress concentration, such as a welding defect, fatigue crack, or stress-corrosion crack. The plane of the fracture is always perpendicular to the principal tensile stress that caused the fracture at that location.

Fig. 3 Fragment of a fractured thick-wall drum. The fracture, which started at the right side of the photo, ran rapidly to the left, resulting in a well-defined chevron pattern. Source: Ref 7

Note that the preceding discussion of the ductile/brittle transition primarily concerns carbon and alloy steels, as well as certain nonaustenitic stainless steels. Other metals with body-centered cubic crystal structures behave similarly but are less common. Most nonferrous metals, such as alloys of aluminum and copper, and austenitic stainless steels have crystal structures (face-centered cubic) that are not susceptible to the ductile/brittle transition characteristic of body-centered cubic metals.

See Chapter 6, "Basic Single-Load Fracture Modes," in this book for other discussion on ductile/brittle fracture.

Microstructural Aspects of Brittle Fracture

Brittle fractures usually propagate by either or both of two fracture modes, cleavage or intergranular. In most cases it is necessary to study the fracture surface with a scanning electron microscope to distinguish the fracture mode.

Cleavage fractures are characterized by splitting of the crystals, or grains, along specific crystallographic planes without respect to the grain boundaries, as shown in Fig. 12. Since the fracture goes through the grains, this type of fracture is frequently referred to as transgranular. Cleavage fractures are the most common type of brittle fracture and are the normal mode of fracture unless the grain boundaries have been weakened by a specific environment or process.

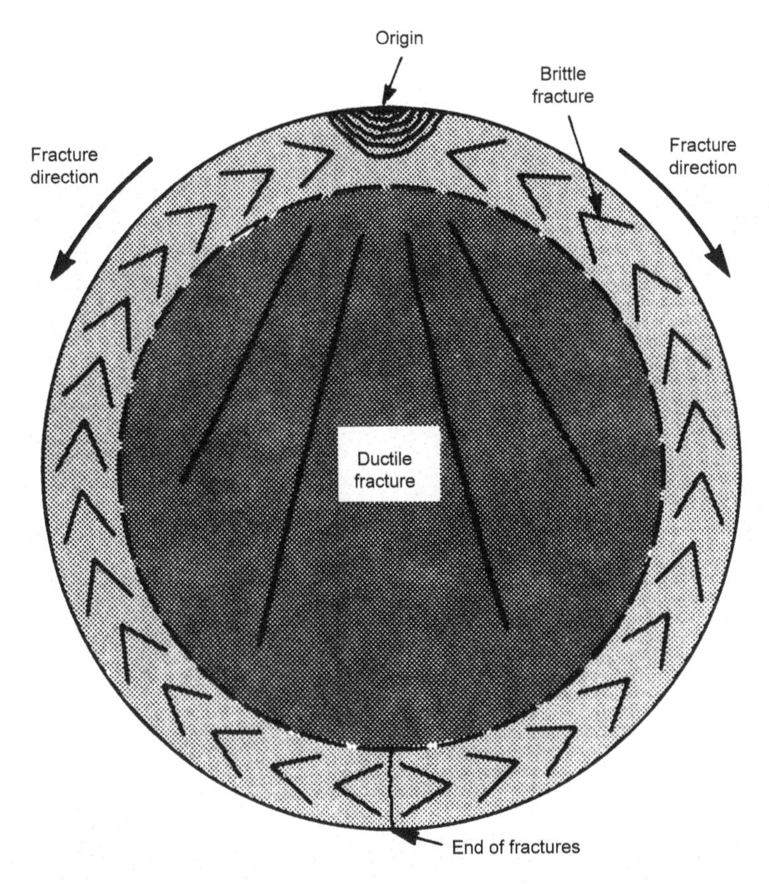

Fig. 4 Sketch of a fractured case-hardened shaft showing chevron marks pointing back toward the fracture origin. If brittle fracture continues completely around the part, the two separate fractures may form a step where they meet opposite the origin. The interior, or core, is likely to have a ductile fracture with dimpled rupture on the microscale.

A typical cleavage fracture viewed by the scanning electron microscope is shown in Fig. 13. Note that the pattern is characterized by the joining together of microscopic ridges, much like the joining of tributaries of a river system to form the main stream of the river. This pattern reveals the direction that the fracture ran; the fracture propagated in the same direction that the water in a river flows: downstream.

Intergranular fractures are those that follow grain boundaries weakened for any of several reasons. An analogy may be made to a brick wall, which fractures through the mortar rather than through the bricks themselves. The mortar is analogous to the grain boundaries, while the bricks are analogous to the metal grains. In steels, intergranular fracture follows the prior austenite grain boundaries. A typical intergranular fracture is shown in Fig. 14.

The reasons for weakened grain boundaries are frequently very subtle and poorly understood. Under certain conditions, some metals are subject

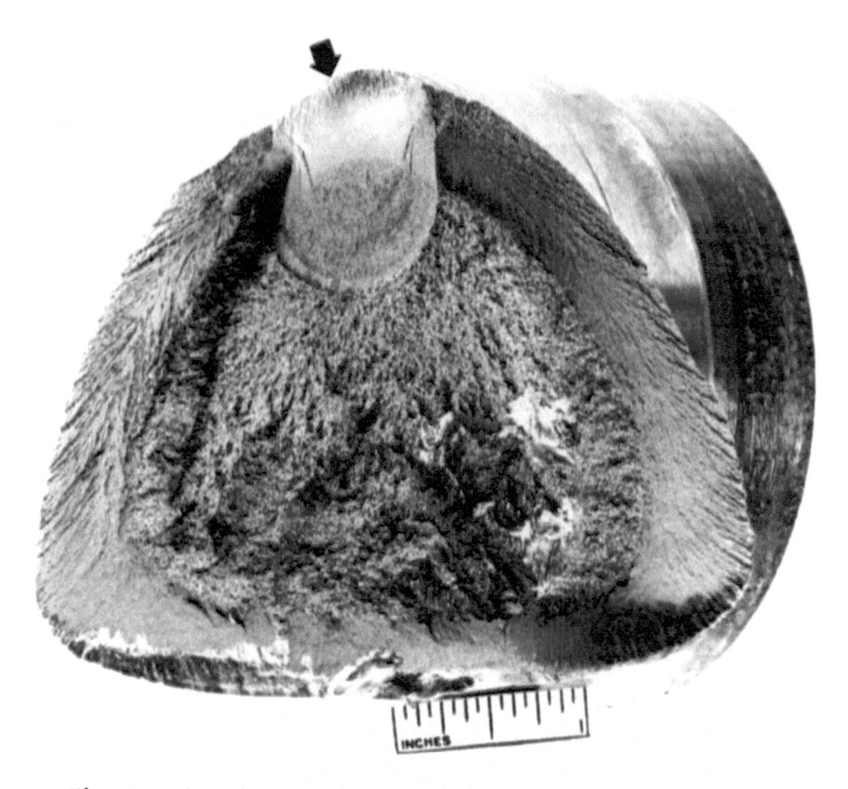

Fig. 5 Surface of a torsional fatigue crack that caused brittle fracture of the case of an induction-hardened axle of 1541 steel. The fatigue crack originated (arrow) at a fillet (with a radius smaller than specified) at a change in shaft diameter near a keyway runout. Case hardness was about 46 HRC at the surface. Note the well-defined chevron marks in the two brittle fractures pointing toward the roughly circular fatigue portion of the fracture. Note also that this fracture is at a 45° angle to the shaft axis, as is typical of high-cycle fatigue and brittle fracture of shafts in torsion.

to migration or diffusion of embrittling elements or compounds to the grain boundaries. The major forms of embrittlement of steel are discussed briefly here but are covered in considerably more detail in various articles in Ref 6 and 9, as well as in other sources. A summary of the types of embrittlement experienced by ferrous alloys is presented in Table 1.

Strain-Age Embrittlement. Most susceptible to the phenomenon of strain-age embrittlement are low-carbon rimmed or capped steels that are severely cold worked during forming processes. Subsequent moderate heating during manufacture (as in galvanizing, enameling, or paint baking) or aging at ambient temperature during service may cause embrittlement.

Quench-Age Embrittlement. Rapid cooling, or quenching of low-carbon steels (0.04 to 0.12% C) from subcritical temperatures above about 560 °C (1040 °F) can precipitate carbides within the structure and also precipitation harden the metal. An aging period of several weeks at room temperature is required for maximum embrittlement.

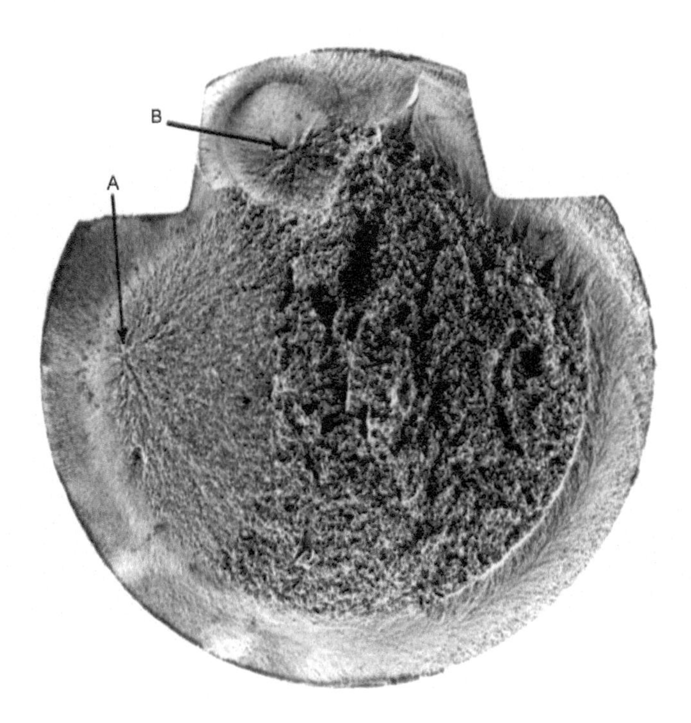

Fig. 6 Fatigue fracture of a 3¼ in. diam induction-hardened shaft of 1541 steel after fatigue testing in rotary bending. Fatigue fracture origins A and B were subsurface due to the steep induction-hardened gradient and lack of an external stress concentration. (See Fig. 9 in Chapter 3 of this book.) Fatigue crack A, the larger of the two, propagated through the core and case regions on the left side; then the shaft suddenly fractured in a brittle manner around the right side of the case. Note the chevron marks on the lower right pointing in a clockwise direction toward fracture origin A.

Blue Brittleness. Bright steel surfaces oxidize to a blue-purple color when plain carbon and some alloy steels are heated between 230 and 370 °C (450 and 700 °F). After cooling, there is an increase in tensile strength and a marked decrease in ductility and impact strength caused by precipitation hardening within the critical temperature range.

Stress Relief Embrittlement. Alloy and stainless steels that undergo precipitation hardening may be susceptible to stress relief embrittlement or reheat cracking. Post-weld stress relief heat treatment can induce reprecipitation of elements dissolved during welding, primarily alloy carbides, in the weld heat affected zone (HAZ) causing strengthening and embrittlement. Residual stresses and grain coarsening from the welding contribute to the embrittlement process. When severe, intergranular brittle cracking or reheat cracking can occur.

Temper Embrittlement. Quenched steels containing appreciable amounts of manganese, silicon, nickel, or chromium are susceptible to temper embrittlement if they also contain one or more of the impurities phosphorus, antimony, tin, and arsenic. Embrittlement of susceptible

Fig. 7 Surface of a brittle fracture in a cold-drawn, stress-relieved 1035 steel axle tube. Fracture originated at a weld defect (arrow) during testing in very cold weather. Note the well-defined chevron marks located clockwise from the arrow, pointing back toward the origin. Note also that the steel around the access hole below the grease fitting is actually necked down along the tube axis. The upper side of the tube is deformed and torn in a ductile manner.

steels can occur after heating in the range of 375 to 575 °C (700–1070 °F) but occurs most rapidly around 450 to 475 °C (840–885 °F).

Tempered Martensite Embrittlement (500 °F Embrittlement). High-strength low-alloy steels containing substantial amounts of chromium or manganese are susceptible to embrittlement if tempered in the range of 200–370 °C (400 to 700 °F) after hardening, resulting in tempered martensite. Steels with microstructures of tempered lower bainite also are subject to 500 °F embrittlement, but steels with pearlitic microstructures and other bainitic steels are not susceptible.

400 to 500 °C Embrittlement. Fine-grained, high-chromium ferritic stainless steels, normally ductile, will become embrittled if kept at 400 to 500 °C (750–930 °F) for long periods of time. Soaking at higher temperatures for several hours should restore normal ductility.

Sigma-Phase Embrittlement. Prolonged service at 560 to 980 °C (1050–1800 °F) can cause formation of the hard, brittle, iron-chromium

(a)

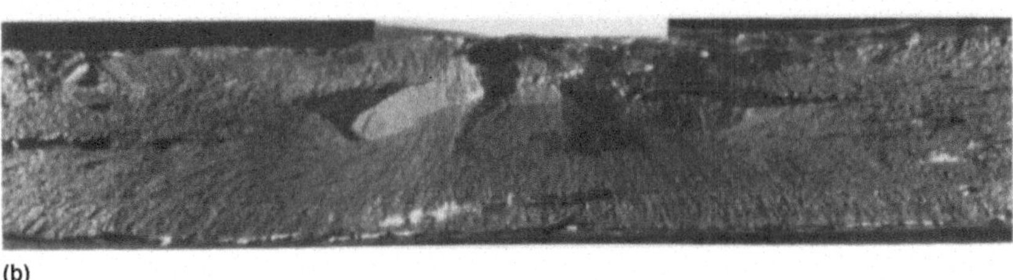

(b)

Fig. 8 (a) Catastrophic brittle fracture of a 260 in. diam solid-propellant rocket motor case made of 18% Ni, grade 250, maraging steel. The case fractured at a repaired weld imperfection during a hydrostatic pressure test. Fracture occurred at about 57% of the intended proof stress. All welds of the case had been carefully inspected by x-ray and ultrasonic inspection. Arrows indicate the origin of fracture. (b) Light fractograph of the crack origin, at $1\frac{1}{3}\times$. A crack was found beneath a gas tungsten-arc repair weld on the inner surface of the case (at the top in this fractograph) in the heat-affected zone of a longitudinal submerged-arc assembly weld. The crack was about 1.4 in. long, parallel to and 0.47 in. beneath the outer surface of the case (at the bottom). Radial marks are visible that confirm that fracture proceeded to the left and to the right. Source: Ref 8

intermetallic sigma phase in both ferritic and austenitic stainless steels and similar alloys. Impact strength is greatly reduced, particularly when the metal has been cooled to about 260 °C (500 °F) or less.

Graphitization. Pearlite transforms to nodular graphite and ferrite in carbon steels with 0.5% or less molybdenum after exposure to temperatures in the range of ≥425°C (>800 °F) for an extended period of time.

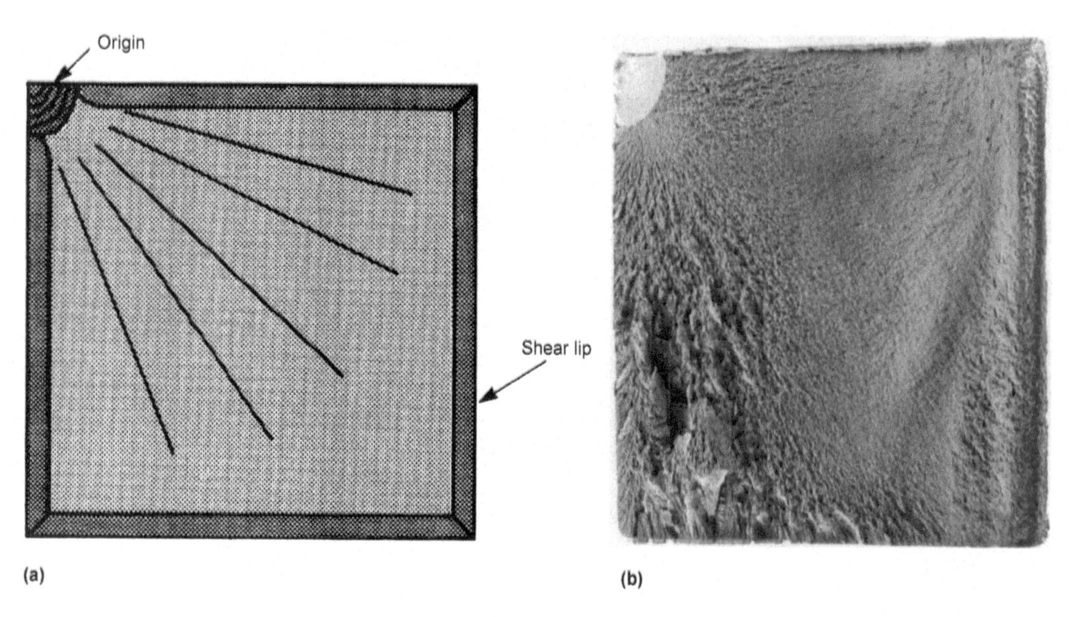

Fig. 9 (a) Sketch of pattern of brittle fracture in a moderately hard, strong metal. The fracture originated at a sharp stress concentration that grew to the critical flaw size for that metal. The sharp stress concentration is frequently, though not always, a fatigue crack or a stress-corrosion crack. Note the fan-shaped pattern radiating from the origin region in the upper left corner. When viewed under the electron microscope, this type of fracture is likely to reveal a cleavage or quasi-cleavage fracture mode. (b) Fracture surface of a large (~5-¾ × 6 in.) equalizer bar made from D6B steel heat treated to a hardness of 45–47 HRC. This bar, which supports the front end of a large crawler tractor, was in service for about 200 h and was then returned to the laboratory, where it was flexed in high-stress, low-cycle fatigue, fracturing at 60,000 cycles. Note that the fatigue-crack zone at the upper left has a very fine texture. Note also the radial, fanlike ridges emanating from the fracture origin and the very small shear lips around the periphery of the brittle fracture. This fracture of a very large, high-strength part released a large amount of energy in a very short time.

Fig. 10 Origin (at arrow) of a single-load brittle fracture that initiated at a small weld defect. Note also a fatigue fracture in the upper right corner. Radial ridges emanate from the origin in a fan-shaped pattern. The brittle part of the fracture is bright and sparkling, in contrast to the dull appearance of the fatigue zone and the thin shear lips at the top and bottom surfaces.

Fig. 11 Three Charpy V-notch impact test specimens of the same metallurgical conditions tested at three different temperatures. At the highest temperature (left), the fracture is virtually all shear. At intermediate temperature (center), the fracture is combined shear and cleavage. At the lowest temperature (right), the fracture is virtually complete cleavage. Note the increased deformation with increased temperatures.

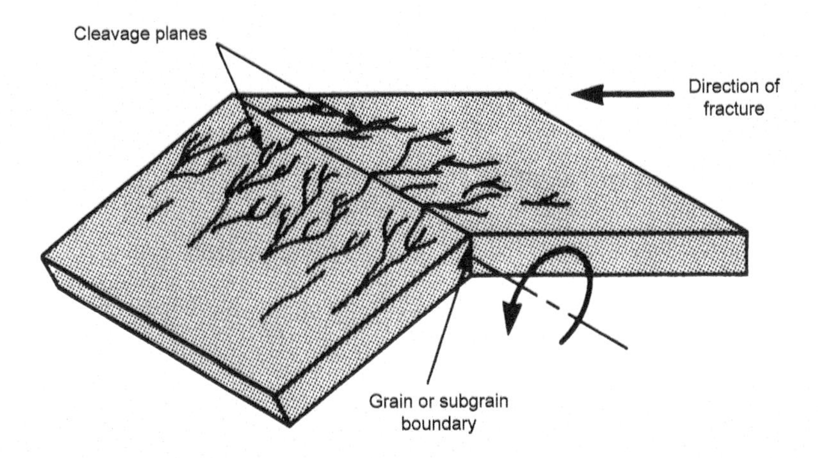

Fig. 12 Cleavage-fracture model showing fracture direction, cleavage planes, and low-angle grain or subgrain boundary. Source: Ref 9

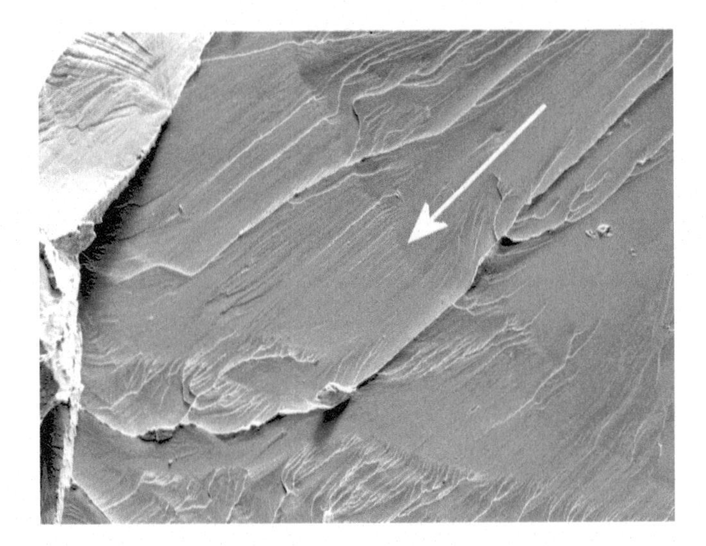

Fig. 13 Cleavage fracture in hardened steel showing numerous "river" marks. The overall direction of crack propagation is in the direction of the arrow (i.e., downstream). New river patterns are created where grain boundaries were crossed. 125×.

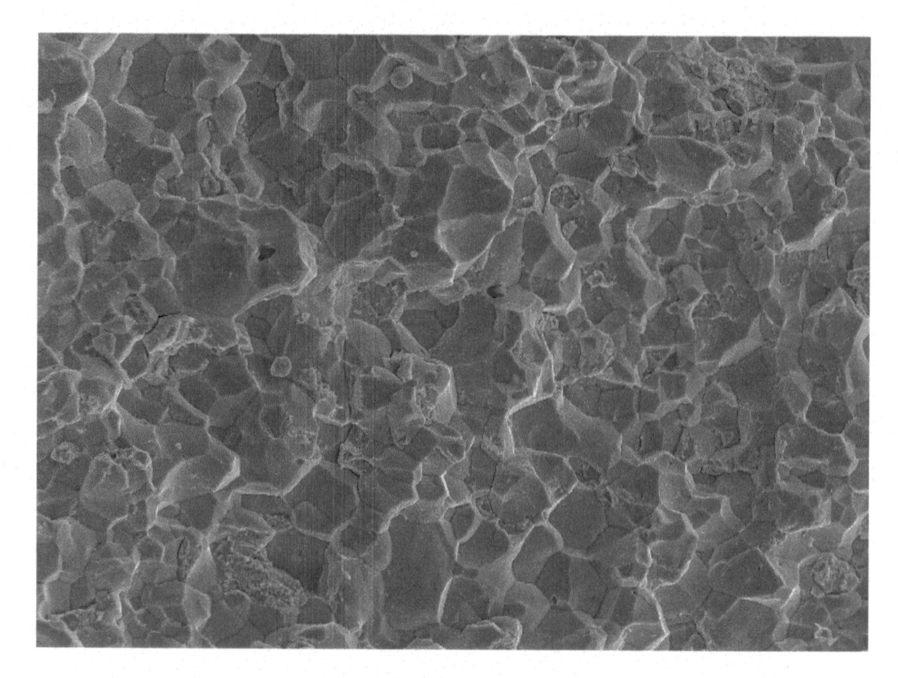

Fig. 14 Intergranular fracture viewed under the scanning electron microscope. Note that fracture takes place between the grains; thus, the fracture surface has a "rock candy" appearance that reveals the shapes of part of the individual grains.

Table 1 Summary of the types of embrittlement experienced by ferrous alloys

Embrittlement type	Susceptible steels	Causes	Result
Strain-age embrittlement	Low-carbon steel	Precipitation after deformation processing	Strength increases, ductility decreases
Quench-age embrittlement	Low-carbon steel	Quenching followed by precipitation	Strength increases, ductility decreases
Blue brittleness	Carbon and alloy steels	230–370 °C (450–700 °F) exposure	Strength increases, toughness and ductility decrease
Stress-relief embrittlement	Alloy and stainless steel	Postweld heat treatment	Toughness decreases
Temper embrittlement	Carbon and low-alloy steel	375–575 °C (700–1070 °F) exposure	Increase in DBTT
Tempered martensite embrittlement	Heat-treated alloy steel	200–370 °C (400–700°F) exposure	Toughness decreases
400–500 °C embrittlement	Stainless steel (>15% Cr)	400–500 °C (750–930 °F) exposure	Strength increases, ductility decreases
Sigma-phase embrittlement	Ferritic and austenitic stainless steel	560–980 °C (1050–1800 °F) exposure	Toughness decreases
Graphitization	Carbon and alloy steel	Exposure >425 °C (>800 °F)	Toughness decreases
Intermetallic compound embrittlement	All steel	Exposure to metal that forms an intermetallic compound	Brittleness
Neutron embrittlement	All steel	Neutron irradiation	Increase in DBTT
Hydrogen embrittlement	Cold-worked or heat-treatment-hardened steel	Processing, service, corrosion, etc.	Ductility decreases, cracking
Stress corrosion cracking	All steel	Prolonged exposure to a material-specific aggressive environment and sustained tensile stresses	Brittle fracture
Liquid metal induced embrittlement	All steel	Molten low-T_M metal exposure	Ductility decreases
Solid metal induced embrittlement	All steel	Near-molten low-T_M metal exposure	Ductility decreases

DBTT: ductile-brittle transition temperature; T_M: melting temperature.
Source: Ref. 9

At grain boundaries, the formation of chains of graphite can result in brittle fracture or cracking.

Intermetallic Compound Embrittlement. Long exposure of galvanized steel to temperatures slightly below the melting point of zinc (420 °C or 787 °F) causes zinc diffusion into the steel. This results in the formation of a brittle iron-zinc intermetallic compound in the grain boundaries.

Other types of embrittlement leading primarily to intergranular fracture are caused by environmental factors. These include:

Neutron Embrittlement. Neutron irradiation of steel parts in nuclear reactors usually results in a significant rise in the ductile/brittle transition temperature of the steel. Metallurgical factors such as heat-treating practice, microstructure, vacuum degassing, impurity control, and steel composition greatly affect susceptibility to this type of grain-boundary weakening.

Hydrogen Embrittlement. Hydrogen atoms diffuse readily into steel during processes such as acid pickling, electroplating, arc welding with moist or wet electrodes, and exposure to hydrogen sulfide. After stressing, delayed brittle fracture may occur, particularly in higher-strength steels.

Stress Corrosion Cracking (SCC). Simultaneous exposure of a susceptible metal to a tensile stress (applied or residual) and to a material-specific aggressive environment may cause brittle fracture that may be either intergranular or transgranular, depending on conditions. If either factor is eliminated, stress-corrosion cracking cannot occur.

Liquid Metal and Solid Metal Induced Embrittlement. Certain low-melting-point metals can embrittle the solid metals with which they are in contact. Embrittlement can occur above (liquid metal induced embrittlement) or below (solid metal induced embrittlement) the melting point of the embrittling metal. A tensile stress is also required for brittle fracture to occur.

Each type of embrittlement described is the result of exposure to one or several environmental factors during manufacture, storage, or service. Each type is extremely complex. Additional information on these types of environmental embrittlement can be found in Ref 1 to 9.

Mixed Fracture Morphology

It must not be assumed that brittle fracture always occurs solely by cleavage or intergranular fracture as described above. In most cases one fracture morphology predominates but is not necessarily the only morphology. For example, in a predominantly intergranular fracture in steels, there probably will be regions, large or small, on the fracture surface that contain cleavage fracture as well. The reverse also is true. In other words, the fracture morphology that occurs at a particular location depends on the local composition, stress, environment, imperfections, and crystalline orientation of the grains. There may also be regions of tough, fibrous, dimpled-rupture fracture, particularly away from the origin of the major fracture (i.e., the shear lip or final fast overload).

It should be noted that cast metals generally tend to be less ductile (or more brittle) than wrought metals of the same composition under the same conditions. The reason is that various types of casting imperfections—particularly shrinkage porosity, gas porosity, and certain types of inclusions—act as internal stress concentrations in castings. As a rule of thumb, wrought metals generally contain fewer inherent imperfections.

Summary

The subject of brittle fracture is exceedingly vast and complex, with many interrelated factors of material, design, manufacture, quality control, and environment, both thermal and chemical. Study of the references cited is urged as well as those in the more recent literature.

REFERENCES

1. E.R. Parker, *Brittle Behavior of Engineering Structures,* Wiley, 1957
2. *Symposium on Effect of Temperature on the Brittle Behavior of Metals with Particular Reference to Low Temperatures,* ASTM, STP 158, 1954, p 3–110, 116–118
3. R.W. Hertzberg, *Deformation and Fracture Mechanics of Engineering Materials,* Wiley, 1976, p 229, 544, 569–573
4. J.A. Bennett and H. Mindlin, *J. Test. Eval.*, 1973, p 152
5. D.B. Ballard and H. Yakowitz, *Scanning Electron Microscopy,* Illinois Institute of Technology Research Institute, 1970, p 321
6. *Properties and Selection: Irons, Steels, and High-Performance Alloys,* Vol 1, *ASM Handbook,* ASM International, 1990, p 689–736
7. M.E. Shank, Ed., *Control of Steel Construction to Avoid Brittle Fracture,* Welding Research Council, 1957
8. J.E. Srawley and J.B. Esgar, *Investigation of Hydrotest Failure of Thiokol Chemical Corporation 260-Inch-Diameter SL-1 Motor Case,* NASA TM X-1194, National Aeronautics and Space Administration
9. *Failure Analysis and Prevention,* Vol 11, *ASM Handbook,* 10th ed., ASM International, 2002, p. 690

SELECTED REFERENCES

- C.D. Beachem and W.R. Warke, Ed., *Fractography: Microscopic Cracking Processes,* ASTM, STP 600, 1976
- J.E. Campbell, J.H. Underwood, and W.W. Gerberich, *Application of Fracture Mechanics for Selection of Metallic Structural Materials,* American Society for Metals, 1982

- *Cracks and Fracture,* ASTM, STP 601, 1976
- *Fatigue and Fracture,* Vol 19, *ASM Handbook,* ASM International, 1996
- *Fractography,* Vol 12, *ASM Handbook,* ASM International, 1987
- *Impact Testing of Metals,* ASTM, STP 4669, 1970
- G.W. Powell, S.-H. Cheng, and C.E. Mobley, Jr., *A Fractography Atlas of Casting Alloys,* Battelle Press, 1992

CHAPTER 9

Ductile Fracture*

BEFORE STUDYING this chapter on single-overload ductile fracture, it is recommended that the reader review the first three paragraphs, at least, of Chapter 8, "Brittle Fracture," in this book. This will give an overall perspective of the very different but closely related subjects of brittle and ductile fracture.

Ductile fracture results from the application of an excessive stress to a metal that has the ability to deform permanently, or plastically, prior to fracture. Thus, the property of ductility is simply the ability of the material to flow or deform, which may or may not lead to fracture, depending on the magnitude of the stress applied.

The property of ductility is somewhat related to the property of toughness, although the latter is usually measured in the presence of a notch or other stress concentration. The Charpy V-notch impact test is commonly used as a measure of toughness. However, the ability to absorb energy and deform plastically prior to fracture is characteristic of both ductility and toughness.

This ability to absorb energy is a valuable property of ductile metals. For example, many modern automobiles are designed so that in the event of a front-end collision, the car will not stop instantly but will fold like an accordion. This absorption of energy serves to slow down the rate of deceleration of the passenger compartment and lessen the impact on the occupants of the car.

The property of ductility permits metallic and nonmetallic parts to be formed without fracture and to be adjusted in shape. A simple example is familiar to all persons who wear eyeglasses: periodically it is necessary to have the frames adjusted because of the various stresses that are applied to

**Understanding How Components Fail, Second Edition*, by Donald J. Wulpi, Chapter 9, Ductile Fracture, ASM International, 1999, reviewed and revised by Craig J. Schroeder, Element Materials Technology.

the frames during service. The adjustment is possible because of the ductility, or formability, of the material in the frame. If the material was brittle, adjustment would not be possible, because the frames would fracture. Actually, the ductility works both ways: it enables adjustment of the frames, but it also permits the distortion that makes adjustment necessary. Somewhat higher yield strength would prevent initial deformation.

Characteristics of Ductile Fracture

Ductile fractures have characteristics that are different from those of brittle fractures. However, many fractures contain some of the characteristics of both types. Ductile fractures have these characteristics (Ref 1):

- There is considerable gross permanent or plastic deformation in the region of ductile fracture, as shown by the bolt on the right side of Fig. 1. In many cases, this may be present only in the final rupture region of a fracture that may have originated with a fatigue or brittle fracture.
- As pointed out in Chapter 6, "Basic Single-Load Fracture Modes," in this book, ductile fractures are those in which the shear stress exceeds the shear strength before any other mode of fracture can occur. Therefore, the micromechanism of fracture is in the shear direction, but this is not always obvious on macroexamination. Scanning electron microscopy inspection is typically required to determine the shear direc-

Fig. 1 Brittle versus ductile fracture in two 1038 steel bolts deliberately heat treated to have greatly different properties when pulled in tension. The brittle bolt (left) was water quenched with a hardness of 47 HRC but had no obvious deformation. The ductile bolt (right) was annealed to a hardness of 95 HRB (less than 20 HRC) and shows tremendous permanent deformation.

tion. The surface of a ductile fracture is not necessarily related to the direction of the principal tensile stress, as it is in a brittle fracture.

- The characteristic appearance of the surface of a ductile fracture is dull and fibrous. This is caused by deformation on the fracture surface, which is discussed in "Microstructural Aspects of Ductile Fracture," below.

The classic example of a ductile fracture is a tensile specimen that has "necked down" or deformed to form a "wasp waist" prior to fracture. A typical fracture of this type, shown in Fig. 2, is the so-called cup-and-cone fracture characteristic of ductile metals pulled in tension. It is instructive to study this type of fracture in some detail:

- The narrowing, or "necking," indicates that there has been extensive stretching, or elongation, of the grains of metal in the reduced area, particularly near the fracture.
- As noted above, shear stress dominates deformation and ductile fracture. In most cases, the 45° plane of maximum shear stress components is not obvious or readily observed. Figure 3, however, shows a low-carbon cast steel tensile specimen with numerous casting discontinuities that tend to emphasize the 45° shear aspect of tensile fracture. The diagonal ridges, frequently called "Lüders lines" or "stretcher strains," are easily visible in this photograph.
- A tensile cup-and-cone fracture originates with many tiny internal fractures called "microvoids" near the center of the reduced area. These voids occur after the tensile strength has been attained and as the stress (or load on the test machine) is dropping toward the fracture stress, as shown in Fig. 4, region (c) (Ref 2).
- A ductile fracture starts near the center of the reduced section in tensile loading and then spreads outward toward the surface of the necked-

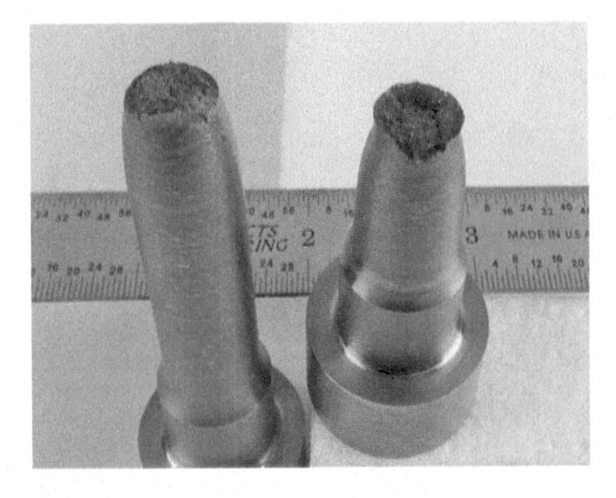

Fig. 2 A fractured tensile specimen with the typical cup-and-cone fracture characteristic of ductile metals fractured in tension.

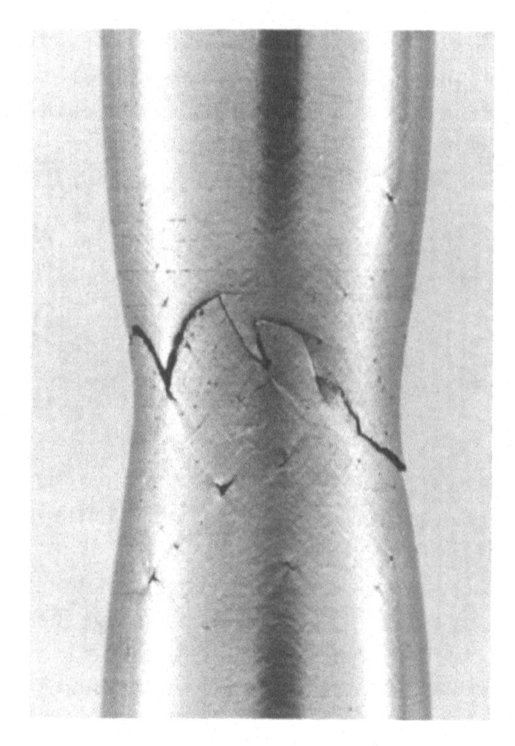

Fig. 3 Low-carbon cast steel test specimen emphasizing 45° shear aspect of tensile fracture of a ductile metal. Diagonal ridges are "Lüders lines" or "stretcher strains"; porosity in the steel shows many localized fractures.

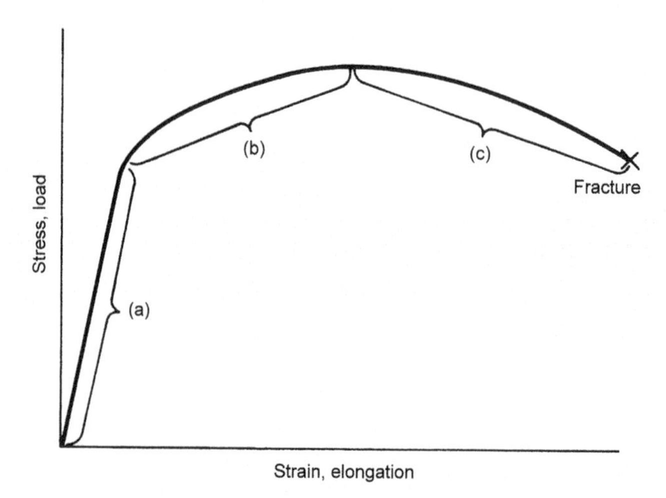

Fig. 4 Typical stress-strain diagram showing different regions of elastic and plastic behavior. (a) Elastic region in which original size and shape will be restored after release of load. (b) Region of permanent deformation but without localized necking. (c) Region of permanent deformation with localized necking prior to fracture at the "X."

down area. Before the fracture reaches the surface, however, it suddenly changes direction from generally transverse to about a 45° angle. It is this slant fracture—frequently called a "shear lip"—that forms the cup-and-cone shape characteristic of many tensile fractures of ductile metal. This slant fracture is useful for study of many fractures, for it represents the end of the fracture process at that location. Tensile fracture of a relatively thin section of a ductile metal may be entirely slant fracture. As the thickness increases, however, the percentage of slant fracture around the central origin area will decrease, sometimes resembling a "picture frame," on a relatively thick rectangular section. The reasons are discussed in Ref 1 and other references.

Note that *shear* is defined as "that type of force that causes or tends to cause two contiguous parts of the same body to slide relative to each other in a direction parallel to their plane of contact" (Ref 3). This term has been used frequently in this work, but it is necessary to point out that shear can be considered on both a macroscale and a microscale. *Macroshear* involves lateral cutting, or shearing, as pointed out in Chapter 3, "Stress versus Strength," in this book, as in direct, or transverse, shear. The cutting action of a pair of scissors (shears) is a common example. *Microshear* involves microscopic sliding on shear planes of the atomic crystals. Microshear causes plastic, or permanent, deformation, as discussed in Chapter 6, "Basic Single-Load Fracture Modes," in this book. Ductile fracture may or may not occur, depending on the relative stress and strength involved. Ductile fractures can appear differently in high-strength, high-hardness, high-alloy metals compared with low-strength, low-hardness, low-alloy metals. This is illustrated later in this chapter.

Microstructural Aspects of Ductile Fracture

It is on a microscopic scale that the characteristics of ductile deformation and fracture really become unique. It is necessary to examine both the changes that are visible with a light microscope and those that can be examined only with an electron microscope at magnifications not attainable by a light microscope.

Figure 5 shows the behavior that occurs within a typical ductile metal immediately prior to fracture due to a tensile force. Detailed study of these three photographs is instructive (Ref 2).

Figure 5(a) is a photograph at low magnification (6×) of the necked-down portion of a tensile specimen made from hot-rolled 1020 steel immediately prior to fracture. In order to make this study, it was necessary to stop the test after the maximum load had been attained, in the region of the stress-strain curve just prior to the fracture at the "X" in Fig. 4. The specimen was sectioned longitudinally and metallographically prepared by polishing and etching to reveal the grain structure. Note that the major crack

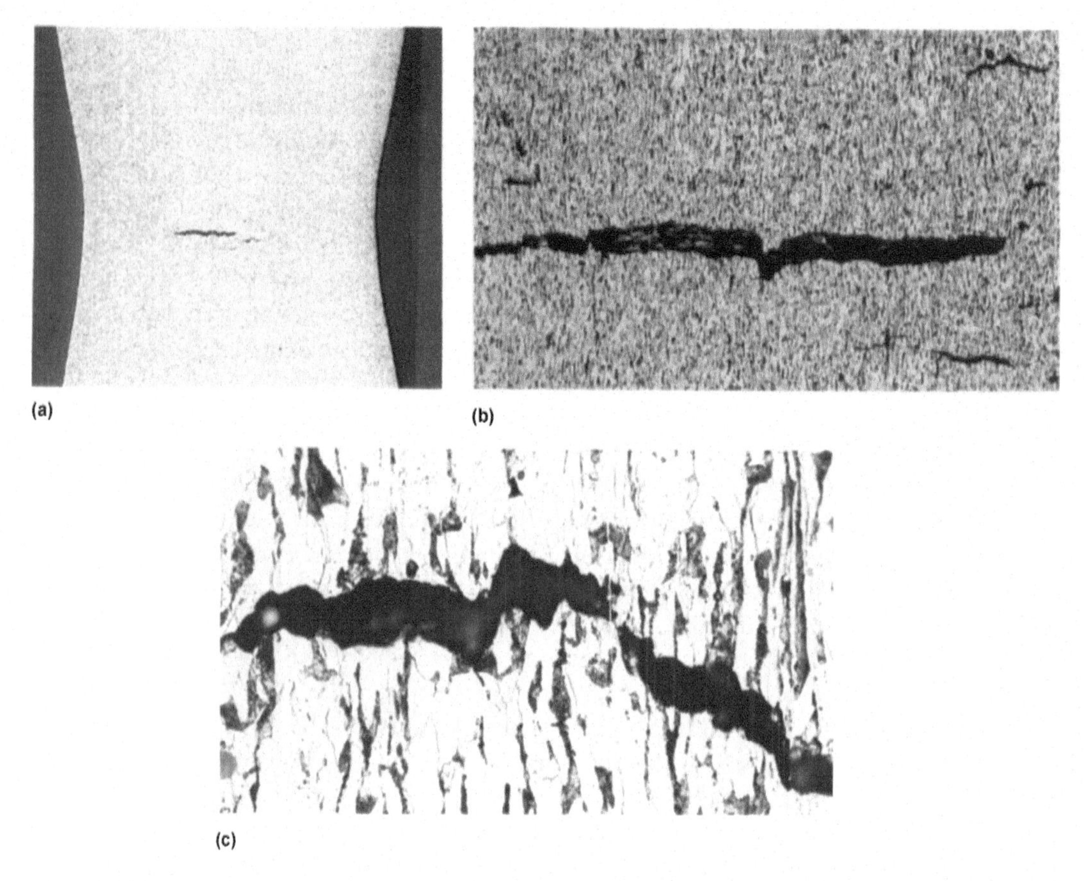

Fig. 5 Photographs at three different magnifications (shown here at 75%) of longitudinal section of tensile test specimen of hot-rolled 1020 steel. The test was stopped immediately prior to fracture (region (c) of Fig. 4). (a) At 6× magnification, internal cracking, reduced section, and severely deformed grain structure can be seen. (b) At 50× magnification, one large crack and several smaller cracks and elongated grains are apparent. (c) At 250× magnification, lack of fit due to distortion is clear, as well as elongated ferrite (light) and pearlite (dark) areas. Specimens were prepared with 2% Nital etching.

is near the center of the reduced section and that there is a smaller crack close to it nearer the center.

Figure 5(b) is a higher-magnification (50×) view of the center area, showing a large crack and many smaller cracks in the same general area. Several smaller cracks have joined together to form the major cracks. There is also tremendous vertical (axial) elongation of the original equiaxed (nondirectional) grains of this hot-rolled steel. Grain distortion of this kind is characteristic of ductile deformation and fracture and is frequently observed in examination of damaged and failed parts, provided the plane examined is in the direction of deformation, as in this photograph.

Figure 5(c) is a view at still higher magnification (250×) of part of a crack that had not yet separated completely when the test was stopped.

Note that the opposite sides of the crack do not fit each other because of the distortion. This is also characteristic of ductile fracture: the opposite surfaces cannot be fitted closely together as they can in a truly brittle fracture. Again, note the axial elongation of the grains; the white areas are soft ferrite, and the dark areas are harder pearlite.

The magnification of these three photographs is too low to reveal the mass of internal fractures or cracks that developed on the fracture surface during the fracturing process. The analyst must visualize a large number of individual microareas in the section being pulled apart under tensile stress. Obviously, the weakest areas, or at least those that contain some minute imperfection or inclusion, separate first into microvoids, or tiny cavities. As soon as the first internal fracturing occurs, the cross section at these locations is slightly reduced, thereby putting a higher local stress on the surrounding areas, which are then likely to form their own microvoids. These microvoids coalesce rapidly, and the process continues until fracture occurs. For this reason, the term frequently used for this process of formation of tiny voids and their growing together is called *microvoid coalescence*.

The process of microvoid coalescence continues to form larger cracks, which have approximately half of each cavity on each side of the fracture surface as a tiny cup or "dimple." The actual fracture surface of a ductile metal, therefore, is essentially nothing but a mass of dimples, or half-voids, which usually can be seen only with the aid of an electron microscope: this is termed a *dimpled-rupture* fracture surface, as shown in Fig. 6. Examination of these dimples is exceedingly useful in studying fractures because the dimples are extremely sensitive to the direction of the stresses that formed them.

A fairly accurate, but exaggerated, analogy to the phenomenon of microvoid coalescence and plastic deformation on a microscopic scale involves pizza on a macroscopic scale. When a slice of hot pizza is pulled away from its neighbor, the hot cheese—an extremely ductile material—is

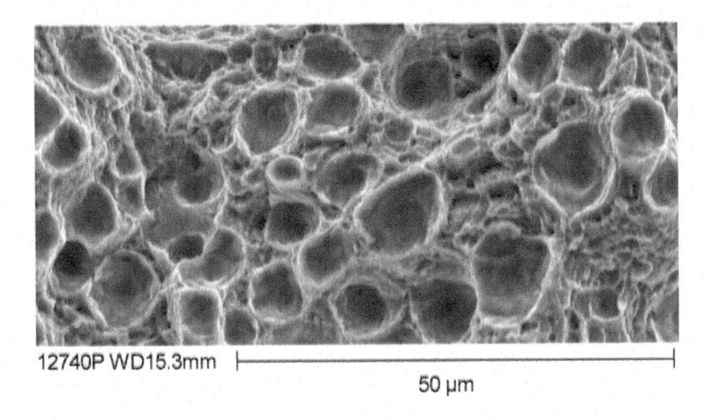

Fig. 6 Typical dimpled-rupture fracture surface of a ductile metal viewed with a scanning electron microscope at a magnification of 1000×.

easily distorted, forming long strings with holes between them that can be pulled or moved laterally until separation, or fracture, occurs.

Two-dimensional pizza holes can be compared to the three-dimensional microvoids that form in fractures of ductile metal, while the strings are analogous to the sides of the dimples. Lateral movement causes formation of elongated holes (dimples) when separation (fracture) occurs.

Dimple shapes have been studied extensively and are shown schematically in Fig. 7:

1. **Tension.** As shown in Fig. 7(a), the tensile force at the center of the specimen causes microvoids first to form near the center and then to spread to nearby areas that must then carry more stress because the cross section is now smaller in area. Under an axial force as shown, the microvoids are not skewed, or tilted, in any particular direction; thus, the fracture surface consists of equiaxed dimples when viewed perpendicular to the surface. The angle of view is particularly important when

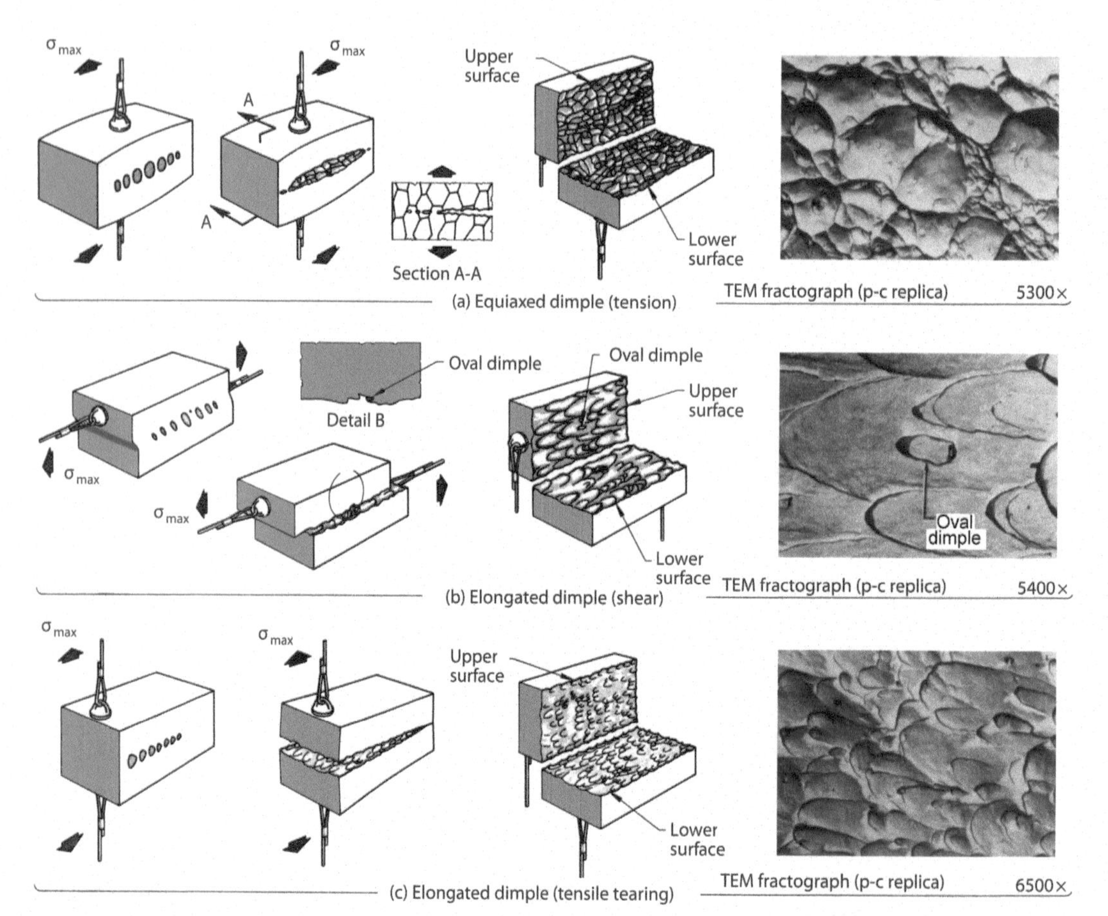

Fig. 7 The influence of the direction of principal normal stress on the shape of dimples formed by microvoid coalescence. Transmission electron microscopic (TEM) fractographs are phase-contrast (p-c) replicas of fracture areas. Source: Ref 2

the dimples are studied in a scanning electron microscope; if they are viewed from an angle, they can appear foreshortened and not equiaxed.

2. **Shear.** As shown in Fig. 7(b), the dimples become elongated at a pure shear fracture surface, such as that of a torsional fracture of a ductile metal. The dimples can become so elongated that they no longer closely resemble dimples except for their C-shaped ends. Also note that the "C" shapes are in opposite directions on the opposed fracture surfaces because each represents one end of a dimple that was pulled in opposite directions. On torsional fractures, these dimples are frequently damaged by rubbing from the opposite sides so that the surface observed is simply a mass of circular rub marks. (These marks must not be confused with fatigue striations, which are discussed in Chapter 10, "Fatigue Fracture," in this book.)
3. **Tensile Tearing.** As shown in Fig. 7(c), this mechanism of fracture is somewhat similar to the pure tension mechanism shown in Fig. 7(a) except that the fracture actually originates at the edge of the metal rather than at the center. This is due to a bending force on the part that causes tensile fracture with equiaxed dimples at the region close to the origin, while the actual tensile tearing causes the C-shaped dimples to form near the end of the fracture, opposite the origin.

In addition to these three basic modes of microvoid coalescence and fracture, several others are described in Ref 2 and shown in Fig. 8. The three modes just discussed appear at the upper left of Fig. 8.

The effect of the direction of fracture can be seen dramatically on many bending fractures of ductile metals, even if the exterior has been case hardened. Close to the surface on the convex side of the fracture (the origin surface), the dimples are essentially round and equiaxed because of the tensile stress at that location, as shown at the bottom center portion of Fig. 9(a) and (b). In pure bending, the fracture starts on this convex side and progresses across the part to the opposite side, which acts as a hinge as the fracture opens up.

When the fracture approaches the hinge (concave) side of the fracture, the local stress is no longer pure tension but is in the tensile tearing mode, as shown in Fig. 7(c). At high rates of loading, the dimples can become extremely elongated, as is shown in the center and upper portions of Fig. 9(b) and (c). Also, if there is a tensile force on the part in addition to the bending force, the dimple elongation will not be as dramatic as it is in this sequence of a rapid, pure bending fracture. Note that the closed end of the elongated dimples points back toward the origin, as is also shown in Fig. 9(c).

The fractograph in Fig. 9(b) is a view of the fracture surface in the interior of the part where the transition between equiaxed and elongated dimples occurred. Note that on the bottom side of the fractograph the dimples are equiaxed, while on the top side, they start to become elongated.

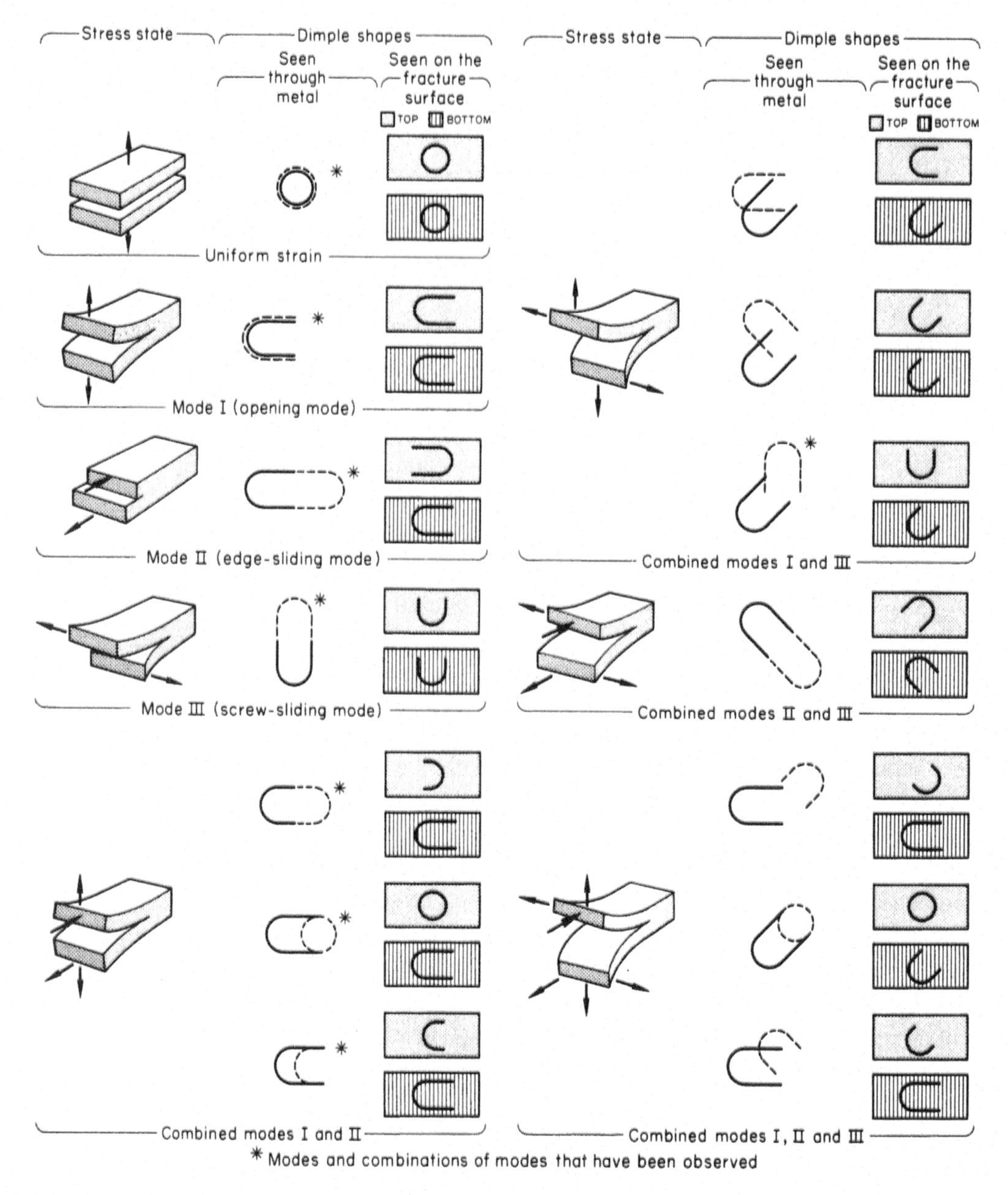

Fig. 8 Fourteen probable combinations of mating dimple shapes, resulting from different stress states that caused metal at the crack tip to deform by various modes. Source: Ref 2

Cautions in Interpretation

At this point, it is prudent to insert some cautions about interpretation of the evidence visible with this type of microscopic examination. The analyst must be aware of complicating factors extraneous to the fracture itself:

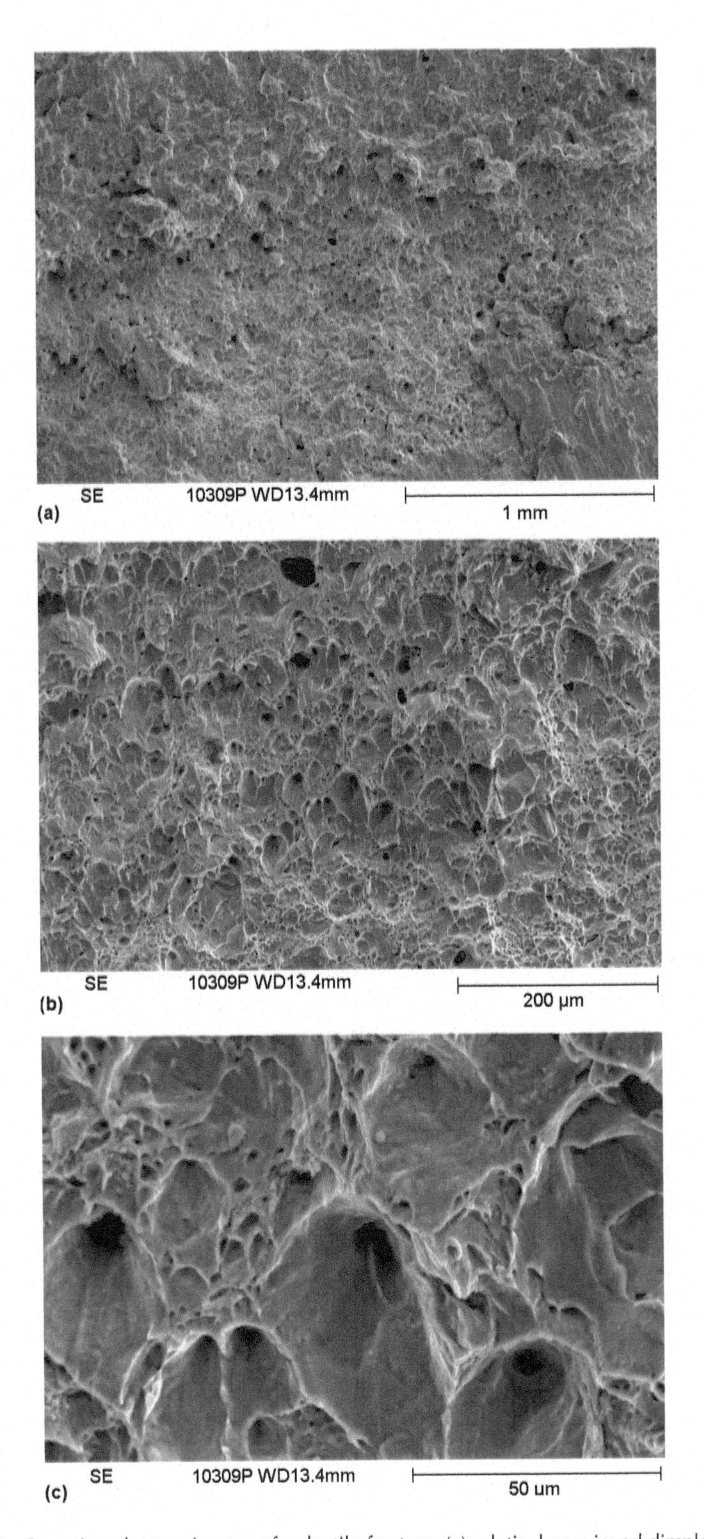

Fig. 9 Scanning electron images of a ductile fracture: (a) relatively equiaxed dimples at the center portion of the image, with some postfracture mechanical rubbing damage at the bottom portion of the image (50×); (b) elongated dimples indicating ductile rupture with a shear component are shown at the center portion of the image (200×). (c) The elongated dimples shown at the center portion of (b) are presented at higher magnification (1000×). Finer dimples are present along the ridges between the larger dimples.

- *Mechanical damage:* The fracture of a metal part is always a violent event because of the sudden release of energy. Fracture surfaces frequently are rubbed or pounded against each other, causing rub marks, abrasion, dents, smearing, or other scars of postfracture damage. It is common for much of the fracture surfaces to be obliterated so that microscopic examination as discussed above is difficult or impossible. The microscopic features on fracture surfaces are fragile and are easily damaged; when this occurs, one must rely on the macrofeatures of the part or parts for useful evidence in analyzing the fracture. In some cases, examination of relatively nondisturbed features between islands of damage is possible.
- *Chemical damage:* Surface pits resembling the dimples of ductile fracture may instead be the result of severe etching with corrosive acids (or bases, in some metals) or of cleaning techniques used to remove atmospheric corrosion. In other cases, a corrosive film, or layer, may have formed on the fracture surface and completely obliterate the underlying character of the surface.
- *Cavities may not be dimples:* In certain metals, cuplike cavities may be present on a fracture surface, but the cavities may not be the dimples characteristic of ductile fracture. In cast metals and weldments, particularly, this type of rounded cavity may actually be evidence of gas porosity. Gas porosity occurs because the inevitable gases in castings and weldments (which are actually localized castings) cannot escape to the atmosphere before the metal solidifies and traps the gas in tiny, smooth-walled bubbles within the metal. This is particularly true in welds in metals such as aluminum, which has a high rate of thermal conductivity. In effect, large, cold members having high thermal conductivity act as heat sinks, causing rapid solidification of the weld metal and entrapping gas bubbles before they can escape. In ductile fractures of metals with extensive gas porosity, it may be quite difficult to determine which cavities are true dimples and which are evidence of gas porosity.
- *Mixed fracture modes:* During study of a fracture surface, one should not be surprised to find different fracture modes. An example of a fracture through a metal injection molded specimen with a mixed fracture mode is presented in Fig. 10. The fracture consists predominantly of dimples, indicative of ductile rupture, along with flat features that were present due to grain boundary separation. Other cases of mixed fracture modes may include something like a predominantly ductile fracture surface with isolated regions of cleavage or intergranular fracture because of differences within the metal. These differences may result from crystalline orientation with respect to the fracture stress, or from differences in the microstructure of the metal.

A steel with a partly pearlitic, partly ferritic microstructure, for example, may fracture with dimple rupture in the ferrite regions but frac-

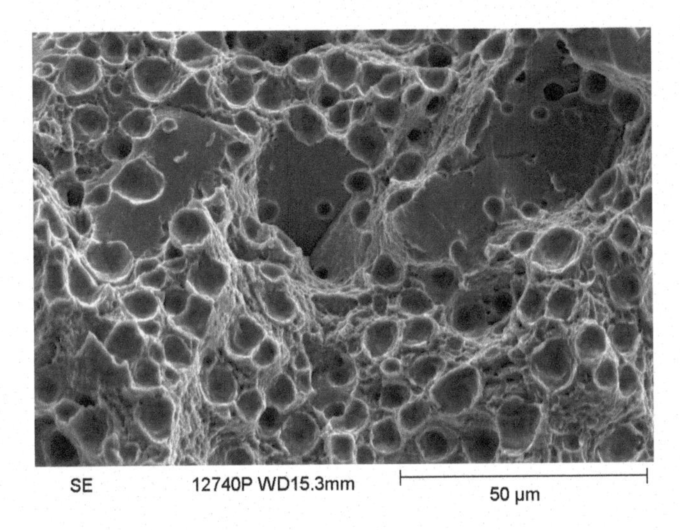

Fig. 10 A scanning electron micrograph of a mixed fracture mode from a metal injection molded (MIM) steel. The dimples indicate ductile rupture. The flat features indicate separation of nonfused particles. The circular features on the flat fractures are porosity.

ture by cleavage in the pearlite regions. A notched-bar impact test specimen may fracture by cleavage in some areas and with dimple rupture in others. A hydrogen-embrittlement fracture may have certain areas with intergranular fracture and others with dimple rupture. No true fatigue striations are possible in a single-overload fracture; however, there may be somewhat similar-appearing parallel ridges resulting from fracture through pearlite or other lamellar structures, or resulting from mechanical rubbing either during or after the fracture. These spurious marks can resemble striations that can confuse the analyst, who has a difficult task even without these complications. Of course, if cyclic loading has in fact occurred, true fatigue striations may also be present in addition to any or all of the other fracture modes. See Chapter 10, "Fatigue Fracture," in this book for more information.

Ductile fractures may not always have the classic cup-and-cone or dimpled structures shown in this chapter. High-strength, high-hardness, high-alloy metals may possess a rough texture with very fine dimples, as shown in Fig.11. Knowledge of the material, loading conditions, and the texture of the remainder of the fracture surface, along with the experience of a knowledgeable failure analyst, are often required to correctly interpret the nature of the fracture features.

Fractures through porosity can be misinterpreted as ductile rupture. Such a case in an aluminum fatigue bar specimen is presented in Fig. 12. Locations 1 and 5 in Fig. 12(a) indicate regions of macroporosity on the fracture surface. A scanning electron micrograph of the top center portion

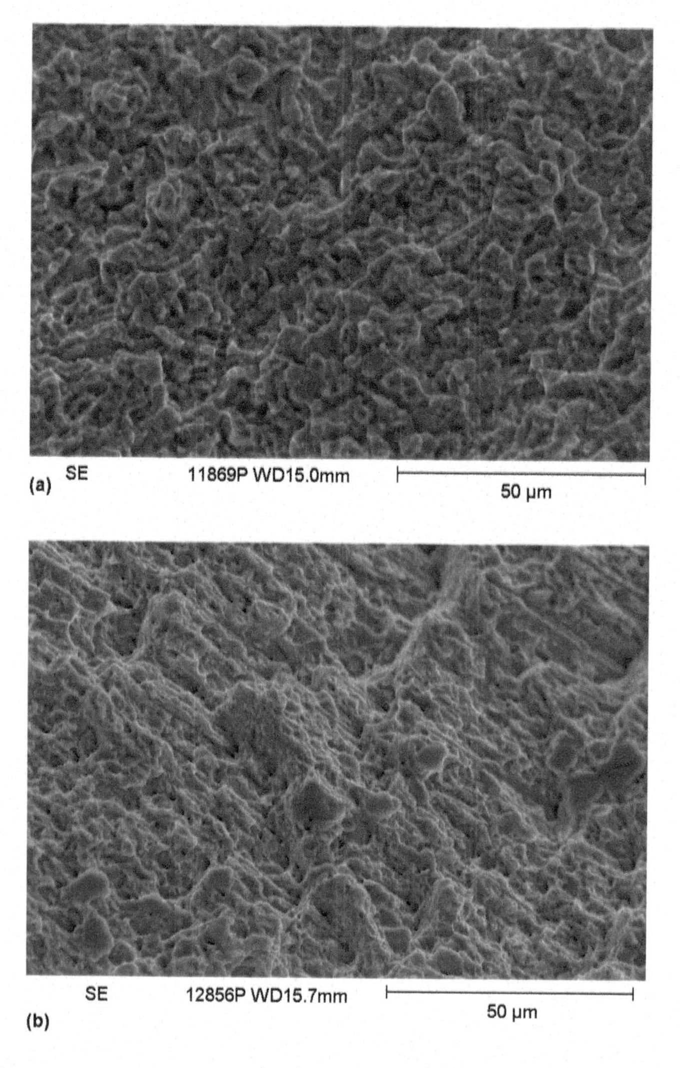

Fig. 11 (a) Ductile fracture through type 420 stainless steel hardened to 51 HRC equivalent (1000×). (b) A woody texture, indicative of ductile fracture, through grade 9254 spring steel.

of Fig. 12(a) is presented in Fig. 12(b); the smoothness of the macropores is evident in this view. Higher magnification of location 4 in Fig. 12(a) is presented in Fig. 12(c), showing a spherical pore. The texture surrounding the pore is indicative of fatigue-related cracking. An inexperienced failure analyst may incorrectly interpret the pore as an isolated ductile rupture.

Summary

Ductile fracture occurs when the shear strength is the limiting factor. Permanent (plastic) deformation is inherent in ductile fracture and is usually obvious. The micromechanisms of dimple formation and distortion

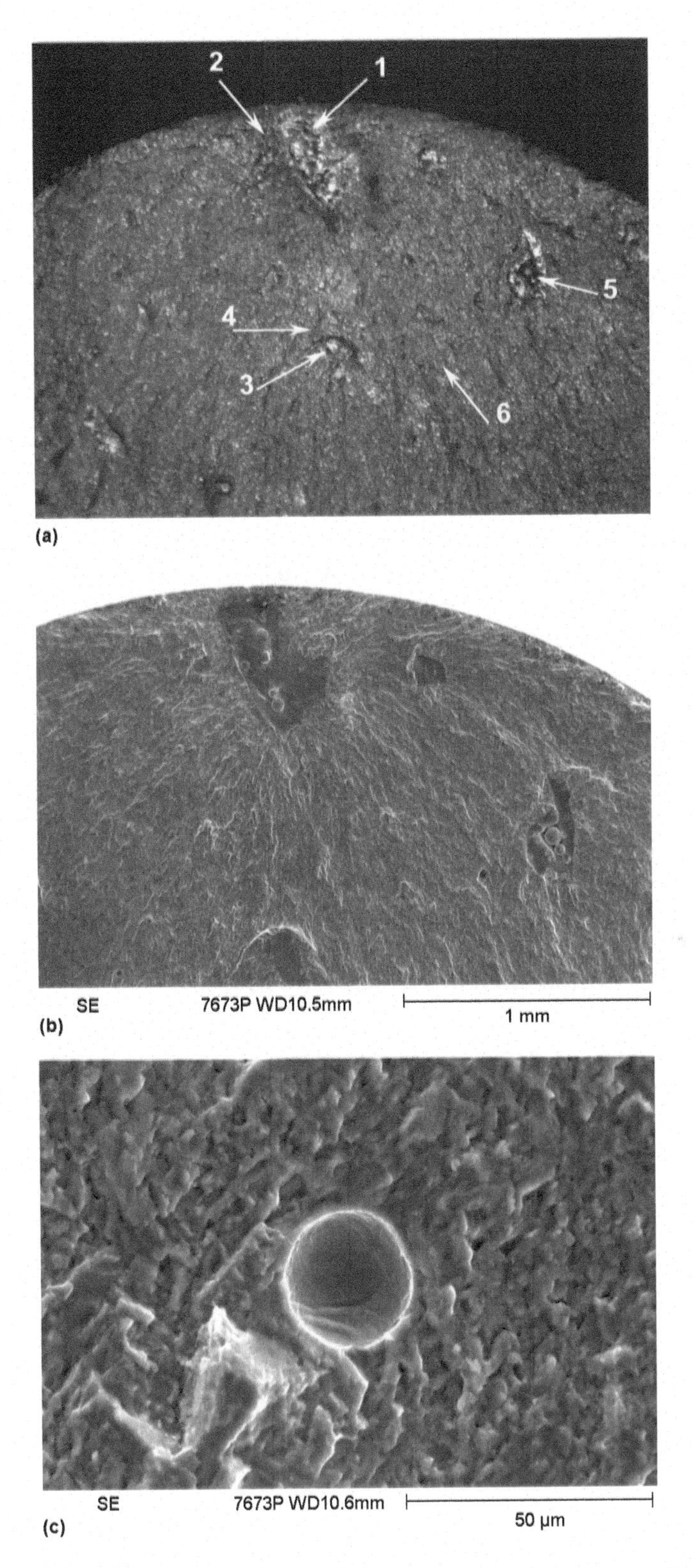

Fig.12 (a) Fracture through an aluminum fatigue bar specimen (33×). Locations 1 to 6 were selected for closer inspection. (b) A scanning electron micrograph of the fracture (50×). (c) Higher magnification of the feature shown at location 4. The feature is a gas pore.

must be understood in order to study electron microscope views of ductile-fracture surfaces. Careful examination and knowledge of the metal, its thermal history, and its hardness are important in determining the correct nature of the fracture features.

REFERENCES

1. *Failure Analysis and Prevention,* Vol 11, *ASM Handbook,* ASM International, 1986, p 20, 82–101
2. *Fractography,* Vol 12, *ASM Handbook,* ASM International, 1987, p 12–15, 173–174, 284
3. *ASM Metals Reference Book,* 3rd ed., ASM International, 1993, p 77

SELECTED REFERENCES

- S. Bhattacharyya, V.E. Johnson, S. Agarwal, and M.A.H. Howes, Ed., *IITRI Fracture Handbook: Failure Analysis of Metallic Materials by Scanning Electron Microscopy,* Metals Research Division, Illinois Institute of Technology Research Institute, 1979
- *Fatigue and Fracture,* Vol 19, *ASM Handbook,* ASM International, 1996
- R.W. Heitzberg, *Deformation and Fracture Mechanics of Engineering Materials,* 4th ed., Wiley, 1995
- H. Liebowitz, Ed., *Fracture: An Advanced Treatise,* Vol 1–8, Academic Press, 1968
- J.L. McCall and P.M. French, Ed., *Metallography in Failure Analysis,* Plenum, 1978
- G. W. Powell, S.-H. Cheng, and C.E. Mobley, Jr., *A Fractography Atlas of Casting Alloys,* Battelle Press, 1992

CHAPTER 10

Fatigue Fracture*

FATIGUE FRACTURES are generally considered the most serious type of fracture in machinery parts simply because fatigue fractures can and do occur in normal service, without excessive overloads, and under normal operating conditions. For example, in the aerospace industry, fatigue accounts for upwards of 90% of all service failures. Fatigue fractures are serious because they are insidious; that is, they are frequently "sneaky" and can occur without warning. Obviously, if service is abnormal as a result of excessive overloading, corrosive environments, or other conditions, the possibility of fatigue fracture is increased.

Let us consider the definition of *fatigue* that is commonly accepted:

> *The phenomenon leading to fracture under repeated or fluctuating stresses having a maximum value less than the tensile strength of the material. Fatigue fractures are progressive, beginning as minute cracks that grow under the action of the fluctuating stress. (Ref 1)*

This definition refers to fracture "under repeated or fluctuating stresses having a maximum value less than the tensile strength." This is one of the most interesting—and puzzling—aspects of fatigue fracture. Many submicroscopic changes take place in the crystalline structure of the metal under the action of relatively low-level, repetitive load applications. These minute changes in the structure, as suggested in the second part of the definition, may progress gradually to form tiny cracks that may grow to become large cracks under continued cyclic loading, and that can then lead to fracture of the part or structure. Once initiated, the fatigue fracture can propagate by high stresses and low cycles or by low stresses and high cycles. The final rupture may have characteristics of brittle and/or ductile

***Understanding How Components Fail*, Second Edition, by Donald J. Wulpi, Chapter 10, Fatigue Fracture, ASM International, 1999, reviewed and revised by Nicholas E. Cherolis, Rolls-Royce, and Wesley D. Pridemore, GE Aviation

fracture modes, depending on the metal involved and the circumstances of stress and environment.

Stages of Fatigue Fracture

The preceding discussion suggests that there are three stages of fatigue fracture: initiation, propagation, and final rupture (overload). Indeed, this is the way that most authors refer to fatigue fracture, for it helps to simplify a subject that can become exceedingly complex.

Stage 1: Initiation. Initiation is the most complex stage of fatigue fracture and is the stage most rigorously studied by researchers. Obviously, if this stage can be prevented, there will be no fatigue fracture. The submicroscopic changes referred to previously are difficult to visualize, difficult to describe, and difficult to understand.

The most significant factors about the initiation stage of fatigue fracture are the irreversible changes in the metal caused by repetitive shear stresses, or slip, typically occurring at 45° to the principal stress axis. Figure 1 shows schematically these microstructural changes, or dislocation movements, that occur in the crystal lattice because of stresses repetitively imposed on the structure. A single load application makes little change, but the cumulative effect of thousands or millions of load applications makes many microchanges and in some alloys can lead to formation of intrusions and extrusions on the surface, as shown in Fig. 2. Although each is insignificant by itself, they accumulate to cause very significant alterations in a few or many grains that can lead to fatigue fracture. Indeed, the accumulation of microchanges over a large number of load applications, called *cumulative damage*, has been the subject of study over the years (Ref 3). Study of the cumulative damage concept seeks to predict or estimate fatigue life of a particular part or structure but is handicapped by complications that are discussed throughout this chapter.

Stage 1 fractures typically consist of crystallographic facets commonly oriented at 45 degree angles to the applied stress. The stage 1 fatigue initiation site of a given fatigue fracture is very small, seldom extending for more than two to five grains around the origin (Ref 1); an example of such a fracture is shown in Fig. 3. Under higher-stress conditions, such as at the location of a severe stress concentration (notches, inclusions, etc.), the faceted stage 1 fatigue region may be extremely small or even absent and difficult to differentiate from the succeeding stage of propagation, or crack growth. It must also be recognized that there can be any number of these initiation sites. The number depends on the geometry of the part, as well as on environmental, stress, metallurgical, and strength conditions, as described in this chapter.

Stage 2: Propagation. As repetitive loading continues, the direction of the crack changes from parallel to the shear stress direction to perpendicu-

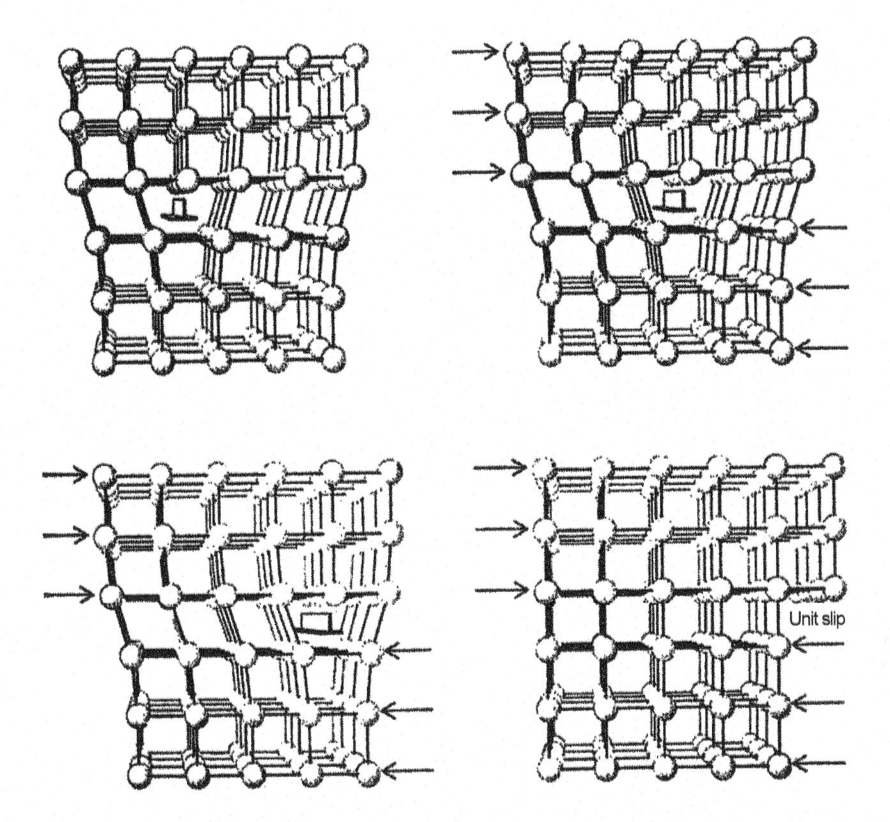

Fig. 1 Schematic sketch of microstructural changes in crystal structure due to repetitive shearing forces. Spheres represent atoms, and lines represent attractive and repulsive interatomic forces. An edge dislocation, represented by the inverted T-shaped symbol, is an imperfection in the crystalline structure. When repetitive shearing forces (top right) are applied, the dislocation slips one cell to the right. Continued repetitive shearing forces (bottom) make the dislocation move to the edge of the crystalline structure. When this action is going on in many other crystals, the dislocations eventually join to form microscopic cracks. This is stage 1 in the fatigue process.

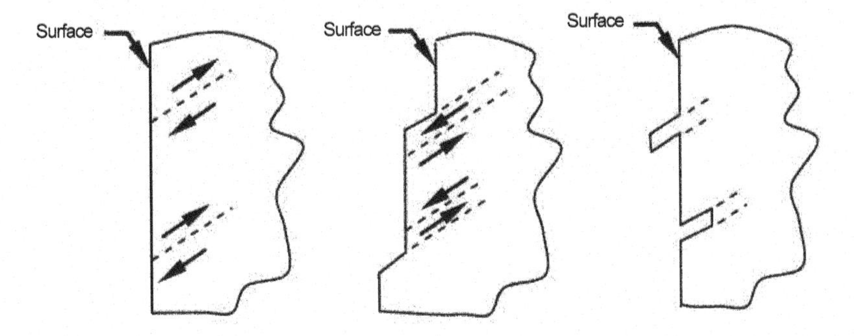

Fig. 2 Schematic illustration of the formation of extrusions and extrusions in stage 1 fatigue. Source: Ref 2

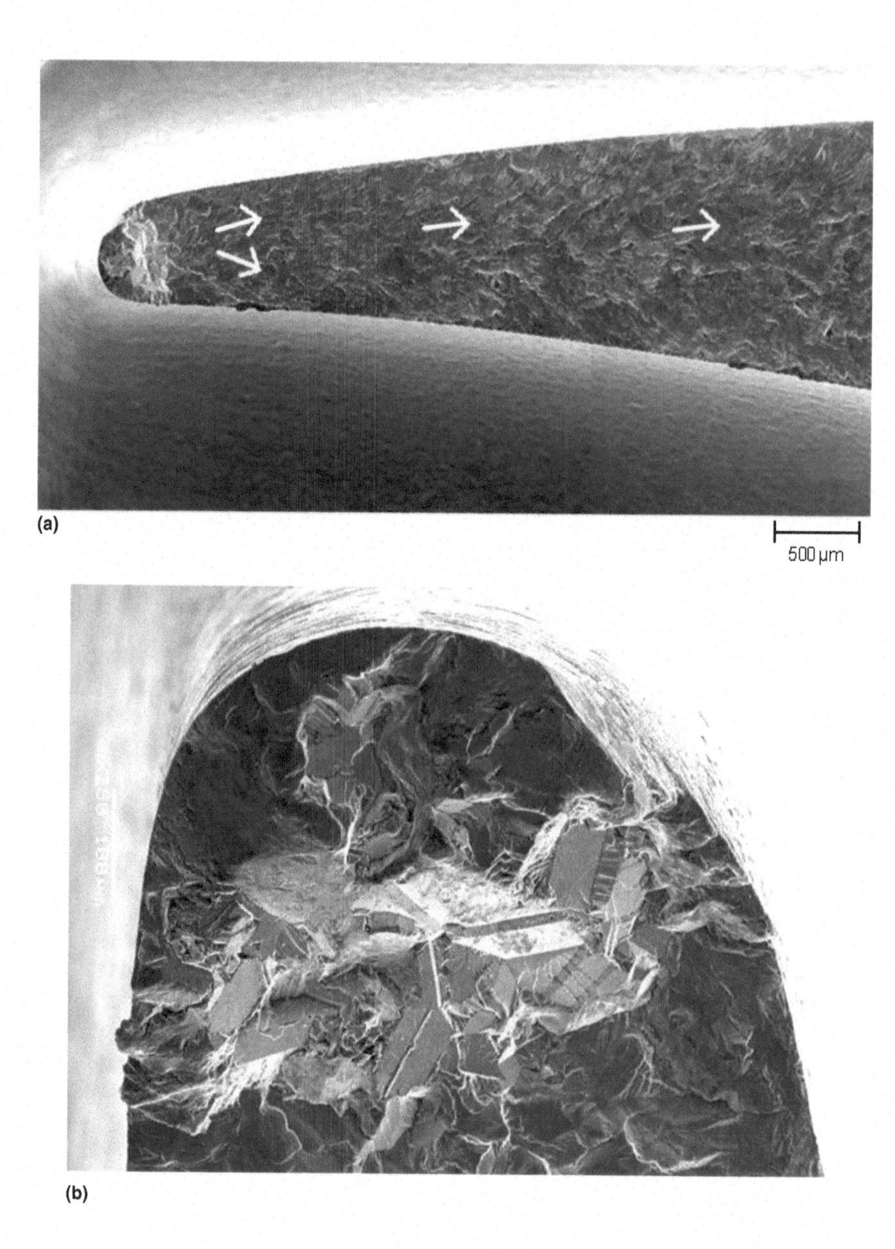

Fig. 3 (a) Example of faceted stage 1 fatigue initiation in a cast nickel-base superalloy turbine blade. Arrows denote crack propagation direction. (b) Overall blade fracture and continuation of flatter stage 2 crack propagation emanating from stage 1 region.

lar to the tensile stress direction. It may be necessary to review Chapter 7, "Stress Systems Related to Single-Load Fracture of Ductile and Brittle Metals," in this book, which discusses both the shear stress and the tensile stress directions for different types of loading.

Stage 2 fatigue propagation is usually the most readily identifiable area of a fatigue fracture. As the fracture progresses across the section, it acts as an extremely sharp stress concentration that tends to drive the crack ever deeper into the metal with each tensile stress application, assuming that the maximum cyclic stress is of a magnitude high enough to propagate the crack. The local stress intensity at the tip of the crack (see Chapter 15, "Fracture Mechanics," in this book) is extremely high because of the sharp "notch," and with each crack opening, the depth of the crack advances by one "striation" under many (but not all) circumstances. Striations are microscopic, closely spaced ridges that identify the tip of the crack at successive points in time. Striations are normally visible only with the aid of an electron microscope; they are discussed in the section "Microscopic Characteristics of Fatigue Fracture" in this chapter. In the later portion of the stage 2 region, some materials tend to exhibit cyclic tensile morphology where individual striations are not formed, with only tensile dimples being observed.

Stage 3: Final Rupture (Overload). As the propagation of the fatigue crack continues, gradually reducing the cross-sectional area of the part or test specimen, it eventually weakens the part so greatly that final, complete fracture can occur with only one more load application. This is stage 3, or the final rupture. From a technical standpoint, this is the stage in a fatigue fracture that is easiest to understand. Actually, stage 3 is not fatigue at all because the fracture mode may be either ductile (with a dimpled fracture surface), brittle (with a cleavage, or perhaps even intergranular, fracture surface), or any combination thereof, depending on factors such as the metal concerned, the stress level, and the environment. Stage 3 represents the "straw that broke the camel's back."

However, the failure analyst must pay careful attention to the size, shape, and location of the final rupture area because it can greatly assist in understanding the relation between the stresses on the part and the strength of the part; they can also indicate imbalance and nonuniform stresses.

Microscopic Characteristics of Fatigue Fracture

Striations are the most characteristic microscopic evidence of fatigue fracture, although interpretable striations are not always present on fatigue fracture surfaces.

During stage 2, the primary crack propagation stage of a fatigue fracture, the tip of the crack is an extremely sharp notch, or stress concentration. Actually, it can be considered to be infinitely sharp, for it basically has zero radius at the tip. However, each time the crack is opened by a tensile stress of sufficient magnitude, the crack tip deforms plastically, extending the crack and slightly blunting the tip on a microscopic scale, as in Fig. 4. On crack closure, the crack tip sharpens and the process repeats, creating a series of striations across the fracture. The exact process is still being studied, but for practical purposes, the above description is adequate. Electron micrographs of striations in forged titanium alloys are shown in Fig. 5 and

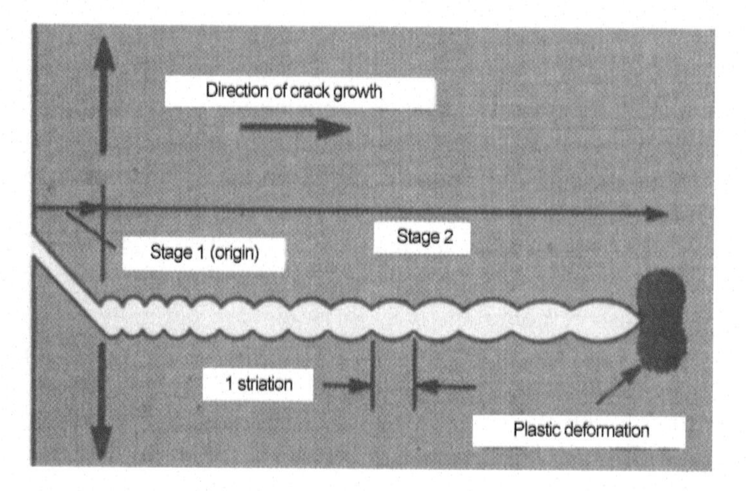

Fig. 4 Highly enlarged schematic cross-sectional sketch of stage 1 and stage 2 fatigue. The edge of the metal is at left. When tensile forces repeatedly act on the surface, the microstructural changes of stage 1 cause a submicroscopic crack to form. With each repetitive opening, the crack jumps a small distance (one striation). Note that the spacing of each striation increases with the distance from the origin, assuming the same opening stress. The metal at the tip of the fatigue crack (right) is plastically deformed on a submicroscopic scale.

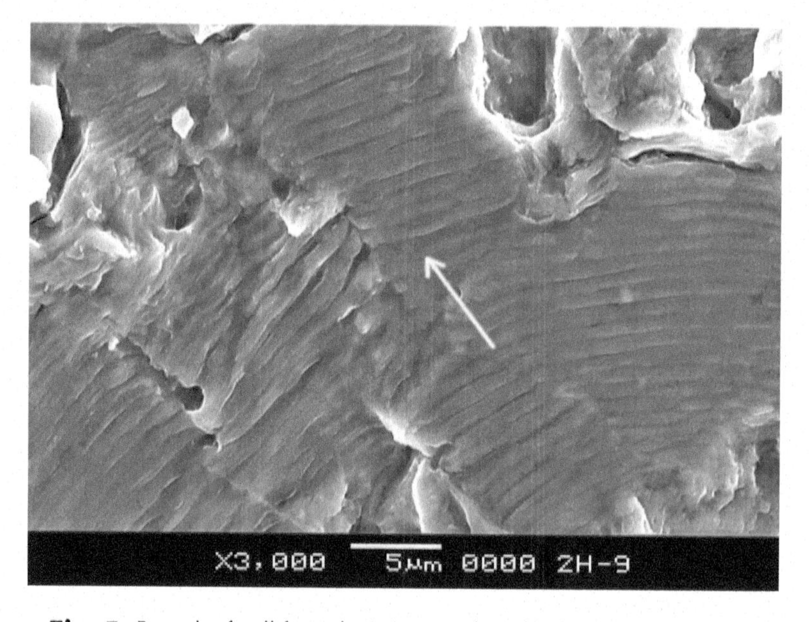

Fig. 5 Example of well-formed striations in a forged high-pressure compressor blade made of titanium alloy. The striation density is approximately 30,000 striations/in. (~3.3×10^{-5} in./striation). The arrow denotes the crack propagation direction.

6. Note the range of striation appearance and spacing and the slight curvature of the striations as opposed to straight linear features, such as slip lines or other features related to microstructure.

If the maximum cyclic load remains constant, the striations near the fatigue origin are extremely fine and closely spaced; the crack grows at a slow rate because the part's cross section has not been significantly reduced. However, as the crack gradually propagates, the spacing between striations increases, and the crack grows at an increasingly rapid rate because the crack greatly weakens the section, eventually leading to final fracture (stage 3) and separation if left unabated. Examples of variable load conditions are shown in Fig. 7 and 8. Figure 7 shows a test article loaded with a spectrum of various stress levels in a repeating pattern. This test condition produced a fracture exhibiting successive bands of coarser and finer striations. Figure 8 shows a band of 10 higher-load cycles inserted in a group of lesser-load cycles.

Unfortunately, striations are not always visible on fatigue-fracture surfaces, for a variety of reasons:

- Near the origin, striations tend to be either absent or too fine to resolve in the scanning electron microscope because of the lower stress intensity at this location. In these areas, structure-related fatigue features tend to dominate, as shown in Fig. 9. In lower-stress fatigue, striations

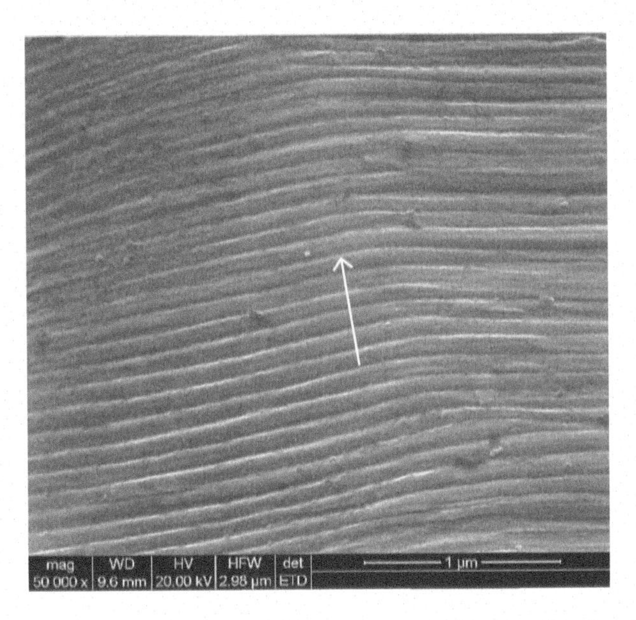

Fig. 6 Example of well-formed fatigue striations in titanium alloy (R = 0.05; maximum alternating stress, 105 ksi). (R is the minimum stress divided by the maximum stress.) The striation density is approximately 263,000 striations/in. (~3.8×10^{-6} in./striation). The arrow denotes the crack propagation direction.

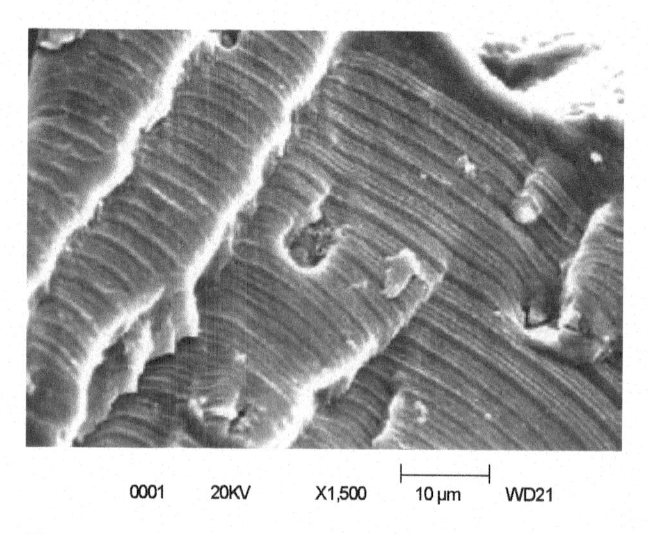

Fig. 7 Striations generated on a spectrum-loaded test article. Source: Ref 2

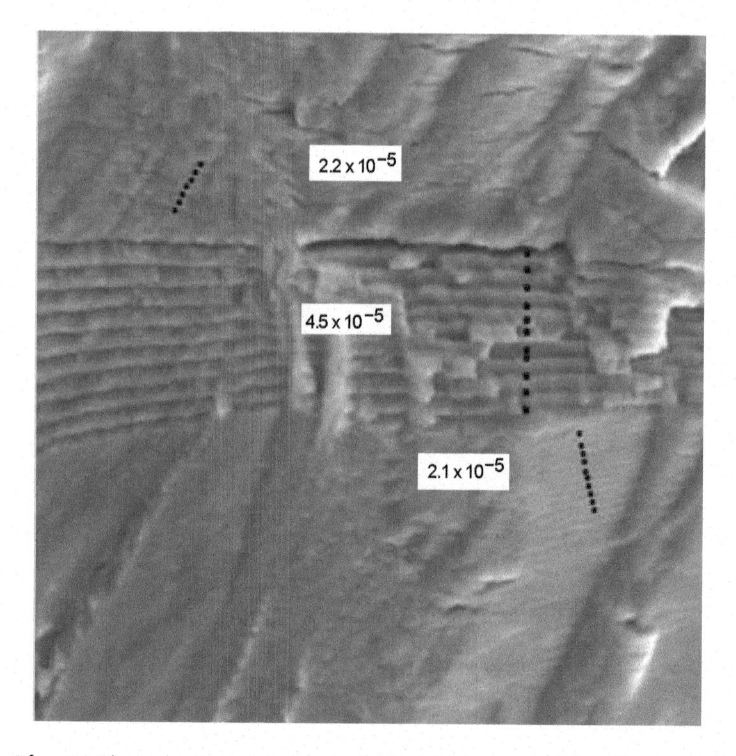

Fig. 8 Electron micrograph showing a block-loaded test article with a band of 10 cycles of a higher load. Source: Ref 2

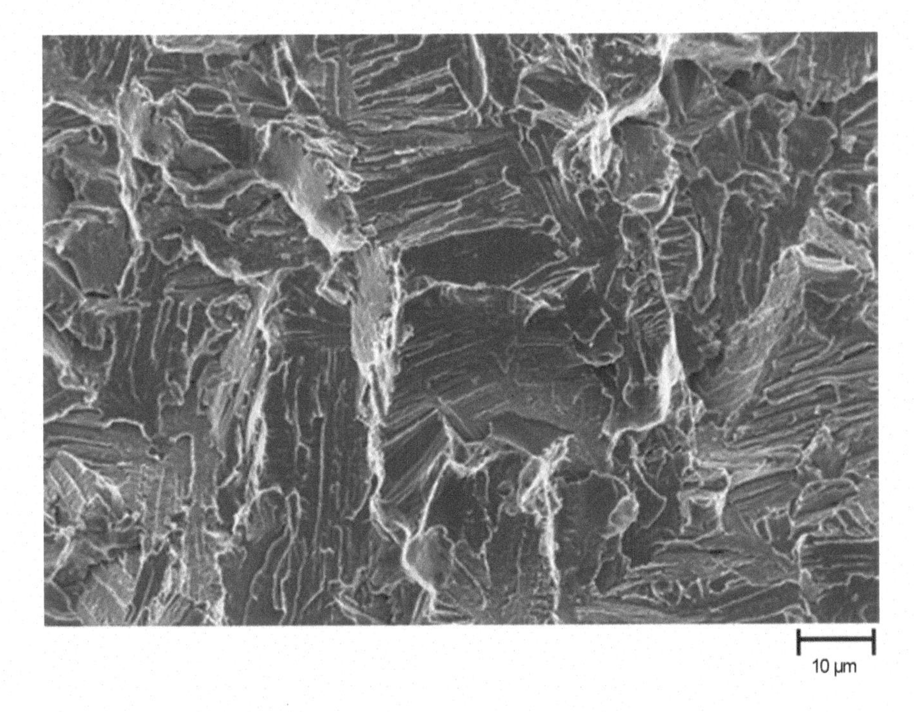

Fig. 9 Scanning electron micrograph of surface structure near the origin of a low-cycle fatigue-loaded test article

may also not be visible over a larger portion of the fracture for the same reason they are not visible near the origin: lower stress intensity.

- Striations are usually not present on very hard or very soft metals. Hardened steels over approximately 50 HRC develop either poorly formed striations or no striations, probably because of their lack of ductility. As hardness increases, striations are more difficult to observe, often resulting in fracture surfaces that appear to be featureless. At the opposite end of the hardness scale, extremely soft and weak metals do not form good striations either, probably because their low strength makes them so vulnerable to smearing damage.
- Quite often, areas exhibiting striations will be adjacent to areas that do not exhibit readily observable striations, sometimes due to differences in local material orientation.
- Impact damage after the fracture occurs can produce parallel smear lines that may resemble striations. This kind of damage also may occur when opposite sides of the fatigue fracture rub against each other during compression or shear loading of a crack. For example, a shaft in rotating bending has a tensile stress at one position (convex side) of the rotating member, while on the opposite (concave) side the compressive stress tends to rub any existing crack surfaces together.
- Loose particles from the fracture surface or externally generated

sources can "walk" across a fracture as it cycles, producing repetitive features called *tire tracks*.

- Certain lamellar microstructures in metals, such as pearlite in steel and cast iron and beta phase in titanium, as well as in eutectic alloys, may have fracture surfaces that can also somewhat resemble fatigue striations. However, careful study with the electron microscope will reveal that the orientation of the platelets varies randomly from one location to another, whereas true striations are generally concentric around the origin, typically possessing some curvature. Also, the platelets may appear to penetrate into the surrounding structure, or they may appear to cross, while true striations do not cross each other.
- Another feature mistaken for striations is wavy glide. These parallel lines can sometimes be seen in overload regions or within individual ductile dimples.

Macroscopic Characteristics of Fatigue Fracture

A tremendous amount of information can be learned about a fatigue fracture with only macroscopic examination. That is, study with the unaided eye and relatively low magnification—up to perhaps 25 to 50×—is usually the most important single way to study and analyze fatigue fractures. In most cases, it is not necessary to exceed 15 or 20× magnification. After all, fatigue fractures were successfully diagnosed for many years before electron microscopes were invented. However, these instruments are of great assistance in understanding the micromechanisms of fractures and in analyzing very difficult cases.

Lack of Deformation

Because initiation of fatigue fracture does not require a high stress, there is usually little or no deformation in a part or specimen that has fractured by fatigue. If the maximum stress did not exceed the yield strength (actually the elastic limit), there can be no gross plastic, or permanent, deformation, although the final rupture (overload) region may have some obvious macroscopic deformation. However, if the part has been subjected to repeated high-stress, low-cycle load applications, there may be deformation depending on the stress-strength relationship. An example of this is the permanent deformation in a paper clip or wire coat hanger fractured by "fatigue" in a relatively small number of extremely high-stress bending applications. Strictly speaking, this is fatigue because more than one stress application was required for fracture. However, this is not the typical fatigue fracture that occurs in most load-bearing parts, which commonly have relatively low-stress, high-cycle loading.

There is no exact dividing line between extremely high-stress, low-cycle fatigue (such as in the paper clip) and the more common low-stress, high-cycle fatigue. The change may occur at any number of load applications up to perhaps 100,000, although this is an arbitrary number (Ref 1). A practical measure of the relative stress level is the final size of the fatigue fracture itself. Low-stress, high-cycle fatigue fractures tend to exhibit smaller final overload regions, and conversely, high-stress, low-cycle fatigue exhibits larger final overload regions. In general, higher-stress fatigue fractures are initiated and propagated at lower-frequency loading conditions on the order of one high load cycle per usage cycle. On the other end of the fatigue spectrum, lower-stress fatigue fractures are initiated and propagated by higher-frequency loading events (vibrations, rotating bending, etc.), as illustrated in Fig. 10.

Not just the fracture surface but the entire part should be examined for deformation. For example, if a unidirectional (one-way) bending fracture is observed, it is useful to carefully reassemble the pieces to determine if there was gross deformation in the part prior to fracture. This would show up as a bent part, either generally or locally, depending on the design. Of course, the origin of the fracture would be on the tension side in bending. Extreme care is necessary in this reassembly, particularly if electron microscope examination is planned, because the fracture surfaces are easily damaged.

As pointed out at the beginning of this section, in a "true" high-cycle fatigue fracture, there will be no deformation in the fatigue region, provided that there has been no postfracture smearing/contact damage to the fracture surface. Frequently, mating fracture surfaces come in contact with

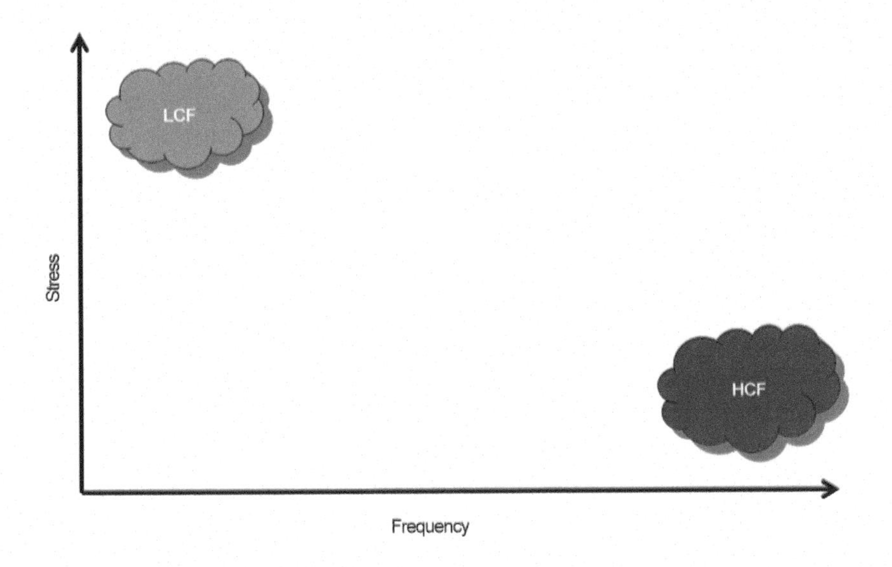

Fig. 10 Schematic showing the relationship between low- and high-cycle fatigue. In systems where significant vibration loads are present, high-cycle fatigue (HCF) tends to be related to high-frequency loading, and low-cycle fatigue (LCF) tends to be related to slowly applied higher-stress loads.

each other and are damaged in the process, especially if the fatigue cycle possesses a compressive component. If the final rupture region (stage 3) is ductile, the resulting deformation will prevent close realignment of the fractured pieces.

Beach Marks

Beach marks are a unique feature found in many fatigue fractures, and their presence is a positive means of identifying fatigue fractures. Beach marks also have been called stop marks, arrest marks, clamshell marks, and conchoidal marks, all in an attempt to describe their origin or characteristic appearance. The term *beach marks* is the most commonly used term but is not really as descriptive as some of the others. At any rate, this term is used to describe *macroscopically* visible marks or ridges that are characteristic of interruptions in the propagation periods (stage 2) of fatigue fractures in relatively ductile metals. Beach marks must not be confused with striations, although they frequently are present on the same fracture surface; there may be many thousands of microscopic striations between each pair of macroscopic beach marks.

Beach marks appear to be formed in several ways:

- *Microscopic plastic deformation at the tip of the fatigue crack from a particularly high load cycle.* A delay in crack propagation can occur following an application of a higher-stress load cycle. This results in a work-hardened zone at the tip of the crack, leading to a temporary cessation of crack propagation and formation of a beach mark. Eventually the crack will resume propagation once it is through the work-hardened zone.
- *Differences in the time of corrosion (or oxidation) in the propagating fatigue crack.* The area near the origin(s) is exposed to the environment longer than any other portion of a progressing fatigue crack. Sometimes these differences in corrosion form concentric rings or lines emanating from the origin(s). In some metals and environments, the corrosion products are of a different color from that of the uncorroded metal near the final rupture (stage 3). This color contrast frequently is of help in determining the location and number of fatigue origins, as shown in Fig. 11. However, if one or both fracture surfaces suffer uniform corrosion after final fracture, then the differences noted above may be obliterated by the presence of more uniform corrosion products over the entire surface. In some cases it may be possible to carefully clean the uniform corrosion products from the fracture surfaces and reveal the underlying contrast; however, the cleaning also may remove the underlying corrosion products if it is performed too vigorously. In any case, compare both fracture surfaces, if both are available, to determine differences in the corrosion patterns.

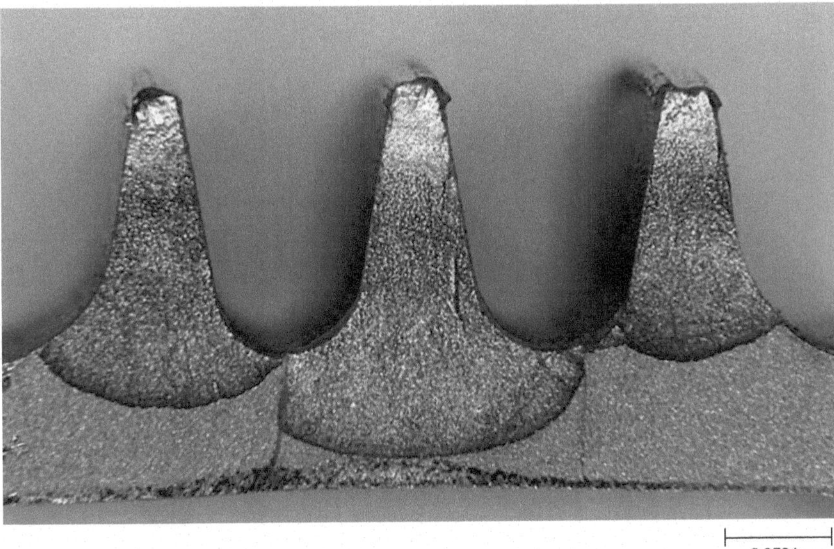

Fig. 11 Fatigue-tested titanium component with cracks initiating in each of three knife seals. The oxide colors reflect the oxide thickness, which would be thicker close to the origin areas.

- *Significant fluctuations in stress magnitude or loading frequency.* These fluctuations result in fractographic texture differences, which can manifest as bands or beach marks on the fracture surface.
- *Changes in direction of loading during crack propagation due to shifting loads within the component or outside influences.* One such change can occur on a rotating shaft. If the torque sense is temporarily reversed, as in during braking on a drive axle, the change in fracture plane will leave a distinct step or beach mark.

Ratchet Marks

The term *ratchet marks* is used to describe features that are very useful in identification of fatigue fractures and in locating and counting the number of fatigue origins. These marks are essentially perpendicular to the surface from which fatigue fractures originate. Therefore, in circular, shaftlike parts, the ratchet marks are essentially radial, pointing toward the center; in flat parts, such as leaf springs, they initially are perpendicular to the surface.

Ratchet marks are formed when several fatigue origins are adjacent to each other; each will start its own fatigue crack propagating, as in Fig. 12. If the origins are in slightly offset parallel planes, the fatigue crack growing from one origin will start to overlap the fatigue crack from another origin. When adjacent cracks overlap, the metal in the overlap area will then fracture across the short distance between the two cracks. This "in be-

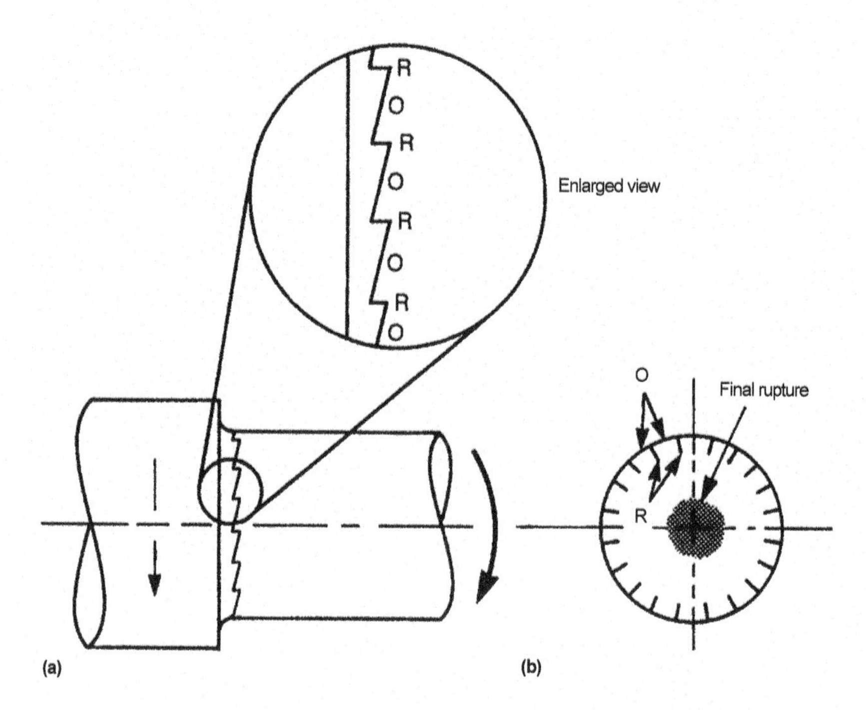

Fig. 12 (a) Formation of ratchet marks in the fillet of stepped shaft under uniform rotating-bending load. (b) Schematic view of fracture surface showing ratchet marks near periphery and central final rupture (stage 3). O: origins of fatigue fractures; R: ratchet marks between fatigue origins.

tween" fracturing creates a corner angle perpendicular to the surface of the part, as shown in Fig. 12. The ratchet marks are not the origins themselves; each ratchet mark separates two adjacent fatigue fractures. As the cracks become deeper, the cracks from each origin tend to grow together and become essentially one fatigue fracture. The number of ratchet marks equals or is one less than the number of origins; thus, recognition of the number of ratchet marks is important in determining the number of origins. Figure 13 shows photos of ratchet marks on a bolt.

Similarities between Striations and Beach Marks

Identification of the Tip of the Fatigue Crack. Both striations and beach marks identify the position of the tip of the fatigue crack at a given point in time. Since fatigue cracks grow over a relatively long period of time, these features can be considered analogous to the growth rings in the cross section of a tree trunk: each ring shows the size of the trunk at a given time; by counting rings, it is possible to determine the age of the tree. Likewise, by counting striations and beach marks, it is possible to estimate the length of time between initiation of the fracture propagation (stage 2) and final rupture (stage 3).

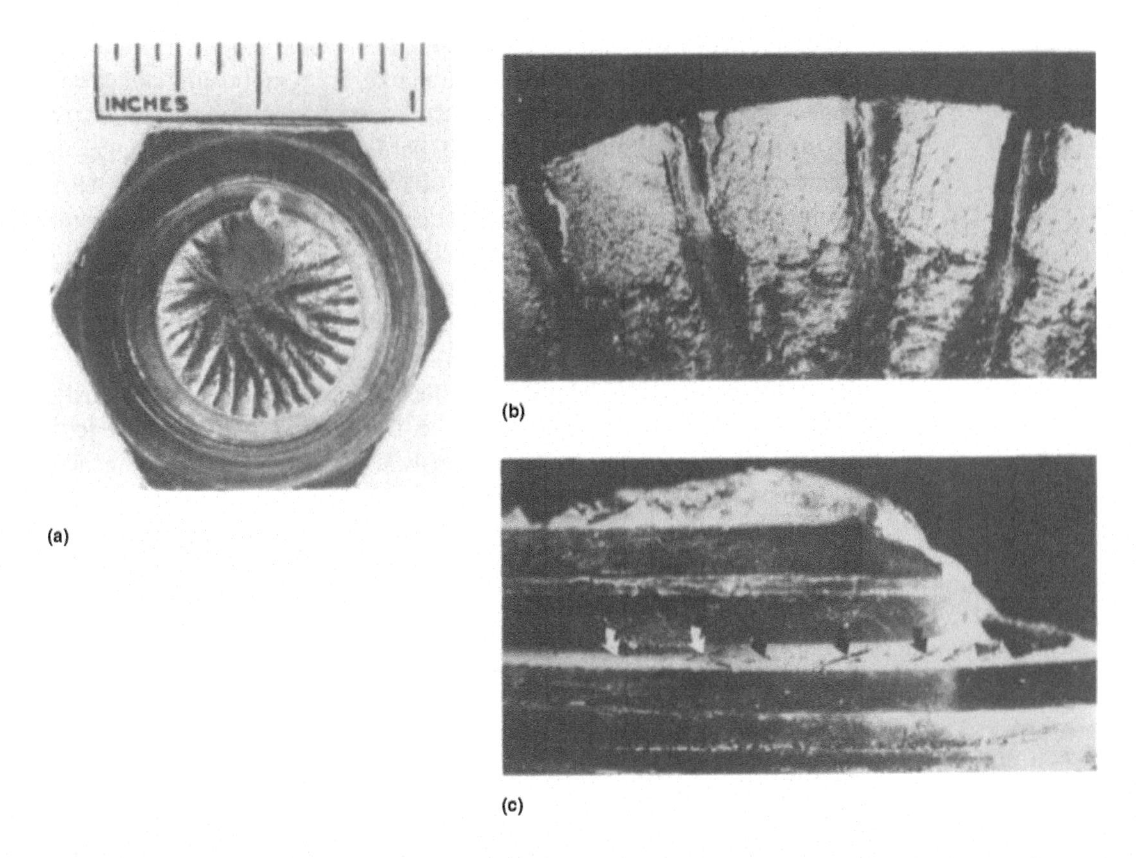

Fig. 13 Multiple-origin fatigue fracture of a short, stubby 5/8 in. bolt that fractured under tensile fatigue. (a) Numerous fatigue origins separated by radial ratchet marks. (b) Close-up of several fatigue origins separated by ratchet marks. Because this bolt was in a continuously operated test machine, there is no evidence of beach marks on the fatigue areas. These areas are at a slight angle to the axis of the bolt, as seen in (c). (c) Fatigue cracks in the root of the thread (arrows) that have not yet grown large enough to come close to the neighboring cracks to form ratchet marks between them. Note the slight angle of fatigue cracks that form a ratchet-like appearance upon fracture.

Expansion from the Fatigue Origin(s). Both striations and beach marks expand from the fatigue origin or origins, often in a circular or semicircular fashion. In addition to the tree growth-rings analogy, this is similar to circular ripples expanding from the location where a stone is thrown into still water. Obviously, the ripples in the water rapidly disappear, whereas striations and beach marks (also tree growth rings) leave a record, unless they are obliterated by contact with the opposite side of the crack either before or after final rupture. Corrosion and oxidation processes can also destroy these fragile features.

Similarity of Appearance. Both striations and beach marks are relatively parallel ridges that do not cross similar features from another origin. Here the ripple analogy breaks down: two or more stones thrown into still water simultaneously form circular ripple patterns that cross each other as they expand. Striations and beach marks are actually the tips of fatigue cracks that gradually progress; as such, they do not cross.

Absence from Certain Fracture Surfaces. Some fatigue fracture surfaces have neither striations nor beach marks. Extremely hard and strong metals are not ductile enough to form striations during stage 2 fatigue. At best, striations are difficult to observe, particularly in the early stages of fatigue-crack growth where striation spacing is extremely small. Also, near the origin, the crack may have been exposed to corrosive atmospheres for some period of time. Another problem is that the two fracture surfaces that form the crack tend to rub against each other on the compressive part of the stress cycle (if there is one) as the crack gets deeper, damaging fracture-surface features.

Beach marks may not be present if the part was operated continuously, or with only brief interruptions in service. Because laboratory fatigue testing is usually performed as continuously as possible to save time in a lengthy test procedure, the fractured test specimen usually will not have beach marks unless there were major changes or interruptions in the cyclically applied load or in the environment during the test. Some types of equipment—such as ventilating and air conditioning systems, power-generating stations, and various types of pumping equipment—frequently run continuously for very long periods of time during normal service. Fractures created under these conditions would be expected to have few, if any, beach marks unless corrosion was a factor.

Confusion by False Features. Artifacts, or false features, can confuse observation of both striations and beach marks. Rub marks from prefracture or postfracture damage can make parallel ridges that somewhat resemble striations when observed in the electron microscope. However, the orientation of true striations is usually predictable, once the origin of fracture is determined; the orientation of rub marks is likely to be different from that of true striations. Also, extremely elongated dimples are present on certain shear fractures, such as the transverse shear fracture of a ductile shaft under a torsional stress. These highly elongated dimples also make parallel ridges that can be mistaken for striations in the electron microscope.

The darkened pattern of a quench crack, seam, or other discontinuity may be mistaken for a beach mark or fatigue area after fracture, thus leading to incorrect diagnosis. Careful study of the surface of the discontinuity, depending on its condition, and also of a metallographic section of the edge of the metal perpendicular to the discontinuity may lead to the conclusion that it is not fatigue but rather a seam or forging lap, for example, with decarburization on its surface. In all such cases, do not leap to conclusions. Make a careful study before deciding the true cause of the problem.

Differences between Striations and Beach Marks

Size. Size is the most obvious difference between striations and beach marks. Striations are extremely small ridges, visible only with an electron microscope. On a given fracture surface, the striation size and spacing are smallest near the origin and gradually increase as the crack depth increases and the part becomes weaker, assuming that there has been no change in the fluctuating load on the part.

Beach marks are much larger than striations. If they are present, they are normally visible to the unaided eye. Beach marks may not be present, however, if there were no significant interruptions of service, as happens frequently in fatigue testing and with some types of continuously operating equipment. Despite their large relative size, proper low-angled lighting might be required to bring out the more subtle beach marks in the macroscope.

Cause. The other difference between striations and beach marks is the factors that cause them. Striations represent the advance of the crack front by one load application in many ductile metals, whereas beach marks locate the position of the crack front when repetitive, fluctuating loading was stopped for a period of time, differential corrosion was present on the fracture surface, or a large increase in load magnitude of direction occurred.

Schematic shown in Fig. 14 highlights typical fatigue fracture surface with several origins, striations, beach marks, and ratchet marks. In this sketch, only a few striations are shown between the beach marks, not the thousands or millions that may exist in an actual fracture surface. The opposite, or mating, fracture surface is a mirror image of this sketch. The same features are present; however, when there is a macroscopic projection on one surface, there is a depression on the opposing surface.

Numerical Analysis of Striations

Striations have been found to be useful in tracking both the number of operating cycles and crack growth rate on high-stress, low-cycle fatigue fractures. The quantity of striations on a typical fracture can range from a couple thousand to approximately 100,000, so direct counting is not practical. In short, striation counting proceeds as follows. The fracture is examined in a scanning electron microscope. Striations are photographed along a propagation path from the origin to the deepest part of the fatigue fracture, usually right down the middle if possible. The distances from the origin to each location are recorded. Striation spacing or, conversely, the striation density (striations per distance) is measured on each photograph and plotted as a function of depth from the origin site, as in Fig. 15. In the simplest of cases, low-cycle fatigue produces a curve that decays rapidly, indicating a commensurate rapid acceleration in crack growth rate. The curve fitted to these data can be integrated to approximate the number of cycles between any two depths. Remember, this approximated cycle number is for propagation cycles only and does not include the number of cycles for initiation. If the total number of cycles on the component is

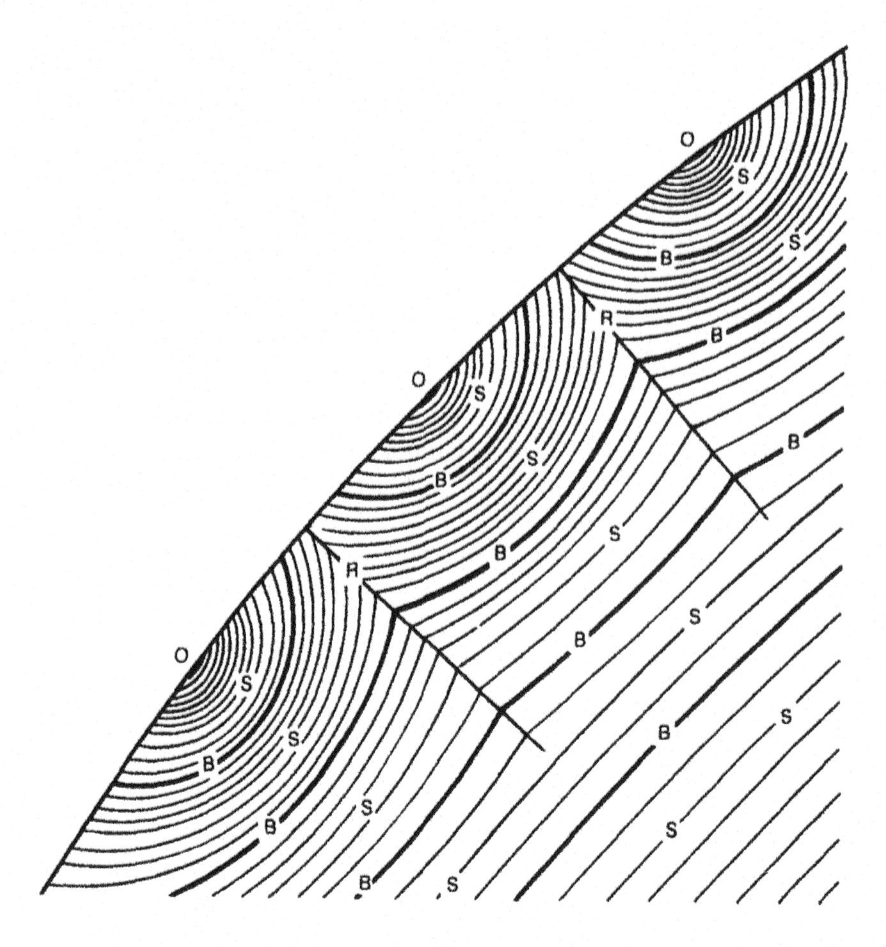

Fig. 14 Schematic, highly enlarged sketch of a typical fatigue-fracture surface. Sketch shows three origins (stage 1) at O; thousands of microscopic, closely spaced fatigue striations (stage 2) at S; a few beach marks, or arrest lines, at B; and two ratchet marks at R where fatigue cracks growing from the origins join as the cracks become deeper.

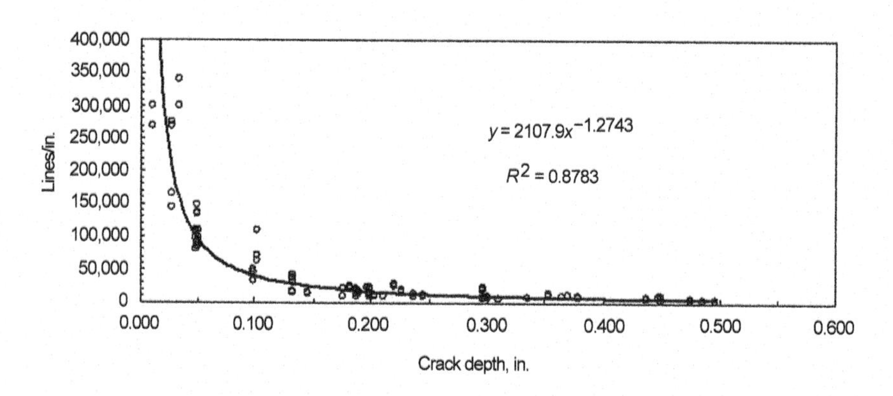

Fig. 15 Plot of fatigue striations as a function of depth, showing the expected increasing crack growth rate with increasing depth.

known, then one can also estimate the number of cycles needed for crack initiation by subtracting propagation estimate from the total since $N_t = N_i + N_p$, where N_t equals the total number of cycles on the component, N_i equals cycles to crack initiation, and N_p equals crack propagation cycles.

Numerical Analysis of Beach Marks

Sometimes beach marks can represent events that are meaningful, especially in the case of spectrum-loaded components, where one recurring event leaves a distinctive mark that can be tracked in the part's operating history. However, the part or test component history must be known in absolute detail. Generally, beach marks are not used to determine or estimate crack growth rates because they are macroscopic in nature.

Relationship of Stress to Strength in Fatigue

In understanding fatigue and in diagnosing fatigue fractures, the relationship between stress and strength assumes utmost importance. This relationship determines the survival or failure of a part. Since the number of fatigue origins in a given fatigue fracture depends on this relationship, the analyst must determine the number of fatigue origins. Typically, for any given surface and geometry condition (in absence of defects), more origins equate to higher operating stress conditions.

Figure 16(a) is a schematic drawing of a two-diameter shaft rotating under a bending stress. This is a very common situation, but the sketch could be of any configuration that has a critically stressed zone. The part shown in Fig. 16(a) has its critically stressed zone in the rather small radius fillet where the small diameter joins the larger shaft.

This critical fillet has a large number of grains, or crystals. For the purpose of this explanation, let us assume that it has one million grains. As the shaft rotates, each grain will successively follow the path of grain number 1, as shown in Fig. 16(a). It starts at the neutral axis where the applied stress is zero, as shown in Fig. 16(b). With the loading as shown, the maximum applied tensile stress is at the top, while the maximum applied compressive stress is at the bottom. When grain number 1 reaches the top, it is stressed in maximum tension; then, as it continues to rotate, it reaches the neutral axis (zero stress) on the opposite side. At the bottom it is in maximum compression; then it returns to the neutral axis where it started. Each grain in succession goes through the same sequence, and each traces out a sinusoidal (or wavelike) curve as shown in Fig. 16(b). In this type of loading, the mean applied stress is zero, and the maximum applied tensile and compressive stresses are equal.

Each of the one million grains in the fillet has a different strength with which to resist the applied tensile stresses (actually shear stress components) that are repetitively loading the individual grains as they progress

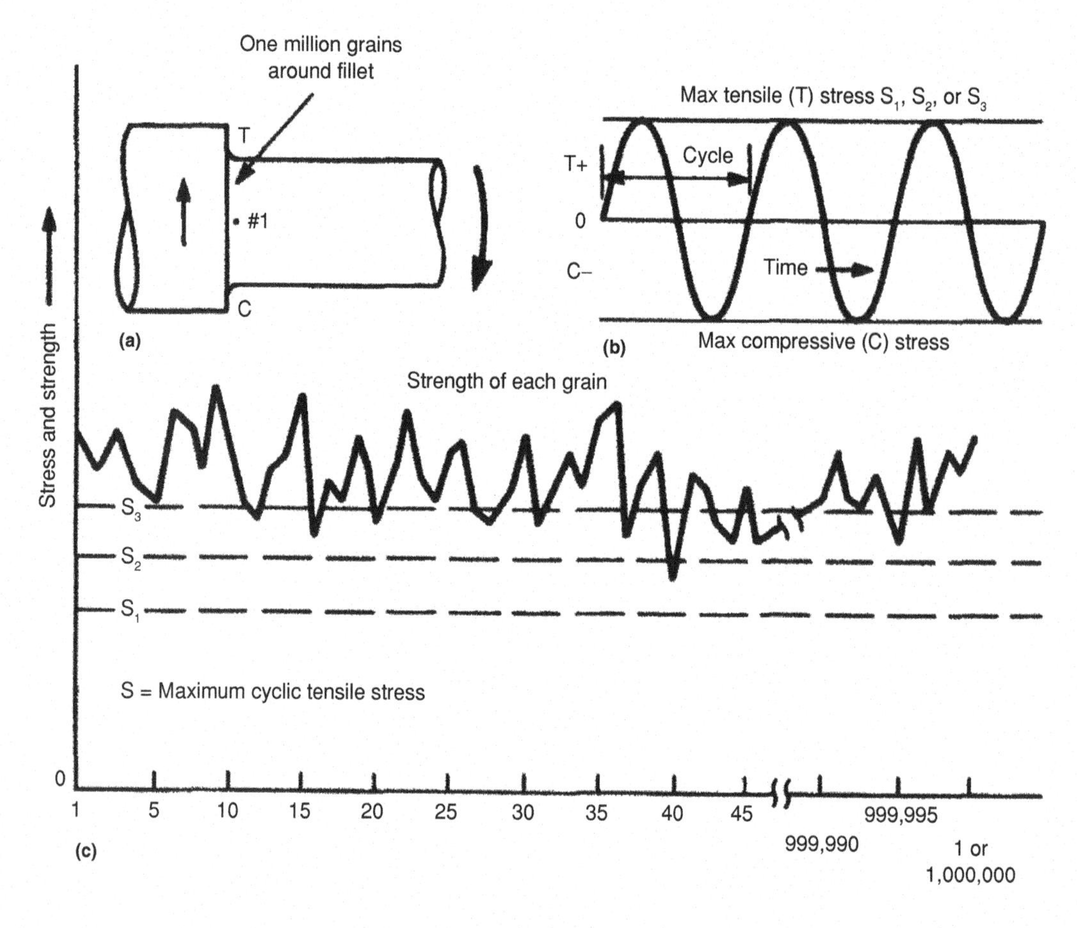

Fig. 16 Schematic sketch of a two-diameter shaft rotating under a bending stress. (a) General shape of a critical stress region in such a part. (b) The stress imposed on each grain, such as grain number 1, as it rotates, with the top of each sine wave representing the maximum tensile stress. (c) Strength of each of a large number of grains in the critical area and three stress levels (S_1, S_2, and S_3), which are the maximum cyclic tensile stress to which the grains are subjected. Successively higher stresses automatically cause more fatigue origins to form.

through their sinusoidal stressing with the rotation of the shaft. The strength of each grain is slightly different due to different orientations of the unit cell structure within the grain, different discontinuities within the grain (Fig. 1), different inclusions in the metal in each grain, and different local microstructures, among other factors.

Figure 16(c) represents some of the one million grains in this critical zone. The jagged, saw-tooth line connects the strength of the first and last grains adjacent to the starting grain number 1. Note that grain number 1 is at both the extreme left and the extreme right, for it has come full circle. The individual points represent the strength of the adjacent grains.

If the shaft rotates with no applied stress, obviously there is no applied stress on the grains. Now, if the maximum applied tensile stress is increased up to level S_1, there still is no effect on the shaft because it is still below the fatigue strength of the weakest grain, number 40 of those shown here.

If the maximum applied tensile stress level is increased to the level S_2, there is the strong possibility that submicroscopic changes (slip) will take place in the crystalline structure of grain number 40 if rotation at this stress level is continued. This could lead to a fatigue fracture with one origin at grain number 40, of those shown here. If this is, in fact, the weakest grain of all of the one million, this would be a single-origin fatigue fracture that would progress very slowly over many millions of stress cycles because the stress level is so low that it affected only this one grain.

However, if the maximum applied tensile stress level is increased to the level S_3, an entirely different, more serious situation arises. Now the stress level is above the fatigue strength of many grains—approximately eleven of those shown here, including the weakest one, number 40, which would start fatigue action first. Because there are so many fatigue origins, the shaft would not survive nearly as long as it would have at level S_2. Examination of the fracture surface would reveal many fatigue origins quite close together, with ratchet marks separating individual fatigue areas that usually grow together to form a nearly circular fatigue crack.

If the load on the rotating shaft is well balanced, the maximum applied tensile stress will be uniform around the fillet. Also, the final rupture region will be located in the center of the shaft and will be quite large because of the high applied stress. If the load on the rotating shaft is unbalanced, the maximum applied stress will not be uniform around the periphery, and the fatigue will start on the side where it is highest, with most of the origins in that region; the final rupture will be offset toward the low-stress side, where there may have been no fatigue action, as in Fig. 17.

Now, the practical question arises: How can the fatigue life (or strength) of this part be increased? Obviously, there are two main options, increase the strength or reduce the applied stress, but there are several ways to do either of these. The strength can be increased by at least two means: increasing the hardness (strength) level by more effective heat treatment or using a higher-quality metal with fewer internal discontinuities, such as a vacuum-treated steel. The resultant stress level can be decreased by redesigning to permit a more generous radius in the fillet or, if that is impracticable, by putting an undercut radius into the shoulder, or by mechanically prestressing the fillet with shot peening or surface rolling. See Chapter 4, "Residual Stresses," in this book for more information. Either shot peening or surface rolling, when properly performed, induces compressive residual stresses at the surface to neutralize the cyclic tensile applied stresses that cause fatigue cracks to form and propagate. This mechanical prestressing can be considered either as increasing the general strength level or as decreasing the maximum cyclic tensile applied stress level. In either case, the result is the same: improved fatigue performance.

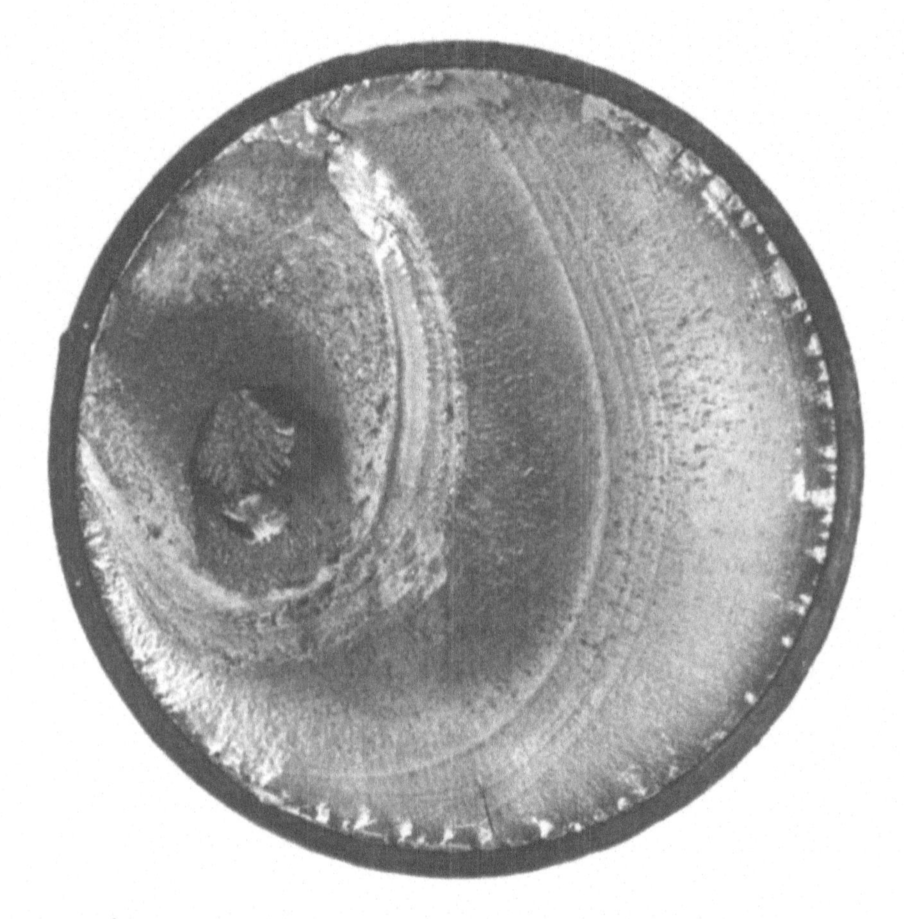

Fig. 17 Surface of a fatigue fracture in a grade 1050 steel shaft, with hardness of about 35 HRC, that was subjected to rotating bending. The presence of numerous ratchet marks (small shiny areas at the surface) indicates that fatigue cracks were initiated at many locations along a sharp snap-ring groove. The eccentric pattern of oval beach marks indicates that the load on the shaft was not balanced; note the final rupture area (stage 3) near the left side.

Laboratory Fatigue Testing

Anyone who has performed laboratory fatigue testing of parts or of test specimens can attest to the fact that this can be a frustrating procedure because there are often not enough test specimens or parts, fatigue test machines, time, and/or money to conduct significant testing. And there are sometimes too many variables to evaluate, too much scatter in test results, and too many specimens that do not fracture in the desired manner because of problems with the test fixtures and/or the specimens themselves.

Study of Fig. 18 and the following discussion may help the uninitiated to appreciate some of the problems of fatigue testing and of the fatigue process itself.

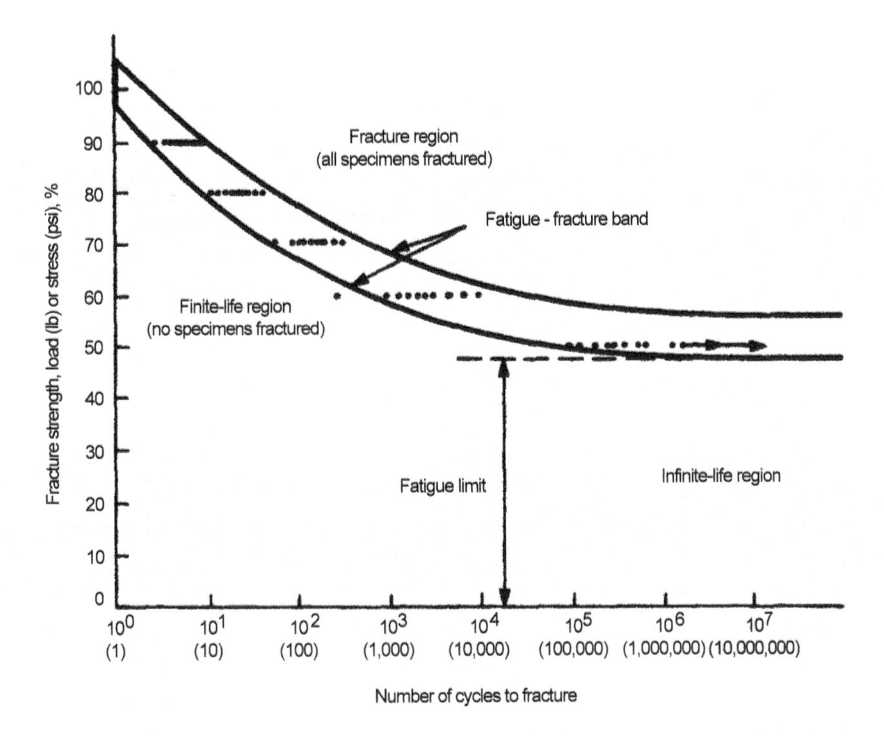

Fig. 18 Typical fatigue (*S-N*) diagram of laboratory fatigue testing of medium-strength ferrous metal.

Note: The test procedure described below usually is not followed in actual laboratory fatigue testing but is shown in this manner for illustrative purposes only. Actual laboratory fatigue test procedures will differ, depending on the purpose of the fatigue test.

First note the scales of the diagram. The vertical axis is percentage of either load or stress, depending on how much is known about the critical location of the test specimens and on the available instrumentation. The load or stress considered is the maximum value attained during the cycle, at the peak of the sinusoidal curve shown in Fig. 16(b). The horizontal axis consists of the number of cycles of load application to cause fracture. Because of the very large numbers that are encountered, this scale is logarithmic, starting with one at the left and going to 1 million, 5 million, 10 million, or even higher, depending on the metal involved and the purpose of the tests. This type of diagram is usually called an *S-N* (for stress-number) diagram, with the resulting curve the *S-N* curve, plotted on a "semilog" scale.

If the single-load fracture strength of the specimens is considered to be 100%, then for purposes of illustration this is the starting place, because the specimens can sustain no higher load without fracture. If 10 specimens are fractured, the results are placed as points at the top of the left axis at one load application.

Intuitively, we know that if the maximum load (or stress) is lowered to 90% of the tensile strength, it will require more than one load application to fracture the specimens. The 10 points shown in the diagram at 90% represent the possible life to fracture of each of the 10 specimens. Because the scale is logarithmic, the points appear to be relatively close, but in fact the scatter in life from longest to shortest is on the order of more than 2 to 1. At this high stress, plastic deformation of the test specimen is likely to be great, such as in bending a paper clip or wire coat hanger to make it fracture. Actual parts are not intentionally designed to operate in this regime, and normal fatigue fractures have no obvious plastic deformation.

If we drop the load to 80% of the single-load fracture strength and test 10 more specimens, we will find that they will run longer, with a fatigue-life scatter of perhaps 3 to 1, which is not unusual, even for theoretically identical specimens (which, of course, they are not). When the load is dropped to 70%, the lives get longer and the scatter in fatigue life increases to perhaps 5 to 1. Again dropping the load, now to 60% of the single-load fracture strength, the fatigue lives again increase, as does the scatter from longest to shortest life. Invariably, in actual fatigue testing, there is at least one specimen that inexplicably fractures far earlier than any of the others in the same group. One such specimen is shown at the 60% level fracturing at about 150 cycles, while the other supposedly "identical" specimens or parts had lives of from about 1,000 to 10,000 cycles. The cause of such an early "anomaly" is often sought in vain, although it is possible that some metallurgical reason, such as an inclusion on the surface, might be found.

Dropping now to 50% of the single-load fracture strength, the fatigue lives increase dramatically, as the *S-N* curve starts to flatten out. This flattening out is characteristic of ferrous metals of low and moderate hardness; many nonferrous metals and some very high-hardness ferrous metals tend to continue their downward path at very large numbers of cycles. Now the problem is when to stop the tests. The test machine will be needed for another test specimen after a very long test time, depending on the rate of loading, or cycles per minute. If 10 million is selected as the end point, the test must be stopped at that figure even if a specimen is unbroken, and the point shown with an arrow pointing to higher values, for it did not actually fracture. Frequently, 5 million or even 1 million cycles is selected as the end point, depending on the metal, purpose, and urgency of the tests. However, 500 million cycles is sometimes used in the aluminum industry.

The region below the lowest portion of the *S-N* curve is called the *infinite-life region*, because specimens that are tested at stresses below the curve should run indefinitely; that is, they should have infinite life. The leveling of the *S-N* curve is the fatigue limit, characteristic of ferrous metals but not of most nonferrous metals. However, the region to the left of the sloping part of the *S-N* curve is called the *finite-life region*, because at the higher stress levels, the test specimens or parts will eventually fracture in fatigue. This is typical of certain structural parts in aircraft that have their histories carefully recorded so that they can be inspected and/or replaced as their fatigue lives are used up in service. Also, growing fatigue cracks must not be permitted to exceed the critical flaw size characteristic of the metal and the stress state. This is the purpose of fracture mechanics: to predict the maximum-size flaw (or crack) that can be tolerated in a given metal part without causing complete fracture. See Chapter 15, "Fracture Mechanics," in this book.

Other Types of Fatigue

Subsurface-Origin Fatigue. Obviously, subsurface origins can result from internal inclusions and other defects under fatigue loading conditions. This subject was also covered in Chapter 3, "Stress versus Strength," in this book and is sketched in Fig. 9 of that chapter. A photograph of a typical example of fracture resulting from subsurface-origin fatigue is shown in Fig. 19. Subsurface fatigue can also originate from rolling con-

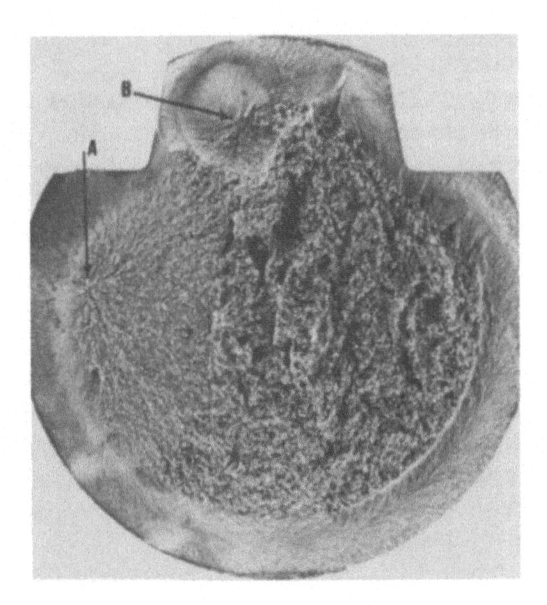

Fig. 19 Subsurface-origin fatigue fracture in an induction-hardened 3¼ in.-diam 1541 steel axle that was continuously tested in rotating bending fatigue in the laboratory. The primary fatigue fracture originated at A, while a smaller crack was progressing at B. Note that no beach marks are visible because of the continuous testing and that both origins are near the inner edge of the induction-hardened zone. The larger fatigue crack (from A) was in the left third of the fracture before it triggered a brittle fracture in the case (notice the chevron marks at lower right) and ductile fracture in the core. Source: Ref 4

tact in gears and bearings, as discussed in Chapter 12, "Wear Failures—Fatigue," in this book.

Fatigue under Compression Forces. A seemingly puzzling type of fatigue crack is one that grows in a part, or a region of a part, that is stressed in compression when the part is under load. However, it is not really difficult to analyze this phenomenon if one understands the mechanical formation of residual stresses. Review of Chapter 4, "Residual Stresses," in this book is recommended, particularly the section "Mechanical Residual Stresses."

It may be recalled that the general principle of mechanically induced residual stresses is that "*tensile* yielding under an applied load results in *compressive* residual stresses when the load is released, and vice versa." Fatigue under nominal compressive loads is explained by the "vice versa" part of the above statement; that is, "*compressive* yielding under an applied load results in *tensile* residual stresses when the load is released." This type of fatigue fracture can be illustrated with the example of the Belleville spring washer, a hardened spring steel washer that is cone shaped in order to give high spring rates in limited space, as shown in Fig. 20.

When this type of spring washer is pressed flat, the upper surface and corner of the hole are heavily compressed in a triangular pattern around the hole, as shown by the shaded area in Fig. 20(b). The compressive stress may be so high that the metal yields compressively around the hole. That is, the circumferential compressive stress on the upper side of the plate at the hole forces the metal to yield compressively when the spring washer is pressed flat. When the fatigue load is released, the washer springs back to its original position, and the upper surface now is stressed in residual tension as a result of compressive yielding when the load was applied. This is in accordance with the "vice versa" part of the mechanical residual stress principle quoted above. Consequently, the fatigue action causes the necessary tensile stress at this location when the load is released, not when it is applied. With continued operation, radial fatigue cracks will form at the upper corner, as shown in Fig. 20(b) and (c). These cracks are innocuous and do not affect operation because they will not progress to cause fracture. In time, however, normal fatigue cracks form at the lower corner of the hole that may eventually cause complete fracture, because this location is stressed in tension when the load is applied.

The solution for this "compression" type of cracking often lies in the field of design, as well as in heat treatment to increase yield strength. In the case of Belleville spring washers, they are frequently made with radial slots from the hole outward, so that a series of "fingers" are the springs that carry the axial compressive force. The slots interrupt the continuity of the circular hole and prevent circumferential compressive and tensile yielding. Shot peening is commonly used on these and many other springs to prevent fatigue fractures.

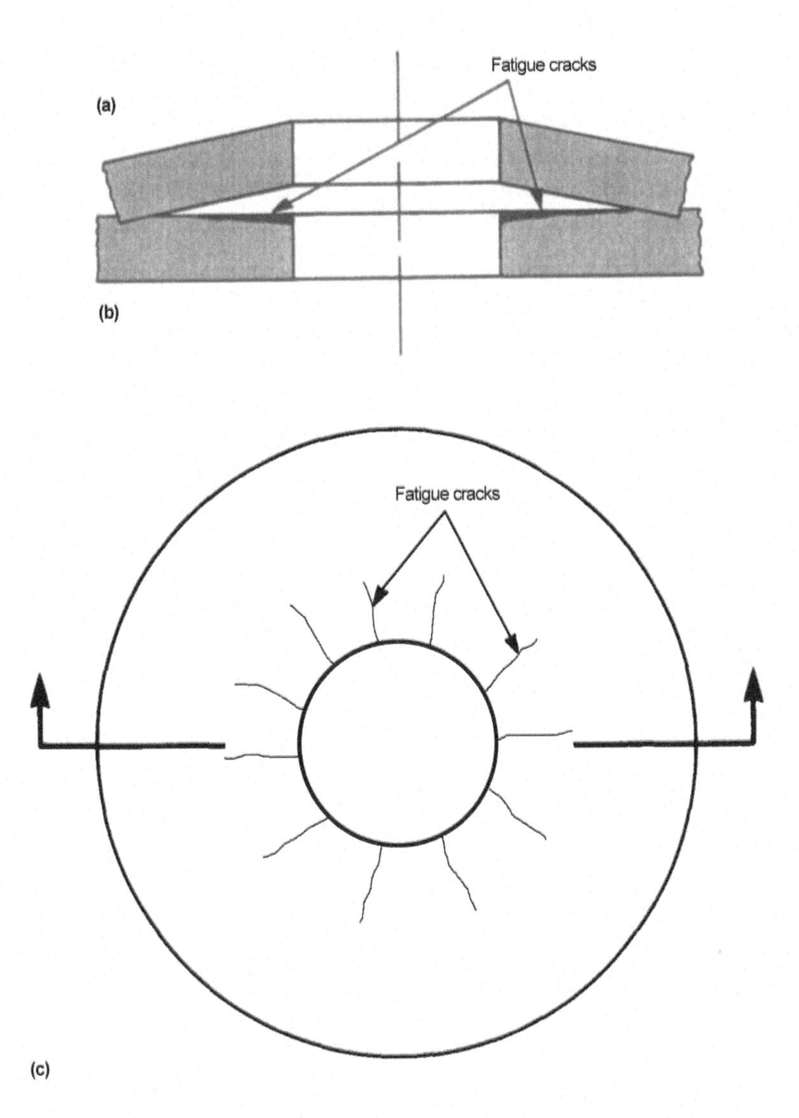

Fig. 20 Sketch of a Belleville spring washer showing how fatigue cracks can form in a nominally compressive stress area. The spring, actually a cone-shaped spring steel washer, is shown in the (a) free condition and in the (b) flattened condition. When flat, the corner of the hole on the convex (upper) side becomes compressively yielded, causing tensile residual stresses when the load is released. Repetitive loading to the flat condition causes radial fatigue cracks to form on the (b) and (c) convex side, but the cracks do not cause fracture. Eventually, independent fatigue cracks may form on the lower (concave) surface that will lead to complete fracture.

Other parts that develop fatigue cracks under nominal "compressive" loads should be studied with an eye for reducing the possibility of yielding when the compressive load is applied. The point to remember is that compressive yielding can result in tensile residual stresses that can cause fatigue cracking when the load is released. The cracks may be harmless to service operation, but they certainly are not desirable and should be prevented.

Thermal Fatigue. A somewhat similar type of fatigue fracture is caused not by repetitive mechanical stresses but by cyclic thermal stresses. Review of the section "Thermal Residual Stresses" in Chapter 4, "Residual Stresses," of this book will help the analyst understand how thermal residual stresses are formed.

The basic principle to recall is that the metal that cools last has tensile residual stresses; also, for thermal residual stress to occur, it is necessary to have both heat and restraint. If, for example, a sharp edge of a part is repetitively heated and cooled while the restraining bulk of the part remains relatively cool, the sharp edge expands when heated and contracts when cooled. That is, the sharp edge develops compressive forces and yields compressively when hot because of the lower strength and lower modulus of elasticity at the elevated temperature. When the edge is cooled, tensile residual stresses form; if this action is repeated many times, thermal fatigue cracks develop and tend to grow each time the metal is cooled from the elevated temperature, for it is at that time that the tensile residual stress is again applied. When the metal is hot, the cracks tend to close, although this is probably resisted by an accumulation of oxide scale or other products of combustion from the high-temperature atmosphere. In fact, these products in the crack can act as a wedge and increase the compressive yielding when hot, so that the tensile residual stress that forms on cooling is increased.

Because both heat and restraint are necessary for thermal residual stresses to occur and repetitive thermal cycling is necessary for thermal fatigue, the methods to prevent such fracture should become obvious. If restraint is internal within the part, as described above with the sharp edge, it may be possible to reduce the restraint by permitting the part to expand more uniformly when it is heated, instead of having just the sharp edge expand. If this is not practical, it may be possible to blunt the sharp edge by changing the shape so that this action is not concentrated on a thin section. Or it may be possible to reduce the thermal gradient within the part by permitting more of the metal to be heated or to design curves into the part so that expansion and contraction forces simply change the shape of the part instead of developing potentially destructive tensile residual stresses. This is the reason that expansion loops are designed into high-temperature piping, such as steam lines. It is also the reason that expansion joints are necessary in bridge and road design. See Ref 5 and Chapter 14, "Elevated-Temperature Failures," in this book for more information on thermal fatigue.

As in so many types of metal failure, once the principles behind a problem are understood, then one can identify the problem and creatively seek ways to eliminate it.

Corrosion Fatigue. As if fatigue itself were not a complicated enough type of failure, it is further complicated by the addition of a corrosive environment. Corrosion fatigue is the combined simultaneous action of repeated or fluctuating stress and a corrosive environment to produce surface-origin fatigue cracking. In one sense, all fatigue fractures should be considered as corrosion fatigue fractures, for the action of the environment within the fatigue crack is critical to the success or failure of the part involved. The fatigue strength of a part or metal in an inert environment, where there is no corrosion, is higher than that of the same part in an aggressive environment where corrosion may play a significant role.

Both corrosion and fatigue are complex subjects that must be understood separately before they can be understood in combination. For this reason, the subject of corrosion fatigue is discussed further in Chapter 13, "Aqueous Corrosion Failures," in this book.

Bolt Fatigue. Assuming that a bolted joint is properly designed and assembled, and that the mechanical properties of the bolt and nut are normal, fatigue fracture of the bolt cannot occur unless the cyclic separating force exceeds the clamping force of the bolt. This can occur if the cyclic separating force is too high or the clamping force is too low, or both. This principle is true no matter how the joint is loaded. It is simplest to illustrate with a model that assumes a tensile separating force, as shown in Fig. 21 and explained in the following discussion.

A bolt is really a very stiff spring, because all metals are elastic. Figure 21(a) shows a typical bolted joint, in which a tensile force in the bolt, O, and tightened nut, P, squeeze parts R and S with a clamping force at the joint, as shown by the arrows. The tensile force in the bolt, caused by tightening of the nut, obviously is equal to the clamping force holding the joint together.

This principle can be demonstrated by replacing the stiff-spring bolt with a soft-spring scale in a jointed frame. Figure 21(b) shows a 25-lb spring scale, T, which can be adjusted with a wing nut, U, to clamp the joint together. The separating force is provided by another spring scale, V, which can be pulled to various steady or cyclic loads by the handle.

If the clamping force is set at 20 lb, the joint cannot separate, nor can the force in the "bolt" (spring scale) exceed 20 lb if the cyclic separating force does not exceed 20 lb, as shown in Fig. 21(c), cycling between 5 and 15 lb.

However, if the cyclic separating force is raised to a range of 15 to 25 lb, the actual force on the "bolt" spring scale increases to 25 lb because the forces are in series, as shown in Fig. 21(d). When this happens, the joint actually separates, and the "bolt" undergoes a cyclic force that could lead to fatigue fracture.

Similarly, if the clamping force drops to 10 lb, cyclic forces in the "bolt" again can occur, although at a lower level, as shown in Fig. 21(e).

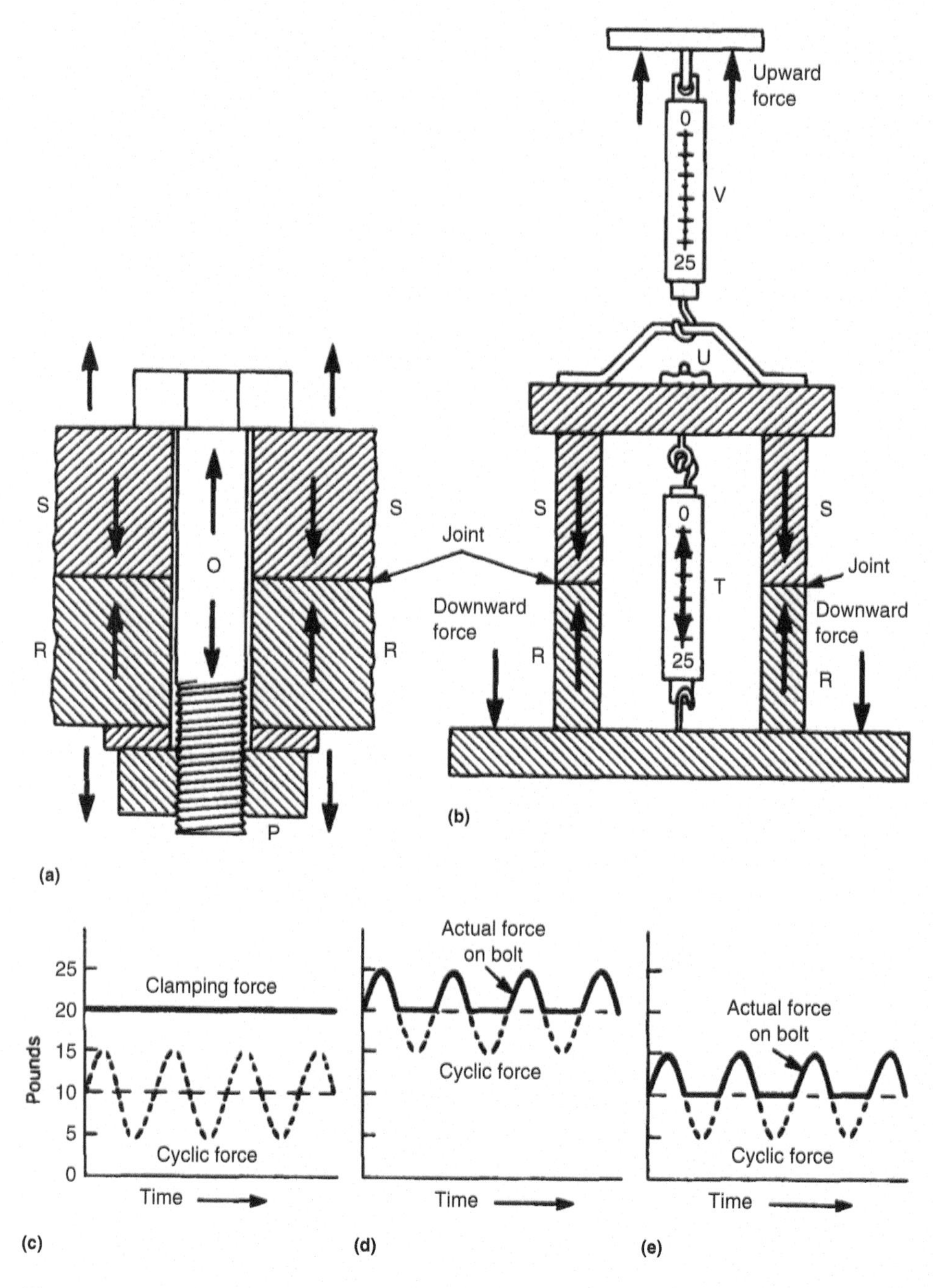

Fig. 21 Demonstration of fatigue of a bolted joint. See text for description.

When the actual force on the "bolt" exceeds the clamping force, the joint separates and the "bolt" again is subject to fatigue fracture. This principle is also discussed in some detail in Ref 1.

Fatigue in Riveted Thin-Wall Pressure Vessels. In riveted thin-wall pressure vessels, such as certain types of tanks and aircraft fuselages, repetitive pressurizing and depressurizing can cause fatigue cracks to develop between the edges of adjacent rivet holes. Each rivet hole is a stress concentration, and the fatigue can be aggravated by corrosion of the metal in many types of environments. In many aircraft fuselages, each flight means another pressurization, and therefore another fatigue cycle. Actually, bonded adhesive sealants are applied between the overlap joints both to distribute the stress by helping the rivets resist the pressurization force, and also to seal the joints.

Statistical Aspects of Fatigue

In discussing laboratory fatigue testing, the term *scatter* was used to describe the variability of results, or the fact that not all specimens tested at a given load, or stress, fracture at the same life or number of stress cycles. Supposedly, the specimens are identical; actually, however, they cannot be identical even though they have the same nominal dimensions, chemical composition, and heat treatment, to name a few factors. The reason is that each specimen is composed of a very large number of different grains, or crystals, each of which has different orientation and imperfections. Before the fatigue test, we have no idea where or when fatigue will originate and propagate to fracture.

Let us try to explain this variability with a simple analogy. Assume that the critical area has a million grains, as was done earlier when discussing fatigue origins, or that we have a population of a million "identical" parts to operate at one load or stress level. Compare these "identical" grains or parts with a million "identical" kernels of popcorn, which are actually miniature pressure vessels. When all kernels are placed in the same environment (hot air or hot oil), the internal pressure in each kernel increases as the moisture turns to steam. Eventually, the internal pressure (or stress) will equal the strength of the hull, the strong exterior of the pressure vessel. When this happens, the hull will fracture, the steam is released, and we say that a kernel has "popped."

However, note the way in which the kernels fracture, or "pop." The million kernels do not go "bang" all at the same time, but instead one pops first, then another, and another, until they are popping furiously. Gradually, the popping tapers off until a few are left that do not fracture (and we throw them away!).

The "random" sequence of popping (actually fracturing) can be explained as follows: The first kernel to pop is the one in which the internal pressure first reaches the fracture strength of the hull. The next one to pop is the next in the sequence, and so on. The kernels that do not pop at all are those in which, for some reason, the moisture content is low, the strength of the hull is high, or the hull had a hole or crack that permitted the steam to leak out without building up pressure to the fracture stress.

The point is that there is no way that we can determine in advance which kernels will pop first, which will be in the main group, or which will not pop at all. The popping occurs in a rough normal (bell-shaped) distribution curve of the entire "pop"-ulation, although there are some "runouts" that do not fracture, as in fatigue testing near the fatigue limit. The variation in time to pop is analogous to the scatter in cyclic fatigue life of a group of specimens that are as supposedly identical as are the popcorn kernels.

Because of the scatter in fatigue-testing results and in parts subject to fatigue loading in service, many statistical methods have been used to try to describe and control this problem. All have limitations in trying to predict the fatigue behavior of a group of parts. Some of the suggested references contain statistical information that is beyond the scope of this work.

Examples of Fatigue Fracture

Because there are a very large number of types of parts and materials that can undergo fatigue fracture, it is impossible to show examples of all combinations. However, a few of the most common types are shown schematically in Fig. 22 and 23 and as photographs in Fig. 24 to 40. Many others are shown in Ref 6 and many of the other publications devoted to the subject of fatigue and fatigue fracture. It is strongly recommended that engineers familiarize themselves with Fig. 22 and 23 and consider using these diagrams for reference.

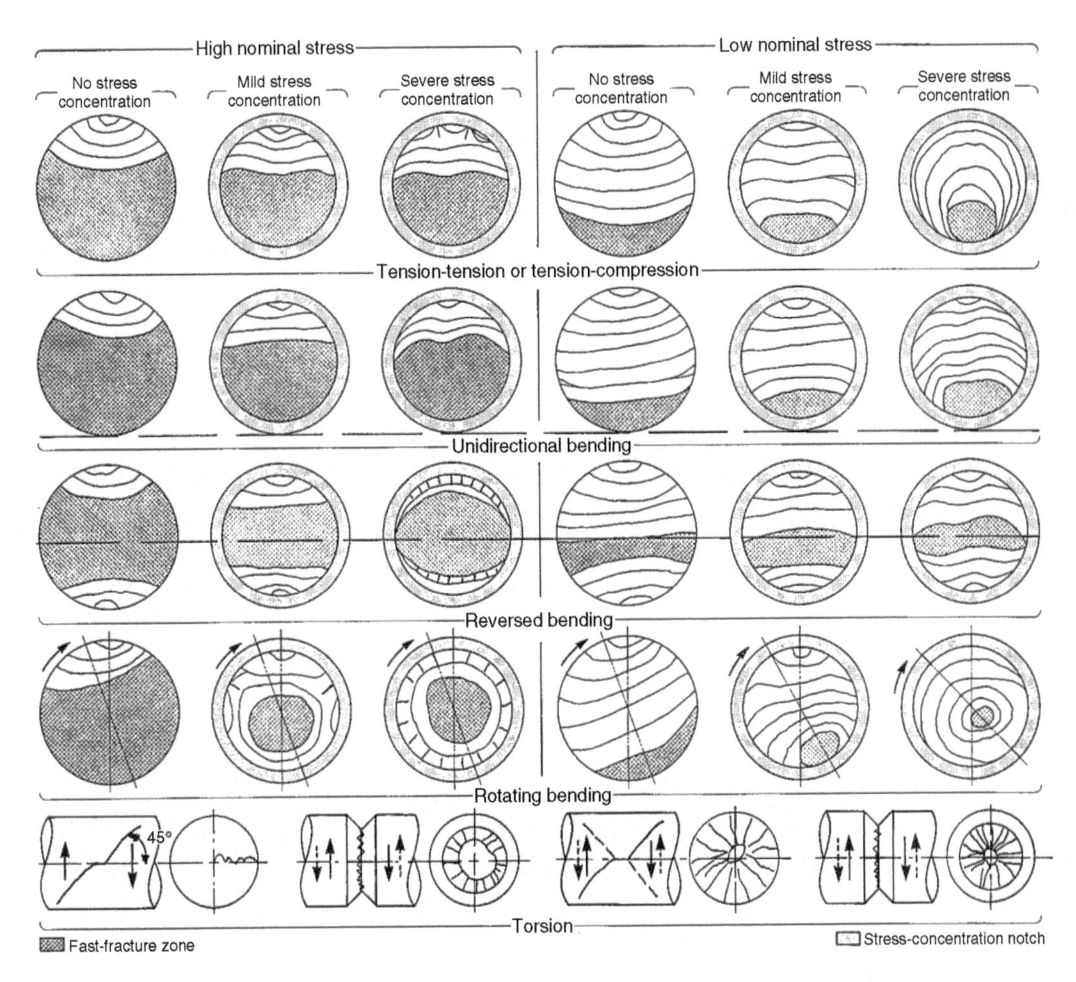

Fig. 22 Schematic representation of fatigue-fracture surface marks produced in smooth and notched cylindrical components under various loading conditions. Note that the final rupture zones (fast-fracture zones) on the left half of the figure, which had a high nominal stress, are considerably larger than the corresponding final rupture zones on the right half, which had low nominal stresses. The individual lines in the sketches represent the general configuration of the progression of the fatigue crack, as recorded by possible submicroscopic striations and macroscopic beach marks. The long horizontal dashed lines in reverse and unidirectional bending indicate the bending axes. Also note the radial ratchet marks between origins of the high nominal stress fractures. In the torsional fatigue fractures (bottom row), note that unidirectional fatigue (left) is at an approximate 45° angle to the shaft axis, while reversed torsional fatigue of a cylindrical shaft (second from right) has an X-shaped pattern with the origin in either the longitudinal or the transverse direction. Torsional fatigue cracks in stress concentrations tend to be very rough and jagged because of many 45° cracks. Modified from Ref 4

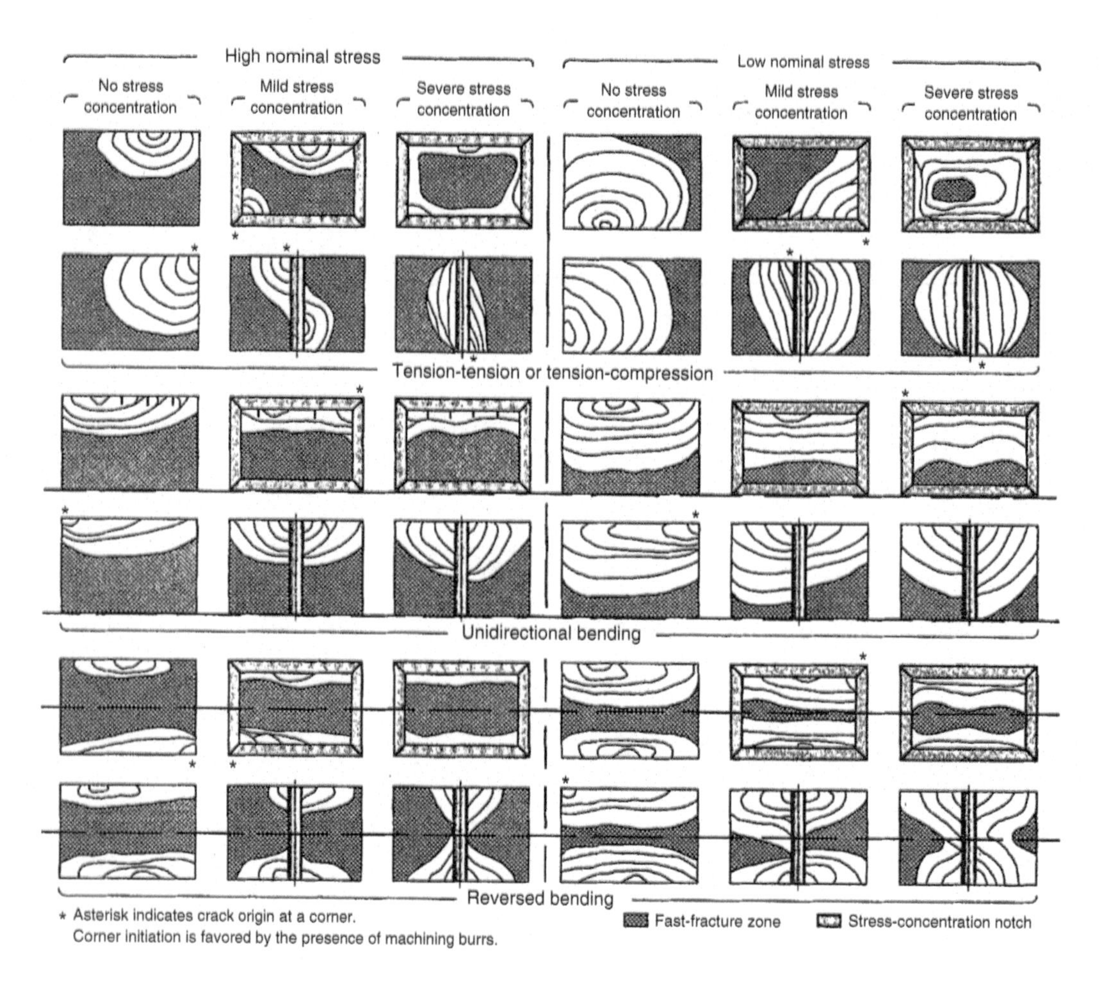

Fig. 23 Schematic representation of fatigue-fracture surface marks produced in components with square and rectangular cross sections and in thick plates under various loading conditions. Note that the final rupture zones (fast-fracture zones) on the left half of the figure, which had a high nominal stress, are considerably larger than the corresponding final rupture zones on the right half, which had low nominal stresses. The individual lines in the sketches represent the general configuration of the progression of the fatigue crack, as recorded by possible submicroscopic striations and macroscopic beach marks. The long dashed lines in reverse and unidirectional bending indicate the bending axes. Note that fatigue origins at corners of rectangular shapes or at ends or drilled holes (asterisks) are quite common in this type of part. Also, multiple fatigue origins are quite common with parts under high nominal stress; ratchet marks perpendicular to the surface separate adjacent fatigue areas. Modified from Ref 4

Fig. 24 Large axle shaft of medium-carbon steel with fatigue fracture across most of the cross section before final rupture. Note the smooth origin region (arrow) and gradually coarsening fracture surface as the fatigue crack progressed. Note that there was a thread groove running around the periphery and that the fracture origin is in the root of the thread. However, the nominal stress was quite low because the part was still intact and operating even though the fatigue crack had gone nearly all the way through the section. This indicates that only a slight improvement in the thread groove would be necessary to prevent this type of long-term fracture.

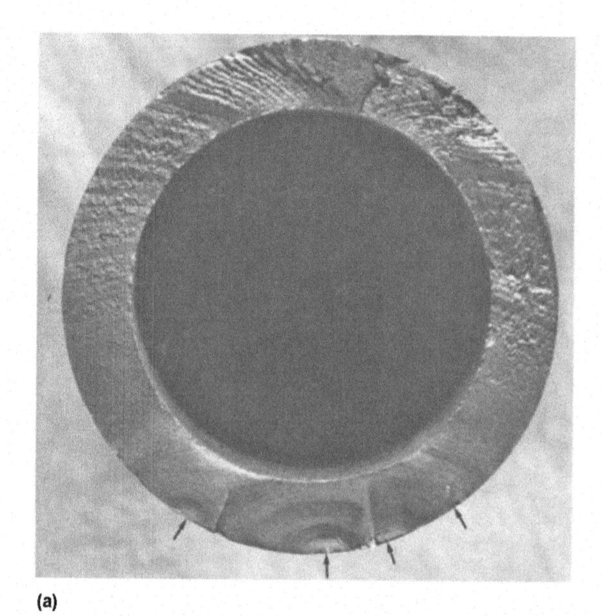

(a)

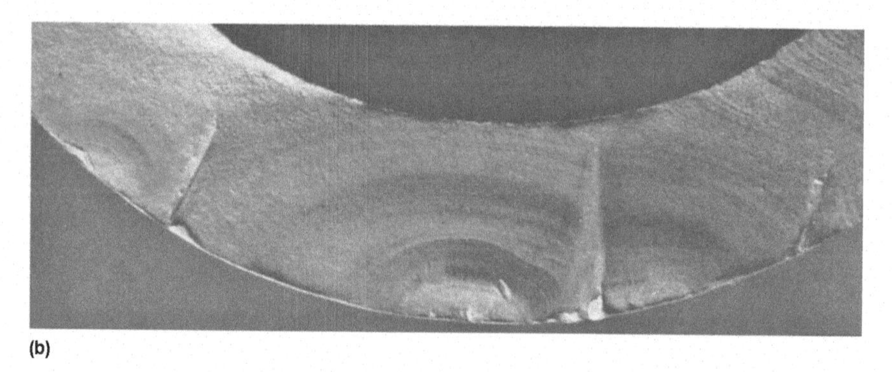

(b)

Fig. 25 (a) Fracture surface of a 3.6-in.-diam axle housing tube showing four major fatigue-fracture origins (arrows) at the bottom. (b) Origin areas at higher magnification. Beach marks are clearly seen. Small areas of postfracture damage are present, but in general, the fracture is in excellent condition after careful cleaning. This part was subjected to unidirectional bending stresses normal for the operation. The metal was a medium-carbon steel with a hardness of 217–229 HB. From the origin areas at the bottom, the fatigue cracks progressed up both sides of the tube and joined at the small final rupture area at the top. Note the increasing coarseness of the fracture surfaces as the cracks grew from bottom to top.

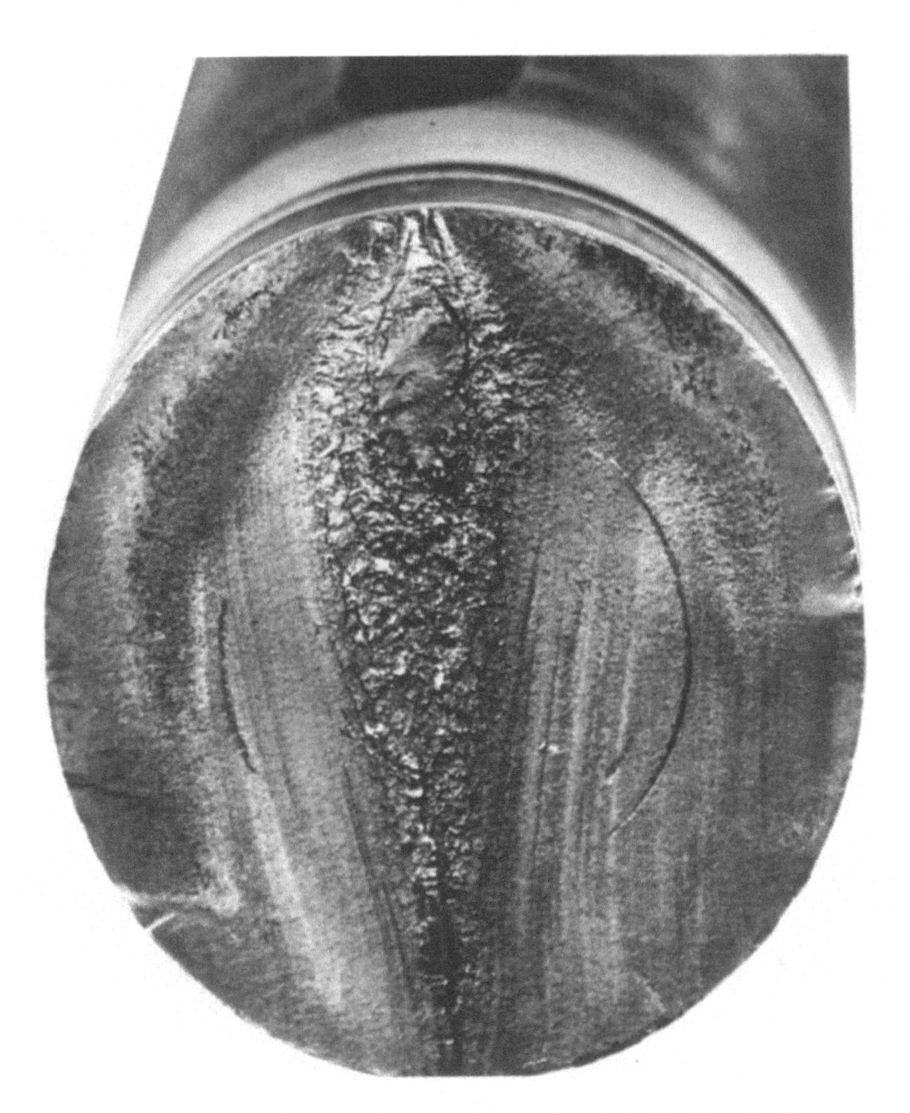

Fig. 26 Reversed bending fatigue of a 1.6-in.-diam shaft of 1046 steel with a hardness of approximately 30 HRC. Note the symmetrical fatigue pattern of beach marks on each side, with the final rupture on the diameter. This indicates that each side of the shaft was subjected to the same maximum stress and to the same number of load applications. The C-shaped marks are postfracture damage.

Fig. 27 Reversed bending fatigue of an alloy-steel steering knuckle at a hardness level of 30 HRC with nonuniform application of stresses. The multiple-origin fatigue at the bottom was caused by the tendency of normal wheel loading to bend the spindle (lower right) of the knuckle upward. The fatigue on the upper side (smaller area) of the fracture was caused by turning maneuvers in which the wheel acts as a lever to bend the spindle downward. Note the many radial ratchet marks on the fatigue surfaces. The part had been overloaded during service.

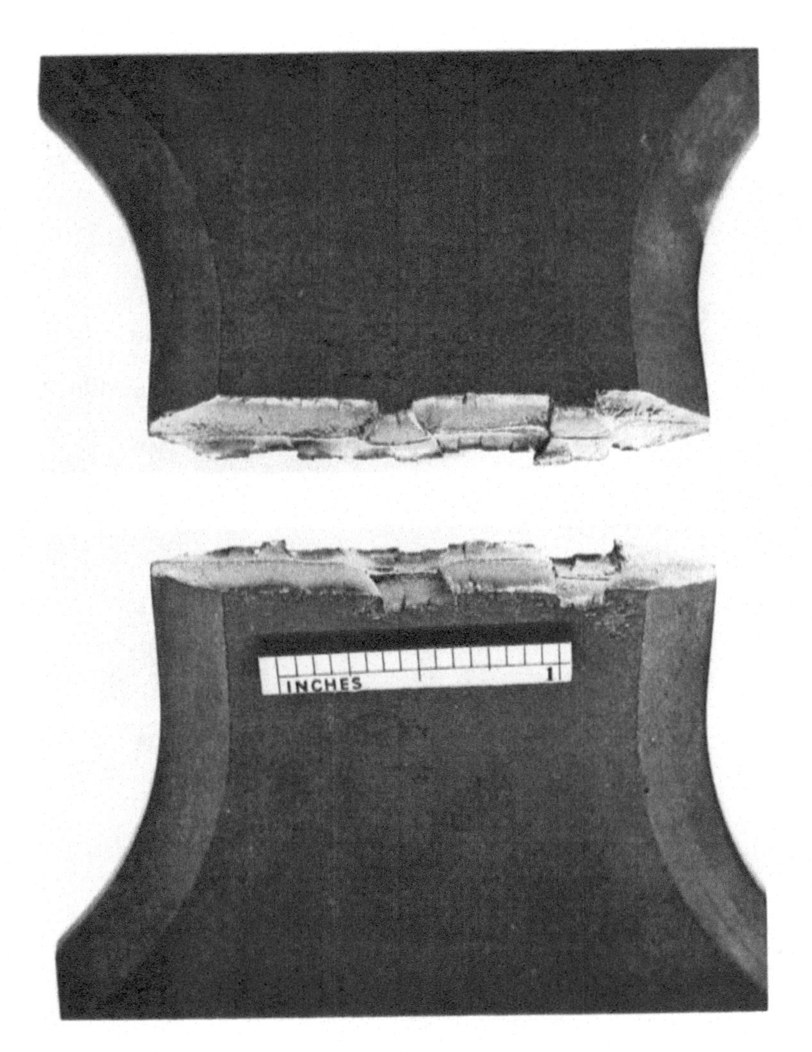

Fig. 28 Reversed bending fatigue of a flat ¼-in. plate of a high-strength low-alloy steel test specimen, designed with tapered edges to prevent fatigue origin at the corners. Note the many separate origins on each side and the very thin final rupture region separating the two fatigue areas on each half. Many other fatigue cracks were present in the reduced section, for this was designed to have a relatively uniform stress in the reduced section.

Fig. 29 Rotating bending fatigue fracture of a keyed shaft of grade 1040 steel, approximately 30 HRC. The fatigue crack originated at the lower left corner of the keyway and extended almost through the entire cross section before final rupture occurred. A prominent beach mark pattern is visible, appearing to swing counterclockwise on this section because the rotation was in a clockwise direction.

Fig. 30 Rotating bending fatigue fracture of a 2-in.-diam grade 1035 steel shaft, hardness 143 HB. The part was designed with a large radius joining the shaft to the shoulder, but it was machined with a sharp tool mark in the fillet. Multiple-origin fatigue around the periphery proceeded uniformly into the shaft, finally fracturing with a final rupture region in the center of the shaft.

Fig. 31 Rotating bending fatigue fracture in a grade 4817 steel shaft, carburized and hardened to a surface hardness of 60 HRC. The fracture started in six fillet areas around the periphery, near the runouts of six grooves. The six fatigue areas penetrated separately, but uniformly, to final rupture at the center. The bright, shiny, flat areas are regions of postfracture rubbing against the opposite surface. Grinding damage in the fillet was the cause of this fracture.

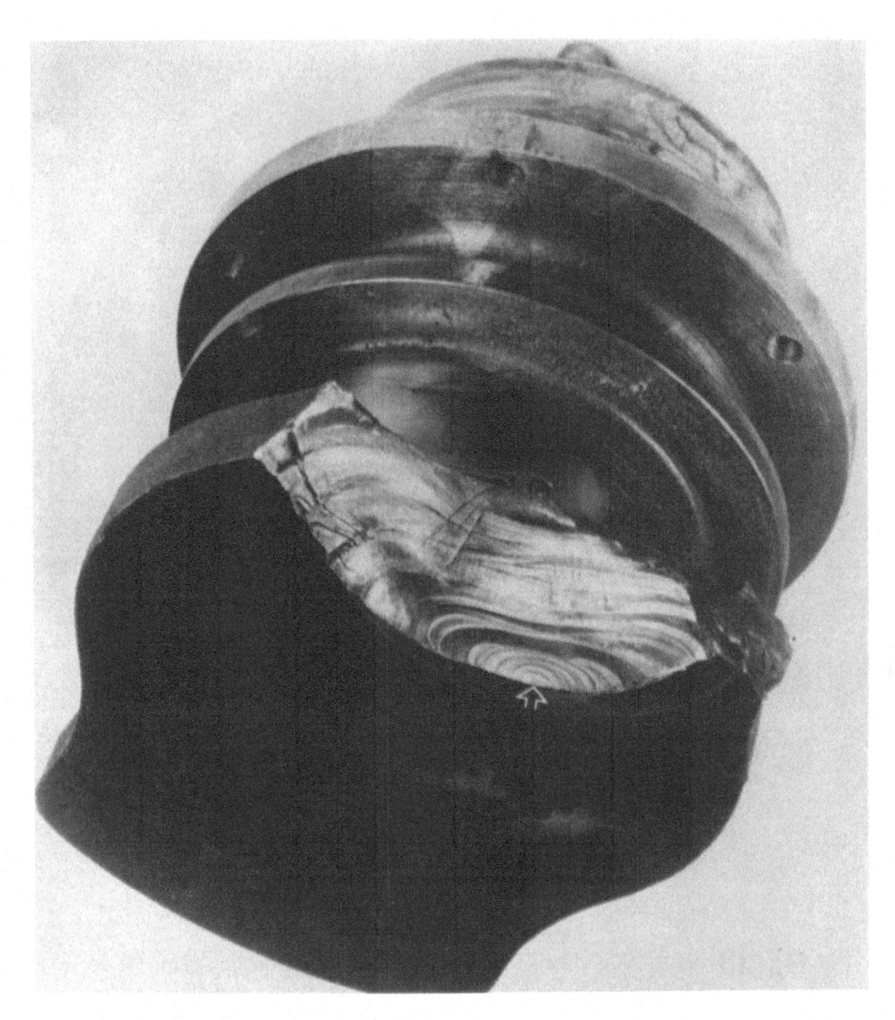

Fig. 32 Bending fatigue fracture through the cheek of a diesel engine crankshaft. The very prominent steps and beach marks were the result of severe overloading during starting and clutching with a very aggressive friction material in the clutch. Though this was a laboratory test, it was not operated continuously and thus more closely resembles fracture from actual vehicle service.

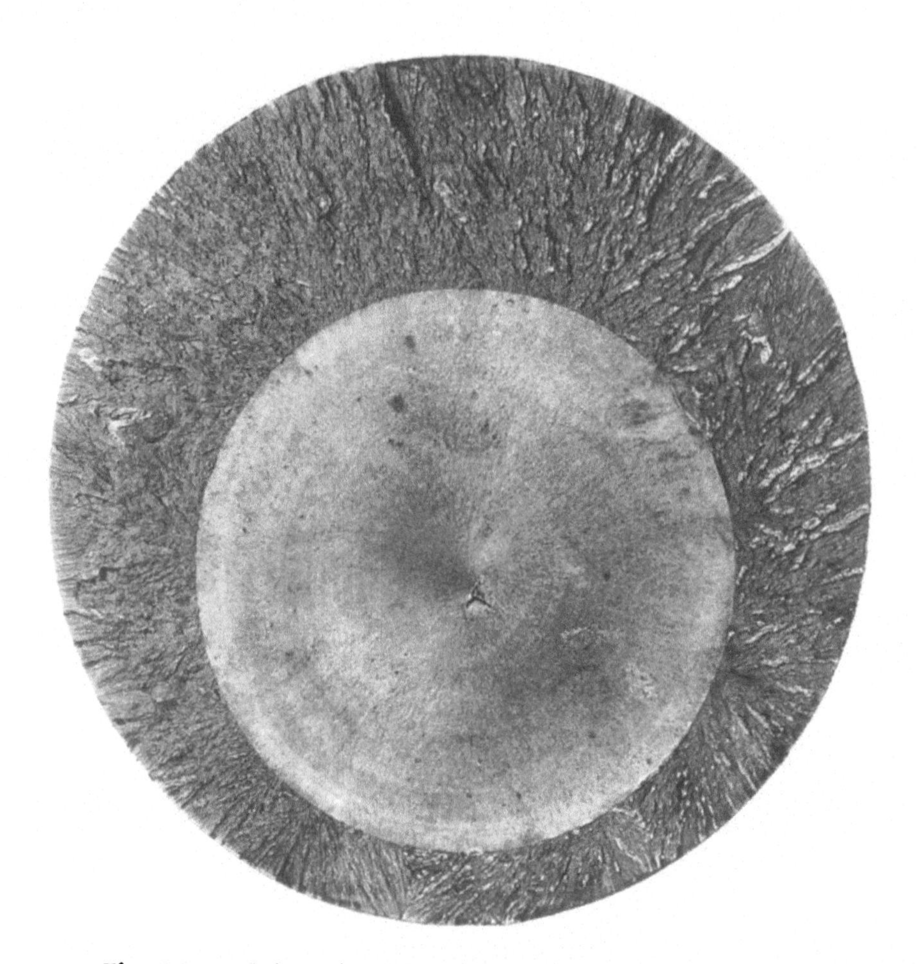

Fig. 33 Tensile fatigue fracture starting near the center of an 8-in.-diam piston rod of a forging hammer, made of low-carbon alloy steel hardened to 24 HRC at the surface and 17 HRC at the center. In an axially loaded part such as this, fatigue fracture can start anywhere in the cross section. In this case, a forging "flake" caused fatigue to start near the center of the section and to grow outward in a circular fashion until it was about 5.2 in. in diameter, at which time final rupture occurred in one load application. At no time did the fatigue crack reach the surface of the large shaft. Use of a vacuum-degassed steel would have prevented the flake that caused this type of fatigue failure, which occurred many years ago.

Fig. 34 Torsional fatigue fracture of a 1050 steel axle shaft induction-hardened to about 50 HRC. The arrow indicates the longitudinal shear fatigue origin, which then changed direction and grew to the small circular beach mark, or "halo." Final brittle fracture (note chevron marks in the case) caused complete separation with the characteristic 45° brittle fracture in torsion.

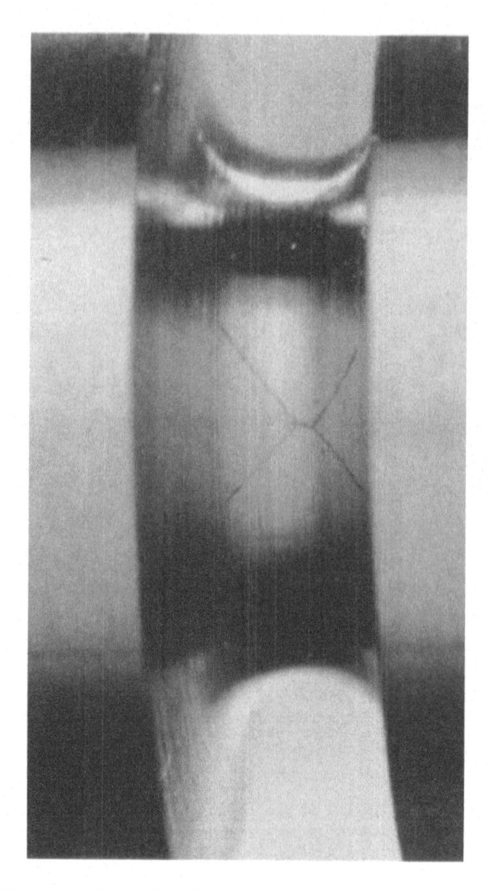

Fig. 35 Close-up of a reduced area on a medium-carbon steel drive shaft showing the X-shaped crack pattern characteristic of reversed torsional fatigue. Reversed torsional fatigue causes approximately 45° spiral fatigue cracks on opposite diagonals. The original shear crack was in the longitudinal shear plane; then each pair of torsional fatigue cracks developed at a 45° angle to the shaft axis.

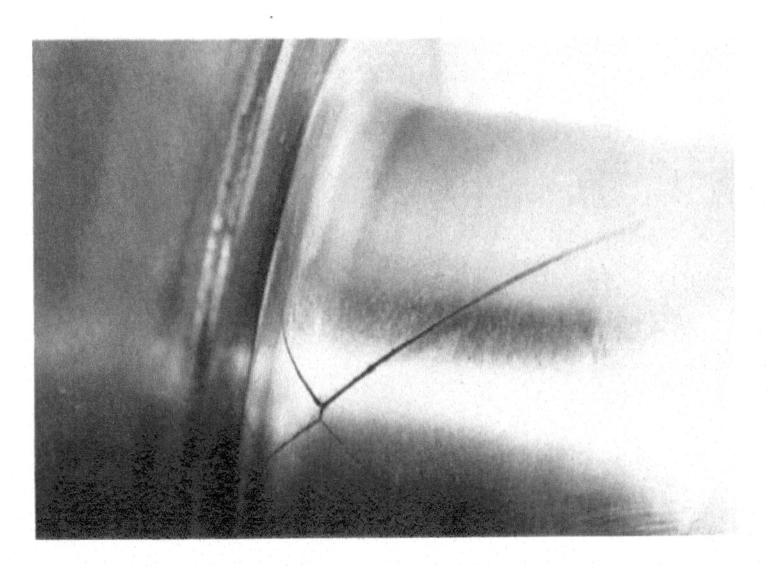

Fig. 36 Characteristic X-shaped crack pattern in a grade 1045 steel crankshaft after testing in reversed torsional fatigue in a special machine, not in an engine. In this case, the original crack was in the transverse shear plane, not in the longitudinal shear plane as in Fig. 35.

Fig. 37 Reversed torsional fatigue of a 6¾ in.-diam spline shaft showing the characteristic "starry" pattern of multiple fatigue cracks. Each of the 32 spline teeth has two fatigue cracks, each at 45° to the shaft axis, that form a V-shaped region. In addition, there are longitudinal radial fatigue cracks that penetrate nearly to the center of the shaft. This type of fatigue progresses at each location where a shaft enters an internal spline and is repeated if there is an internal spline at another location on the shaft. These cracks surround portions of the metal, forming essentially wedge-shaped segments, somewhat like those of an orange.

Fig. 38 A "starry" spline fracture similar to that shown in Fig. 37 due to reversed torsional fatigue on a 1½ in.-diam spline. Torsional fatigue has caused many of the surrounded segments to fall out of the shaft. Note that the longitudinal cracks penetrated nearly to the center of the shaft. This part was made from low-carbon alloy steel with a hardness of 24 HRC in the shaft area.

Fig. 39 Torsional fatigue fracture in a 3-3/8-in.-diam keyed tapered shaft of grade 1030 steel characterized by "peeling" that progressed around the shaft. The fatigue crack originated in the corner (A) of the keyway from pressure of the key aligning a large hub to the shaft. The fatigue crack progressed completely around the shaft, under the keyway, and started around again, all in a clockwise direction, before final separation occurred and nearly broke off the "wing" at the left. The only thing necessary to prevent this type of fracture is to make sure that the large nut that forces the hub to fit tightly against the tapered surface does not come loose. If the nut is loose, the key and keyway, rather than the frictional fit of the conical joint, carry the torsional force.

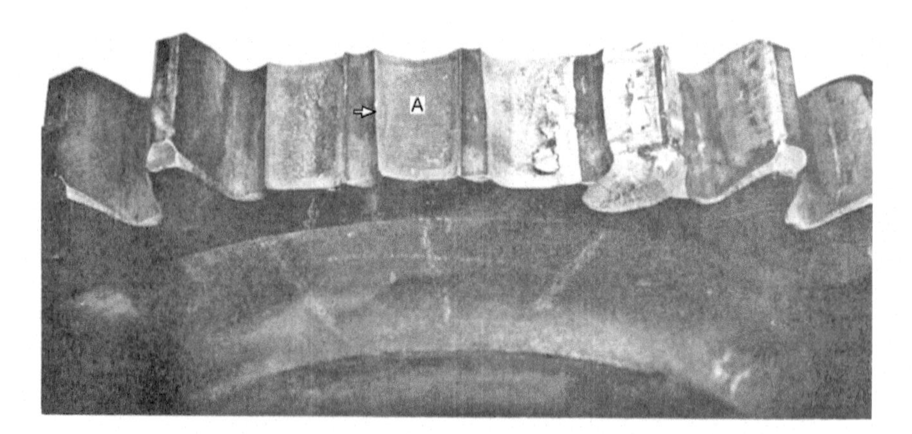

Fig. 40 Bending fatigue fractures in several teeth of a grade 8620 steel spur gear, carburized and hardened to 60 HRC in the case. It can be seen that tooth A fractured first, for it has the largest fatigue area, originating in the fillet on the arrow side of the tooth. Gear teeth are carefully shaped cantilever beams and can be diagnosed in this way. Fracture of the first tooth caused abnormal loading on the adjacent teeth, and they also rapidly fatigued, for the mating gear did not now mesh properly with them. Considerable smashing and battering of adjacent teeth is normal after gear-tooth fracture.

Summary

Fatigue fracture is a progressive type of fracture that can occur under normal service operation in three stages:

1. *Initiation,* by a submicroscopic shear, or slip, mechanism that causes irreversible changes in the crystal structure of the metal
2. *Propagation,* by an increasingly rapid progression of the tip of the fatigue crack in microscopic advances
3. *Final rupture (overload),* which is final separation, or fracture, into two or more parts by a single load application

Fatigue fractures have several microscopic and macroscopic features that may enable them to be properly identified. These include lack of permanent deformation in the origin area, striations, beach marks, and ratchet marks. Unfortunately, these features are not always present or readily observable on the fracture surfaces, depending on the characteristics of the metal itself, the type of operation to which it was subjected, and possible obliteration by pre- or postfracture damage by mechanical and/or chemical action.

Compression fatigue and thermal fatigue are understood most easily with the aid of residual-stress principles.

Fatigue fracture is highly subject to statistical variations because of the submicroscopic origins and internal differences in parts. Thus, there is usually considerable variation, or scatter, in fatigue life of supposedly identical parts. For this reason, fatigue fractures are difficult to predict with any accuracy, except in terms of statistical probability, as discussed in several of the references.

REFERENCES

1. *Failure Analysis and Prevention,* Vol 11, *ASM Handbook,* ASM International, 1986, p 4, 104, 110, 530–533
2. P.H. DeVries, K.T. Ruth, and D.P. Dennies, Counting on Fatigue: Striations and Their Measure, *Journal of Failure Analysis and Prevention,* Vol 10 (No. 2), April 2010, p 120–137
3. M.A. Miner, Estimation of Fatigue Life with Particular Emphasis on Cumulative Damage, *Metal Fatigue,* G. Sines and J.L. Waisman, Ed., McGraw-Hill, 1959, p 278–292
4. *Fractography and Atlas of Fractographs,* Vol 9, *Metals Handbook,* 8th ed., American Society for Metals, 1974, p 68–70, 85–89
5. *Fractography,* Vol 12, *ASM Handbook,* ASM International, 1987, p 111, 133, 266
6. *Fatigue and Fracture,* Vol 19, *ASM Handbook,* ASM International, 1996

SELECTED REFERENCES

- *Achievement of High Fatigue Resistance in Metals and Alloys,* STP 467, ASTM, 1970
- S. Bhattacharyya, V.E. Johnson, S. Agarwal, and M.A.H. Howes, Ed., *ILTRI Fracture Handbook: Failure Analysis of Metallic Materials by Scanning Electron Microscopy,* Metals Research Division, Illinois Institute of Technology Research Institute, 1979
- *Cyclic Stress-Strain Behavior; Analysis, Experimentation, and Failure Prediction,* STP 519, ASTM, 1973
- *Fatigue Crack Growth under Spectrum Loads,* STP 595, ASTM, 1976
- H.J. Grover, *Fatigue of Aircraft Structures,* rev. ed., NAVAIR 01-1A-13, U.S. Government Printing Office, 1960
- *Handbook of Fatigue Testing,* STP 566, ASTM, 1974
- N.R. Mann, R.E. Schafer, and N.D. Singpurwalla, *Methods for Statistical Analysis of Reliability and Life Data,* John Wiley & Sons, 1974
- *Manual on Low Cycle Fatigue Testing,* STP 465, ASTM, 1969
- G.W. Powell, S.-H. Cheng, and C.E. Mobley, Jr., *A Fractography Atlas of Casting Alloys,* Battelle Press, 1992
- B.I. Sandor, *Fundamentals of Cyclic Stress and Strain,* University of Wisconsin Press, 1972

Understanding How Components Fail, Third Edition
Donald J. Wulpi
Edited by Brett Miller

CHAPTER **11**

Wear Failures—Abrasive and Adhesive*

WEAR IS USUALLY DEFINED as the undesired removal of material from contacting surfaces by mechanical action (Ref 1). Although it is not usually as serious a service problem as is fracture, wear is an enormously expensive problem. This problem has both good and bad aspects. Making replacement parts and mechanisms for material that has "worn out" provides employment to millions of people. If all wear could be eliminated, the repercussions in the economy of the nation would be tremendous. The tire, clothing, automotive, and shoe industries, for example, would be devastated, because a large part of their business is to supply replacements. On the other hand, the users of the tires, clothes, automotive vehicles, and shoes would be delighted to have no wear problems and, therefore, no replacement problems. There is no danger of this happening, however, for—like death and taxes—wear will always be with us.

Wear usually is a foreseeable type of deterioration. We expect various rubbing surfaces in any machine—as well as those products previously mentioned—to eventually "wear out." In many cases this type of deterioration can be minimized by proper lubrication, filtering, materials engineering, and design, among other factors. In many respects, wear is similar to corrosion. Both have many types and subtypes, of which usually at least two are progressing simultaneously. Both are somewhat foreseeable unless the environment changes. Both are extremely difficult to test and evaluate in accelerated laboratory or service tests, with rankings of mate-

***Understanding How Components Fail, Second Edition,* by Donald J. Wulpi, Chapter 11, Wear Failures—Abrasive and Adhesive, ASM International, 1999, reviewed and revised by Jake Auliff and Ryan Spotts, Sauer-Danfoss

rials subject to change depending upon seemingly minor changes in the test conditions. Finally, both have enormous economic impact. When studying any failure where wear is known or suspected, it is necessary to have a good understanding of the history and operation of the part or mechanism involved. In many cases, it is not possible to conduct a good investigation by simply examining the worn part itself. Because wear involves the interaction of other parts and/or materials, these must be studied also; because wear is also a surface phenomenon, anything that affects the surface is likely to affect the wear behavior.

Because the definition of wear is quite broad and the various categories or types of wear are not precisely defined, authors on the subject may organize the different types in different ways. This author has chosen to organize the rather overwhelming subject of wear into the following categories, not all of which are conventional:

Chapter 11

- Abrasive wear
  a. Erosive wear
  b. Grinding wear
  c. Gouging wear
- Adhesive wear
- Fretting wear

Chapter 12

- Contact stress fatigue
  a. Subsurface-origin fatigue
  b. Surface-origin fatigue
  c. Subcase-origin fatigue
  d. Cavitation fatigue

Two of the categories, fretting wear and cavitation fatigue, usually are included in the subject of corrosion. There is some justification for this grouping, because both involve some chemical changes. However, since both are primarily the result of "undesired removal of material from contacting surfaces by mechanical action," which is the definition of wear, they are included here with other wear phenomena.

Abrasive Wear

The general category of abrasive wear can be characterized by a single key word: *cutting.* Abrasive wear occurs when hard, typically foreign, free particles or projections from one surface roll or slide under pressure against another surface, as shown in Fig. 1, thereby cutting the other surface. Indeed, machining would fall into the category of abrasive wear except that it is usually not "undesired," which is a condition of the wear definition. Even a hand file, sliding under pressure, will cause microscopic distortion of the surface structure, thereby cutting a softer metal. The cutting action will remove small distorted fragments, or "chips," of the softer metal. Fig. 2 shows a microscopic region of abrasive wear on a relatively soft, low-carbon-steel shaft bearing component. Abrasive wear was intentionally created using a hand file.

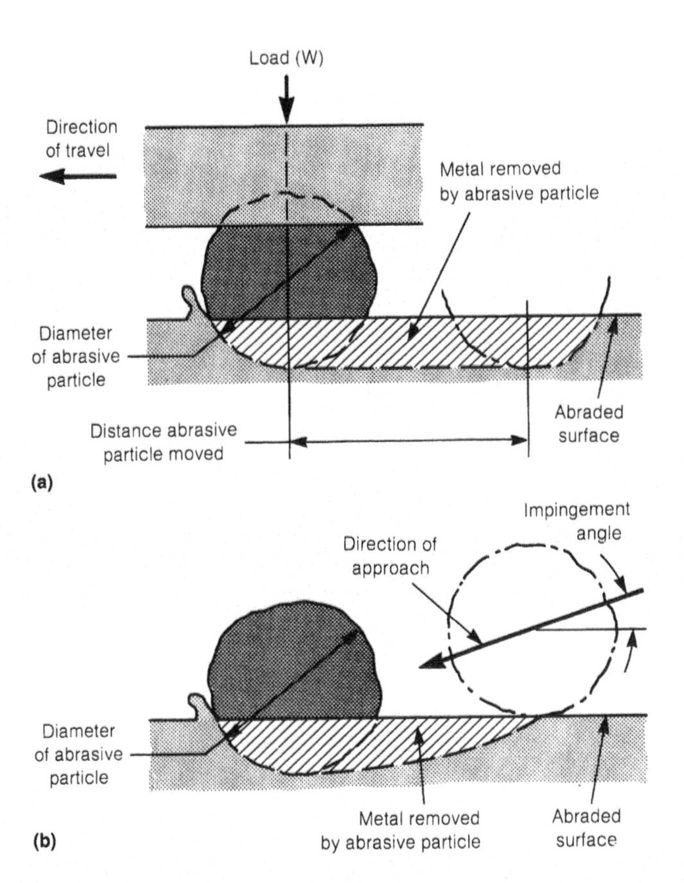

Fig. 1 Idealized representations of the two types of force applications on abrasive wear particles: (a) Cutting or plowing action of a contained particle under pressure. That is, the particle is not free but is under pressure from other particles or a solid object. This is characteristic of grinding and gouging abrasion, in which the hard particles are forced to scratch or cut the metal surface. (b) Cutting or plowing action of a loose particle flowing across the metal surface after impinging upon the surface. This is characteristic of erosive wear, in which free particles strike the surface at an angle and then slide across the surface. Source: Ref 1

Another important characteristic of abrasive wear is the heat that is generated by friction between the two materials. This is especially true with adhesive wear, as described below.

In general, abrasive wear may sometimes be reduced or dealt with by any of several methods, which may or may not be practical in individual circumstances:

- *Increase surface hardness:* This is a rather obvious solution to abrasive wear problems; however, it may not always be the answer to a specific problem (Ref 2). In cutting tools, such as various types of knives, blades, and the like, increasing the hardness may indeed make the cutting tool more resistant to dulling of the sharp edge. However,

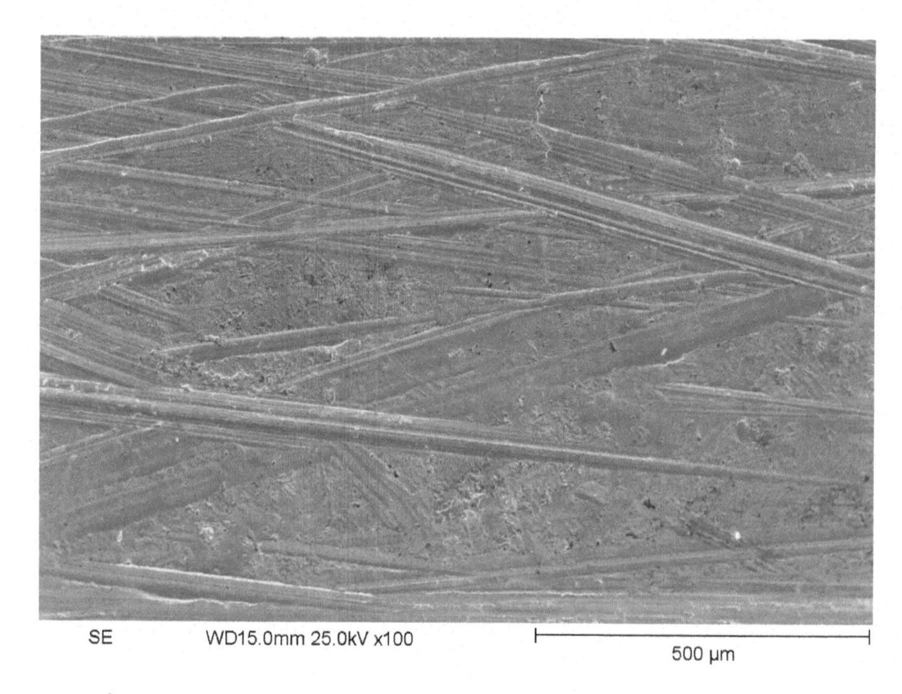

Fig. 2 Scanning electron microscope image of an area of abrasive wear on a soft, low-carbon-steel shaft bearing component, showing classic features of material "cutting" action (100×).

increasing the hardness also increases the chance of brittle fracture of the cutting tool itself. Brittle fracture would be a much more serious problem than the dulling from abrasive wear, for the dull tool can always be resharpened and reused, while the broken blade may cause injury to persons or damage to machines after fracture. The surface can be hardened through a variety of practical methods: using a different material, heat treatment, hardfacing, and other methods that are presented later in this chapter.

- *Remove foreign particles:* If hard, abrasive foreign particles are causing abrasive wear, it again seems obvious that if the particles are trapped and removed, the wear cannot take place. This is exactly the reason that filters for air, water, and oil are used in various types of mechanisms. An automotive engine is a typical example in which air, oil, and fuel filters are always used to prevent entry of external foreign particles and to trap and collect internal foreign particles before they can damage the engine. In heavy-duty engines, filters for the cooling system also are used to minimize abrasive wear of various parts, particularly the impeller that circulates the coolant. In other cases, the foreign particles cannot be removed; this is particularly true with erosive wear, where many high-speed particles slide and roll across a surface (e.g., automobile windshields).
- *Replace worn part:* One of the most practical ways by which we live

with wear is simply to design parts and assemblies that are subject to abrasive wear in such a way that they may be replaced when they are worn out. This is one of the simplest and most common ways of dealing with the problem. However, replacement may not be practical in a given situation because of inaccessibility, excessive labor cost or downtime, unavailability of replacement parts in an emergency, or other problems.

The general solutions mentioned may or may not be effective, depending on the circumstances of the particular wear problem.

Now let us look more closely at the characteristics of the major categories of abrasive wear.

Erosive wear (or erosion) occurs when particles in a fluid or other carrier slide and roll at relatively high velocity against a surface. Each particle contacting the surface cuts a tiny particle from the surface. Individually, each particle removed is insignificant, but a large number of particles removed over a long period of time can produce staggering degrees of erosion. The classic example is the Grand Canyon of the Colorado River. Whenever the dirt particles carried by the river current come in contact with the relatively soft rock of the riverbed, small amounts of the rock are removed. Over millions of years this giant chasm has been cut through the rock by the erosive action of dirt particles in the river, which flows rapidly in some places but is more placid in others.

Erosive wear can be expected in metal parts and assemblies where the above conditions are present. Common problem areas are found in pumps and impellers, fans, steam lines and nozzles, the inside of sharp bends in tubes and pipes, sandblasting and shotblasting equipment, and similar areas where there is considerable relative motion between the metal and the particles.

Erosive wear can be recognized by any or all of these conditions, depending upon the parts involved:

- *General removal of soft surface coatings or material:* This is a common form of wear for fan and propeller blades. In automotive applications, for example, the paint on the rear, or concave, side of the blade is usually removed by the scouring or cutting action of dust and dirt particles in the air. The concave side of a rotating fan blade has a positive pressure, while the convex side has negative pressure; the positive pressure forces the particles against the surface, thus leading to erosive wear.
- *Grooving or channeling of the material:* This type of erosive wear is common in assemblies involving fluids (liquids or gases), where the design of the parts is such that the fluid flows faster or is changed in direction at certain locations. Examples are in pumps or impellers in which vanes push the particle-laden fluid into various passages.

The inside of tubes or pipes is often damaged at curves because the inertia of the particles and the fluid forces them against the outer side of the curve. Obviously, sudden, sharp curves or bends cause more erosion problems than do gentle curves. In textile machinery even high-velocity thread or yarn can cause erosion; a sudden change in direction of yarn caused the grooving and erosive wear in the eyelet in Fig. 3. Grooving and channeling are also quite common in various types of nozzles where high-speed or high-pressure fluids scour and cut their way through the metal (e.g., water jet cutting processes). Liquid droplets can lead to erosive wear, as is frequently seen on the leading edges of high-speed aircraft.

- *Rounding of corners:* Erosive wear can change the shape of impellers, turbine blades, and vanes in such a way as to cause substantially impaired operating efficiency. An example of this type of damage is shown in Fig. 4, which shows a water pump impeller before and after erosive wear. If service had been continued, the vanes would have

Fig. 3 Abrasive wear of a yarn eyelet made of hardened and tempered 1095 steel. Grooving was caused by a sharp change in direction of the yarn as it came out of the hole. Service life was improved by changing the eyelet material to M2 high-speed tool steel, which contains spheroidal carbides in a matrix of martensite. Service life probably could have been improved also by changing the angle of exit or by rounding the corner to make a bell-mouth hole. Source: Ref 1

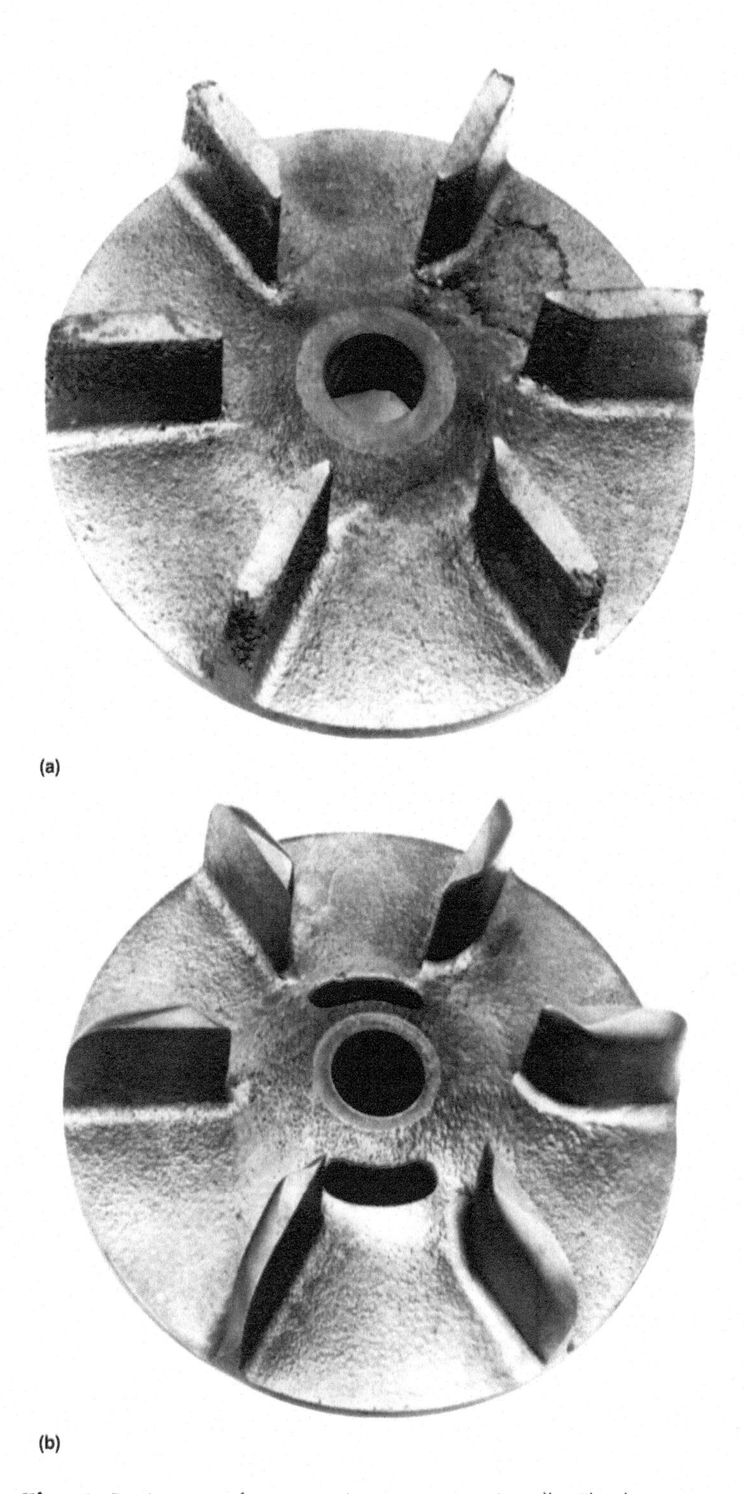

Fig. 4 Erosive wear of a gray cast iron water pump impeller. The sharp corners of the (a) new impeller have been (b) completely rounded off by the abrasive wear of sand in the cooling system. The change in shape of the vanes reduces the efficiency of the pump; if abrasive wear were to continue long enough, the vanes—and efficiency—would be completely gone.

been completely eroded away, leading to zero efficiency in pumping liquid.

For suggested solutions to erosive wear, see the general solutions listed under "Abrasive Wear" in this section.

Grinding wear. The principal characteristics of grinding wear are that it is caused by particles or protuberances under high stress that cut, or plow, many very small grooves at relatively low speed across a metal surface. This high-stress, low-speed operation is characteristic of tillage tools (plows, cultivators, rakes, etc.) and other ground-contact parts, such as bulldozer track shoes, cutting edges of blades, and the like. There are many other operations in other industries that have similar effects on the metal parts. These tend to dull cutting edges, changing their shape to make them perform their function less efficiently or not at all and generally causing unsatisfactory service.

Thus, grinding wear can be recognized if the type of service that caused it is known and if the wear occurs at high-stress locations, particularly points and edges, causing a general change in shape of the part or parts concerned. When two hard metal surfaces slide against each other, frequently in the presence of a lubricant, each may tend to smooth the other, particularly if fine abrasive particles are present. When properly controlled, this process may be very useful as a lapping, or polishing, process in which there is only one metal surface together with the fine abrasive material that is used for the polishing operation, such as in preparation of metallographic specimens for microscopic examination.

When considering means to prevent grinding wear, increasing the hardness is always an obvious thought. As pointed out earlier, however, this may lead to more serious problems in certain types of parts. However, the judicious use of certain hardening methods and of surface coatings may lead to some improvement in managing the wear situation.

Hardfacing by welding, metal spraying, or other means of deposition is frequently used to improve resistance to grinding wear. Usually the deposits contain large quantities of alloy carbides, such as those of tungsten, titanium, chromium, molybdenum, or vanadium. In certain applications, oxides, borides, or nitrides may be more satisfactory—the conditions of the service and the environment are extremely important in selecting the best alternative (Ref 2, 3).

In many cases, diffusion treatments, such as carburizing, carbonitriding, nitriding, chromizing, or boronizing, are sufficient for the purpose. The economics and practical aspects of the situation are usually overriding factors in the choice of a method of improving wear resistance.

Evaluating the best type of coating or diffusion treatment for a given application can be fraught with frustration, because simulated laboratory tests—usually accelerated to reduce testing time—often give misleading results. Actual service testing is usually the best way to evaluate, or rank, wear resistance of various materials or processes. Even this can be misleading, because the combination that is best in one type of abrasive environment may perform poorly in another type of environment. The same abrasive material used under different environmental conditions may produce contradictory results because of differences in packing and thermal characteristics.

Slight, controlled grinding wear may sometimes be used to advantage to maintain a self-sharpening behavior in certain cutting tools. Hardened surfaces may be used with soft surfaces to keep the tool sharp by judicious use of the "rat's-tooth principle": The front tearing teeth of rats (actually all rodents) have very hard enamel on the front convex surface but relatively soft dentine on the rear concave side, as shown in Fig. 5. When the rodent gnaws and tears with its teeth, the high-stress surface is the rear of the teeth, while the front receives little or no stress or wear. Since the rear is the high-stress area, this soft material wears more rapidly than does the hard, low-stress front. Since rodents' teeth grow continuously, the tip of the hard enamel is always sharp because of the gradual wearing away of the softer dentine. The tip of the brittle enamel breaks off, keeping the teeth the proper length. The teeth are, in effect, self-sharpening and cannot become too long, as long as the animal can gnaw.

This same principle can be applied to certain cutting tools. For example, plowshares (the cutting edges of plows) can be made self-sharpening if the front, high-stress surface is soft and the rear, low-stress side is hard faced with an appropriate material, as shown in Fig. 6. During service, the

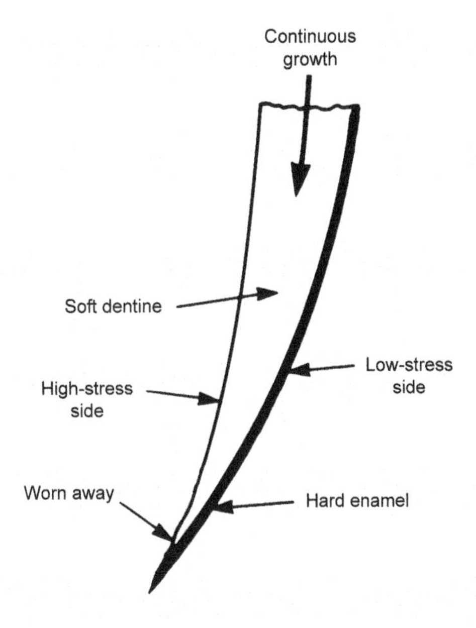

Fig. 5 Diagram of self-sharpening rodent tooth. See text for description. Source: Ref 4

soft, high-stress surface is worn, the hard surface remains relatively undamaged, and the plowshare stays sharp.

Electric carving knives, which have two blades sliding back and forth against each other, are sometimes hard surfaced on the outer, low-stress sides. As the blades slide back and forth against each other while cutting, slight metal wear occurs on the soft, high-stress inner surfaces, keeping the blades sharp.

In the mining industry, digging tools are sometimes hard faced on only one side to maintain the same self-sharpening action, as is shown in Fig. 7 (Ref 5).

Gouging wear. This type of wear is caused by extremely high-stress battering or impact that tends to cut, or gouge, large wear fragments from the surface of the metal. This service is encountered in certain applications in the fields of earthmoving, mining, quarrying, oil-well drilling, steelmaking, cement and clay product manufacture, railroading, dredging and lumbering, and undoubtedly other industries. When hard, abrasive products are crushed, battered, or pounded under extremely high stress, rapid deterioration of the contact surfaces can be expected unless specific steps are taken to prevent this problem.

In certain cases, it may be more economical to use replaceable parts, such as the teeth for backhoe buckets. Fig. 8 shows gouging wear on a tooth made of moderately hard medium-carbon alloy steel on the flat, originally sharp point. Battering against rock in digging eventually changed the shape of the tooth.

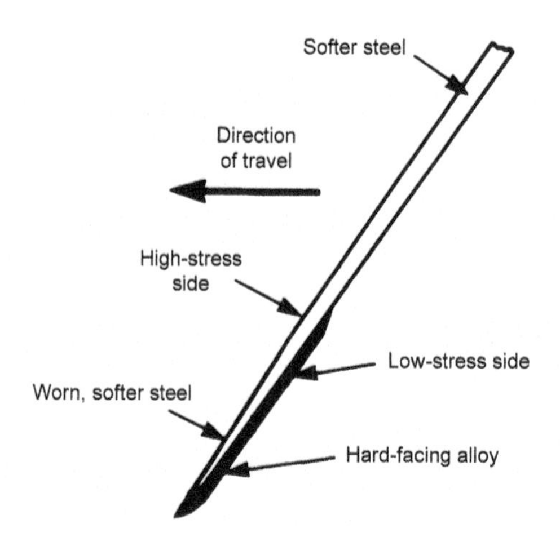

Fig. 6 Diagram of self-sharpening plowshare using the same principle as shown in Fig. 5. As the plowshare cuts through the soil from the right to the left, the relatively soft steel on the forward, high-stress side is gradually worn away, but the hardfacing applied to the rear, low-stress side is continually exposed at the sharp tip. Eventually, of course, the part must be replaced, but service life may be very long in certain types of soil, particularly those without rocks.

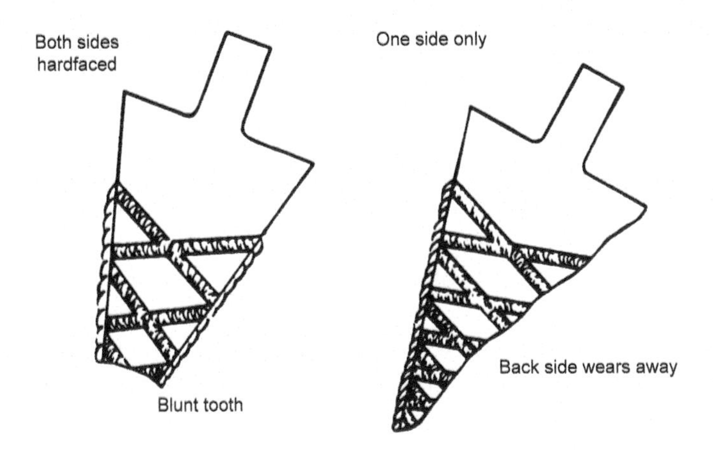

Fig. 7 Self-sharpening of a digging tooth from ground-contact equipment by controlled wear through selective hardfacing. The pattern of hardfacing can be varied to suit the condition, but note that the blunt tooth is hardened on both sides, while the self-sharpening tooth is hardened on only one side. Source: Ref 5

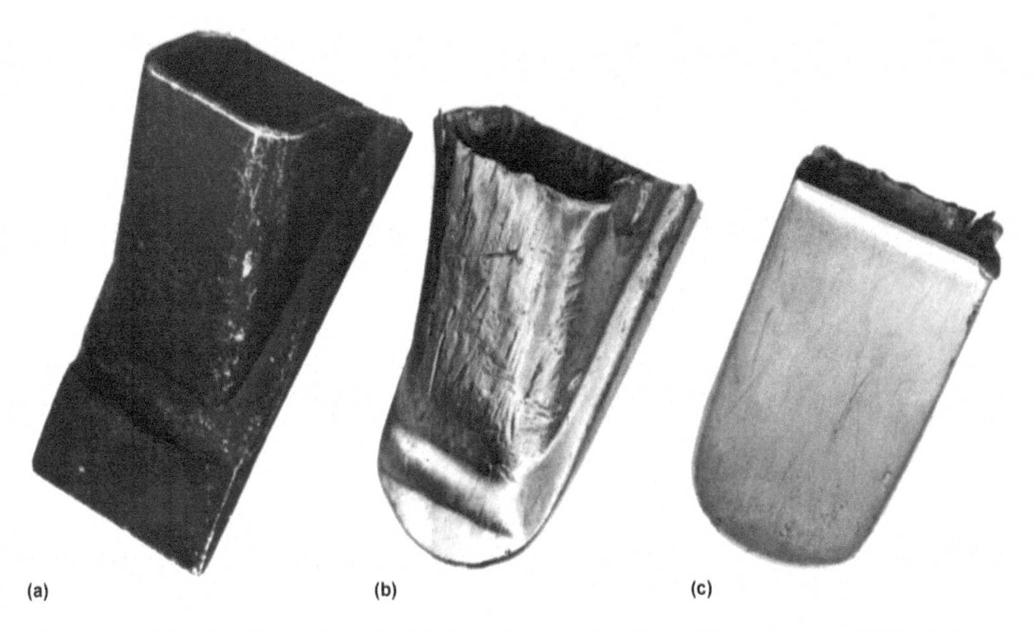

Fig. 8 Tooth for a backhoe bucket: (a) original condition; (b) the soft top of the tooth, made of 1010 steel, wore considerably more during operation in rocky, frozen soil than did (c) the flat opposite side of 8640 medium hard steel. The tooth is a replaceable part that is pinched over a stub to hold it in position.

In other cases, parts are not easily replaceable and must be made from a more resistant material. A type of steel invented over one hundred years ago by Sir Robert Hadfield in England has been used successfully in many applications, such as railway crossing frogs and switches, rock crushers, grinding mills, dredge buckets, power-shovel buckets and teeth, and pumps for handling gravel and rocks (Ref 2, 3).

This specialty steel, usually called Hadfield steel, is an austenitic manganese steel normally either used in the form of castings or welded to a steel base (Ref 2, 3). It is machinable only with difficulty, for the machining operation work hardens the austenitic steel, as does the necessary battering during high-stress service. In some cases, it is possible to preharden this material by submitting it to heavy battering or hammering before putting it into service. This type of steel is not intended for resistance to erosive wear or most kinds of grinding wear.

In any case, the remedies for gouging abrasive wear, as with other types, usually come down to a combination of economics, availability, accessibility, and design. Frequently, there are several ways to improve a product, but only one is chosen because it provides the optimum properties at lowest cost.

Adhesive Wear

Like abrasive wear, adhesive wear can also be characterized by a single word: *welding* or, more precisely, *microwelding*. The actual micromechanism is well described by the term *adhesive wear*. Other terms are sometimes used—for example, scoring, scuffing, galling, and seizing—but these do not accurately characterize the mechanism of failure and should be avoided.

Figure 9 is an exaggerated view of two surfaces that are sliding with respect to each other. They may or may not be separated by a lubricant. When a peak, or *asperity* (from a Latin word meaning "rough"), from one surface comes in contact with a peak from the other surface, there may be instantaneous microwelding due to the frictional heat, as shown in Fig. 9(a). Continued relative sliding between the two surfaces fractures one side of the welded junction, as shown in Fig. 9(b), making the asperity on one side higher and the asperity on the other side lower than the original height. The higher peak is now available to contact another on the opposite side, as shown in Fig. 9(c).

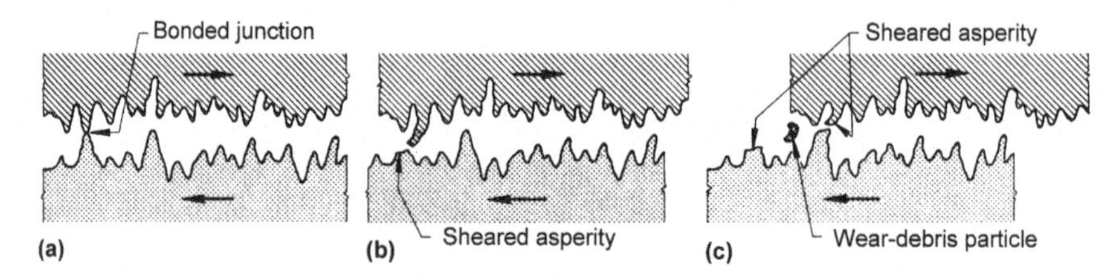

Fig. 9 Schematic illustration of one process by which a particle of wear debris is detached during adhesive wear. As the surfaces slide across each other, (a) a bonded junction is (b) torn from one peak, or asperity, and then is (c) sheared off by impact with a larger, adjacent peak to form a particle of wear debris. The peaks are greatly exaggerated in this sketch, but the principle is accurate; metal also may be transferred from one surface to another by the microwelding process. Arrows indicate direction of sliding. Source: Ref 1

The tip may either be fractured by the new contact or rewelded to the opposite side and the cycle repeated. In either case, adhesive wear frequently starts out on a small scale but may rapidly escalate as the two sides alternately weld and tear metal from each other's surfaces. Also, the debris may be carried by the lubricant, if one is present, to other parts of the mechanism. Figure 10 shows a microscopic region of adhesive wear on a hydrostatic pump piston component made of hardened alloy steel. Material has transferred and has been reapplied between the counterpart and the piston surface. Extreme wear may result, as shown in Fig. 11 and 12, and complete failure of the component may result. In severe adhesive wear, the debris is composed of free metallic particles; in mild cases, the much finer particles can react with the environment to form debris that is largely free oxide particles.

As should be apparent, the interface between two sliding surfaces is an extremely complex system, consisting of two metal surfaces (each with its own metallurgical, mechanical, chemical, and topographical characteristics) and usually a lubricant, which also is an extremely complex blend of physical and chemical characteristics that change with the temperature. In other words, there are both good and poor combinations of metals, and also good and poor lubricants, in a given application. The various modes of lubrication are discussed in Ref 1 and are not covered here except to point out that the ideal situation is for the lubricant to achieve complete

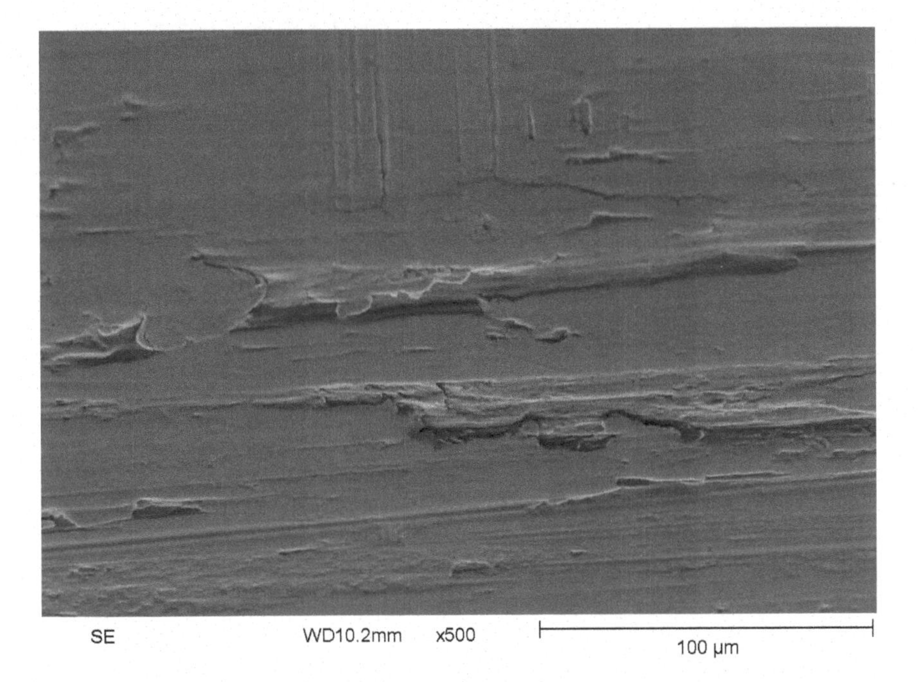

Fig. 10 Scanning electron microscope image of a microscopic region of adhesive wear on a hardened-alloy-steel piston component from a hydrostatic pump (500×). Material has been transferred by microwelding from the adjacent component wall surface during sliding and poor lubrication conditions.

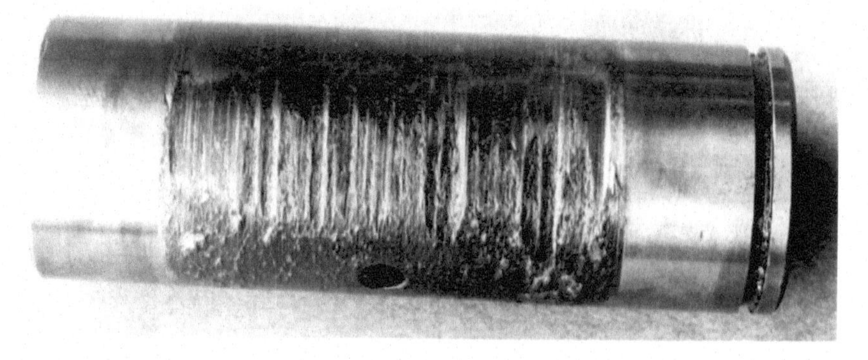

Fig. 11 Severe adhesive wear on a stationary shaft upon which a planetary gear rotated in the presence of inadequate lubricant. Because the radial force was on only one side of the shaft, the adhesive wear was on only one side. However, the entire bore of the gear was damaged by adhesive wear. Both parts were carburized and hardened steel.

Fig. 12 Destructive adhesive wear of a differential cross following fracture of one severely worn trunnion. This vehicle was operated primarily in forward speeds; consequently, only the forward-drive side was damaged severely by the rotation of differential pinions on the trunnions. The lubricant was inadequate for this application. The differential cross, or spider, and the differential pinions were made of carburized and hardened steel.

separation between every part of the two metal surfaces. Because this, unfortunately, does not always occur, there may be problems with adhesive wear.

The heat generated by friction is, locally, high enough to cause microwelding, as described. This means that the temperature is also high enough to cause unintended local heat treatment of the surface metal. Adhesive wear is quite similar to grinding burn in that both can cause tempering of the subsurface regions and actual rehardening of steel microstructures.

This could result in untempered martensite, which is extremely susceptible to cracking because of its brittleness. These cracks can lead to either brittle fracture or fatigue fracture, depending on the part and the application.

A very practical way of checking for the presence of adhesive wear in hardened steels is to use the Tarasov etching technique, as detailed in Ref 6. The two etching solutions described give a high-contrast, nondestructive means of checking for adhesive wear (and grinding burn) in hardened steels. Of course, other etchants, such as the conventional nital (a weak solution of nitric acid in alcohol), also may be used, but the contrast and sensitivity are usually not as good as with the Tarasov etching solutions. Of course, metallographic examination on a cross section of the surface also will show evidence of adhesive wear when studied at moderate to high magnification, depending on the amount of thermal damage.

Prevention of adhesive wear usually can be achieved by use of any or all of these methods:

- *Cool the lubricant:* Because adhesive wear is caused by locally high temperatures, the bulk temperature of the lubricant must be kept relatively cool. This is the reason for the use of transmission oil coolers in racing cars. Obviously, the lower the bulk oil temperature, the lower the interfacial temperature should be.
- *Use contacting metals that are insoluble in each other:* Because adhesive wear is a microwelding process, it follows that if two metals are not weldable to each other, there can be no adhesive wear. This is exactly the principle that is used in developing alloys for sliding bearings: we do not use steel against steel in sliding bearings (at least, not intentionally). However, in many applications steel is used against steel in gears, cams, and the like, and special precautions must be taken against adhesive wear. The antiwelding principle frequently is used in very high-speed gears by plating one of the gears with silver or gold. Since these metals are not soluble in steel, thin plates are usually sufficient to prevent adhesive wear with reasonable lubrication. For obvious economic reasons, use of this type of plating is limited, but it may be extremely useful in preventing downtime in plant equipment.
- *Use smooth surfaces:* If there are no projections to penetrate the lubricant film, there is a reduced probability of adhesive wear. If two smooth surfaces are separated by a thin oil film, they will essentially glide across each other without contact. However, if one of the surfaces has projections to rupture the lubricant film, adhesive wear is more likely. In some cases, some roughness or waviness may be desirable so that the depressions can act as reservoirs to retain lubricant. A word of caution: there is also a level where creating too smooth of a surface will cause severe adhesive wear, depending on surface loading conditions.

- *Contaminate surfaces to keep them chemically "dirty":* Chemical films frequently are used to prevent similar metal-to-metal contact that leads to adhesive wear. Phosphate coatings help separate metal surfaces, particularly during the early phases of operation. During the "wearing in" or "breaking in" period, the projections (or asperities) are removed from mating surfaces; phosphate coating crystals help to retain the lubricant. Eventually, however, the phosphate crystals may be gradually worn, or polished, away, and the parts should enter a long period of service without problems. Special oils and other lubricants have been developed over many years to form monomolecular surface films on steel surfaces. These are the extreme-pressure (EP) lubricants that are used in applications where there are high sliding velocities, such as in hypoid gear sets in automotive axles. These lubricants form extremely thin compounds on the surfaces to prevent metal-to-metal contact.

Fretting Wear

Fretting wear is quite similar to adhesive wear in that microwelding occurs on mating surfaces. The difference is that adhesive wear is related to interfaces that are sliding across each other, while fretting wear is related to interfaces that are essentially stationary with respect to each other. However, when minute elastic deflections or slight motion actually occurs, the cyclic motion of extremely small amplitude is enough to cause microwelding on both surfaces, as is shown in Fig. 13. Fretting wear is also known as fretting corrosion, false Brinelling, friction oxidation, chafing fatigue, and wear oxidation (Ref 1).

Fretting frequently occurs in "stationary" joints that are "fixed" from shrinking or pressing by interference fits or by bolts, pins, rivets, or other mechanisms, and also at the various contact points in antifriction, or rolling-element, bearings. This means that nonrotating antifriction bearings that are subject to vibration over a period of time may have fretting wear wherever a ball or roller contacts a raceway under load. If the bearings subsequently rotate in normal service, they may be noisy because of the wear patterns and small indentations that are present in the raceways and the corresponding flat spots on the rolling elements. The term *false Brinelling* is sometimes used to describe the indentations. However, the mechanism of failure actually is fretting wear. Fretting also is a serious problem on parts such as shafts, where it can initiate fatigue cracking on the contacting surfaces. In fact, many fatigue fractures of shafts are caused directly by fretting. Since fretting is extremely difficult to prevent, special means must be taken to prevent fracture resulting from the fretting, which can occur in the most unexpected and unlikely locations, as shown in Fig. 14.

Because fretting wear is essentially a stationary phenomenon, relative to the interface, the debris that is formed is retained at or near the location

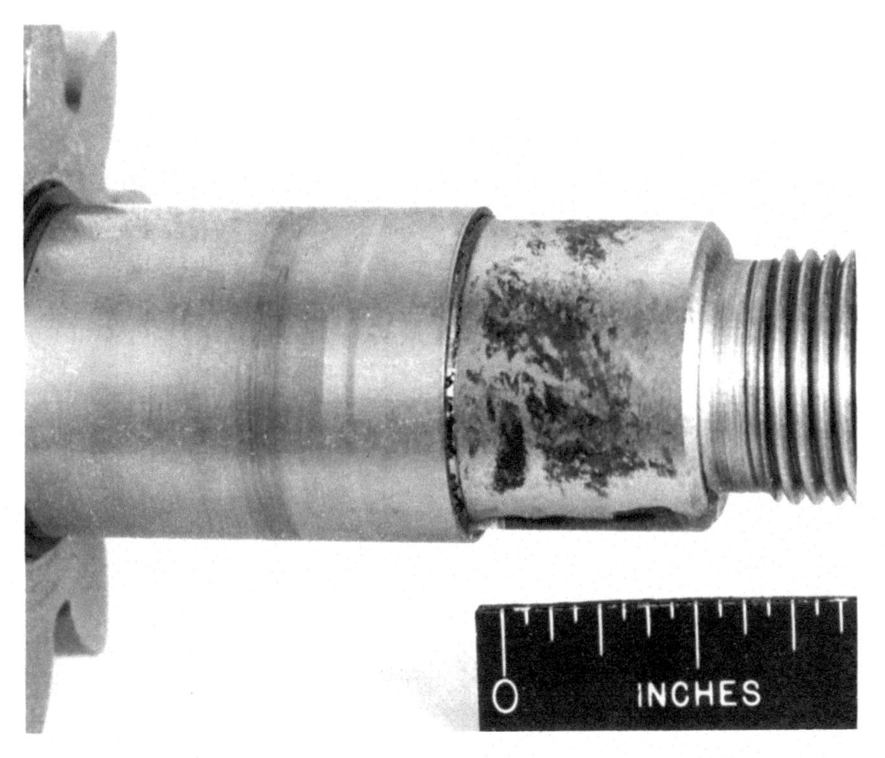

Fig. 13 Fretting wear on a steel shaft at the interface with the hub intended to be a press fit (~2.5×). The same fretting also appeared on the bore of the hub. This is typical of damage in a joint that is nominally stationary but in reality has slight movement between the hub and the shaft.

where it was formed. The debris usually consists of oxides of the contacting metals; with ferrous metals, it is brown, reddish, or black, depending on the type of iron oxide formed. For this reason, with ferrous metals the debris is sometimes called "cocoa" or "red mud" when it is mixed with oil or grease. Aluminum alloys form a black powder when fretting wear is present.

Prevention of fretting wear is not easy. However, its damage can sometimes be minimized with one or more of the following measures:

- *Eliminate or reduce the vibration:* This can sometimes be accomplished with the aid of vibration-damping pads or by stiffening certain members to increase the natural frequency of vibration. Occasionally, however, neither of these measures is effective, and movement must be increased greatly to improve lubrication.
- *Eliminate or reduce the slip at the interface:* This can sometimes be accomplished by trying to "lock" rough mating surfaces together by increasing the pressure between them. However, if the slip is not completely eliminated, the fretting wear may increase because of the increased contact stress between the mating surfaces.

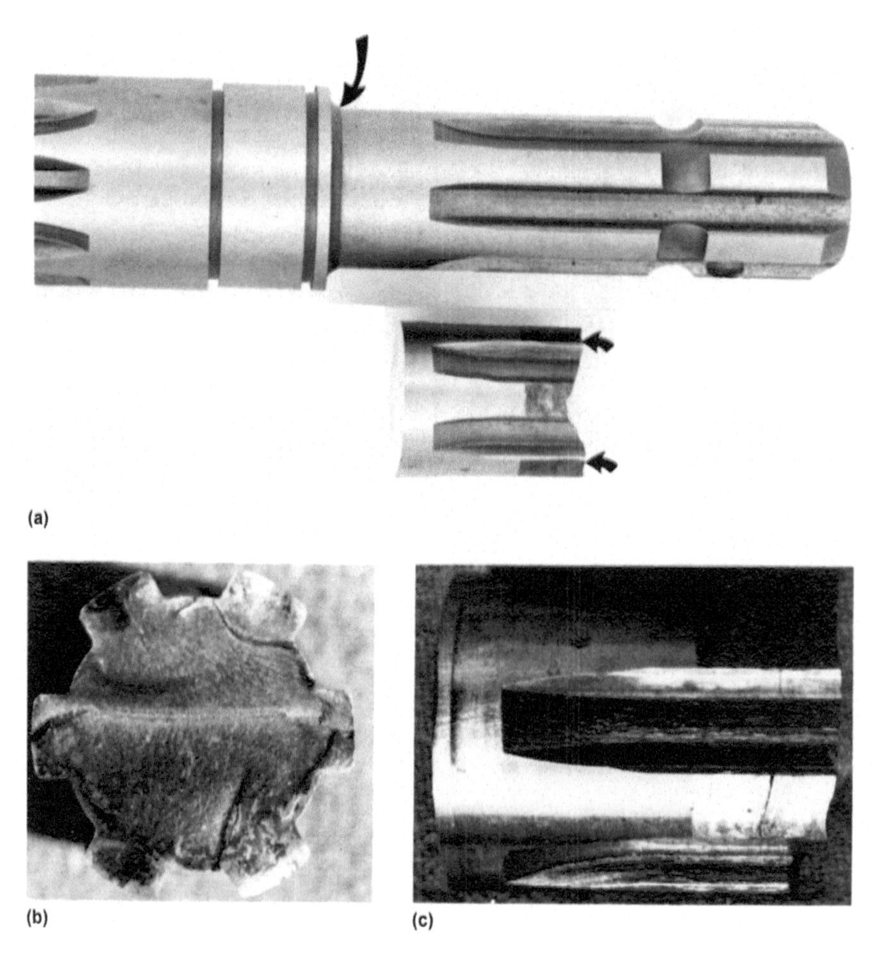

Fig. 14 Severe fretting wear of a splined shaft that led to fatigue fracture. This shaft, though it normally rotates, was being tested in reversed bending by clamped collets on the outside diameter. (a) Dark areas in lower view are areas of severe fretting; fracture occurred at the small arrows. This is an example of the surprises that fretting wear can cause, for the expected location of fatigue fracture was the fillet (large arrow in upper view), shown on a new shaft. However, this area had been strengthened by induction hardening; the fracture location was transferred to the fretting on the outside diameter of the spline. Close-ups of the fracture show (b) the fatigue origins at the outside diameter and (c) multiple fatigue cracks, of which one on each side led to the actual fracture.

- *Use an elastomeric material in the joint:* Complete redesign of the joint to include an elastomeric bushing or sleeve may be necessary. Vibration and minute movement still may be present, but the elastomeric material absorbs the motion and prevents metal-to-metal contact.
- *Lubricate the joint:* Because the joint is essentially stationary, liquid lubricant cannot flow through the interface as it can where there is continual sliding motion. Certain greases, solid film lubricants (e.g.,

molybdenum disulfide), and oils are intended to reduce or delay fretting, as is discussed in Ref 1.

- *Prevent fracture:* Fracture resulting from fretting wear may be prevented by inducing compressive residual stresses or by certain heat treatments, although these probably will not eliminate the fretting. One of the most effective means of preventing fatigue fracture is the use of mechanical prestressing by shot peening, surface rolling, or any other method of prestressing. Because fatigue cracks cannot propagate easily through a compressive residual stress barrier, these methods may be used to prevent fracture unless the part is used at a temperature high enough to stress relieve the material. In another process, the elements diffused into the metal by proprietary salt bath or gaseous nitriding methods form compounds resistant to adhesive and fretting wear. Epsilon iron nitride is one of the most effective surface compounds in preventing fretting wear and possible fatigue fracture.

Summary

Abrasive, or cutting, wear is a normal type of surface deterioration that can be minimized but not completely prevented. The major categories of abrasive wear—erosive, grinding, and gouging—are dependent on variations in stress and relative velocity of the abrading material. Adhesive, or microwelding, wear can occur in sliding joints. Fretting wear in stationary joints is more difficult to eliminate, in most cases, than adhesive wear in sliding joints, where fresh lubricant can be introduced.

REFERENCES

1. *Failure Analysis and Prevention,* Vol 11, *ASM Handbook,* ASM International, 1986, p 145–162, 497–498
2. *Properties and Selection: Stainless Steels, Tool Materials, and Special-Purpose Metals,* Vol 3, *Metals Handbook,* 9th ed., American Society for Metals, 1980, p 563–588
3. *Welding, Brazing, and Soldering,* Vol 6, *ASM Handbook,* 1993, p 789–829
4. H.J. Baker, J.R. Lindsey, and S.H. Weisbroth, *Research Applications,* Vol 2, *The Laboratory Rat,* Academic Press, 1980, p 60
5. C.R. Morin, K.F. Packer, and J.E. Slater, Failure Analysis Associated with Mining and Heavy Mechanical Equipment, *Metallography in Failure Analysis,* J.L. McCall and P.M. French, Ed., Plenum Press, 1978, p 191–205
6. "Injury in Ground Surfaces: Detection, Causes, Prevention," booklet (revised) published by Grinding Wheel Division, Norton Co., Worcester, MA 01606

SELECTED REFERENCES

- R.G. Bayer, *Mechanical Wear Prediction and Prevention,* Marcel Dekker, 1994
- F.P. Bowden and D. Tabor, *Friction and Wear,* Methuen and Co., 1967
- *Erosion, Wear, and Interfaces with Corrosion,* ASTM, STP 567, 1974
- *Friction, Lubrication, and Wear Technology,* Vol 18, *ASM Handbook,* ASM International, 1992
- M.B. Peterson and W.O. Winer, Ed., *Wear Control Handbook,* American Society of Mechanical Engineers, 1980
- E. Rabinowicz, Wear, *Sci. Am.,* Vol 206 (No. 2), 1962, p 127–135
- E. Rabinowicz, *Friction and Wear of Materials,* 2nd ed., Wiley, 1995
- D.A. Rigney, Ed., *Fundamentals of Friction and Wear of Materials,* American Society for Metals, 1981
- D.A. Rigney and WA. Glaeser, Ed., *Source Book on Wear Control Technology,* American Society for Metals, 1978
- J.A. Schey, *Tribology in Metalworking: Friction, Lubrication, and Wear,* American Society for Metals, 1981
- A.Z. Szeri, Ed., *Tribology: Friction, Lubrication, and Wear,* Hemisphere Publ. Corp., distributed by McGraw-Hill, 1980
- *Wear and Fracture Prevention,* American Society for Metals, 1981

CHAPTER 12

Wear Failures—Fatigue*

RECALLING from Chapter 11, "Wear Failures—Abrasive and Adhesive," in this book that *wear* is broadly defined as the undesired removal of material from contacting surfaces by mechanical action, there are certain types of metal removal, or wear, that are not caused directly by sliding action. In this chapter, we consider different reasons for metal removal, such as fatigue fractures that produce cavities, or pits, in either of two surfaces that contact each other primarily by rolling and/or sliding action, or—in the case of cavitation pitting fatigue—in a metal surface subjected to extreme pressure in contact with a liquid.

The cavities themselves are serious because they frequently act as stress concentrations that can cause fracture of the major part. This is particularly true with respect to gear teeth. In bearings, the pits increase the rolling resistance and can lead to overheating of the bearing and increased loads on other components. Also, the metal removed from the cavities usually is very hard and brittle. Thus, it is readily crushed and fragmented into much smaller particles, which can cause abrasive wear and other damage when carried by the lubricant to other parts of the mechanism. Bearings, both the rolling and sliding types; gears; and component parts, such as pumps, impellers, and propellers are subject, to damage by wear fatigue mechanisms.

Fatigue, as discussed in this chapter, is the same mechanism as for fractures that result from cyclic slip under repetitive load applications for many thousands or millions of load cycles. The only difference is that, instead of causing gross fractures of parts, only fragments of the surface are removed, at least initially. These fragments, when lost, result in pits or cavities in the surfaces. This type of pitting wear, or contact stress fatigue, frequently is the limiting factor in the component's load-carrying ability.

***Understanding How Components Fail,* Second Edition, by Donald J. Wulpi, Chapter 12, Wear Failures—Fatigue, ASM International, 1999, reviewed and revised by Larry D. Hanke, Materials Evaluation and Engineering.

The pits that occur on the surfaces of contacting parts as a result of contact stress fatigue seem to have three different types of behavior. Some start as microscopic cavities and may stay microscopic throughout the life of the part. They cause only a dull, frosted appearance on an otherwise bright surface. The second type starts the same way: microscopic. These pits gradually become larger, however, under continued service of rolling and sliding under load. Pits of the third type are large to start with and rapidly become even larger. The latter two types are completely destructive of the surfaces of hardened steel gears, rolling-element bearings, roller cams, and other parts or assemblies where there is a combination of rolling and sliding motion.

The parts subject to wear fatigue failure generally have two convex, or counterformal, surfaces in contact under load. Typical components with this configuration are gear teeth and various types of antifriction bearings. However, the same type of failure can occur where a convex shape fits within a concave shape, such as a shaft within a sliding bearing or balls in a ball-bearing race. These situations are shown in Fig. 1. In these configurations, pitting fatigue occurs on one or both mating surfaces under compressive load with the contact areas concentrated into either a line or point contact, depending on the geometries concerned. Actually, there will be either a broadened line or an elliptical contact area. Since the instantaneous-contact areas may be quite small under heavy loads, the compressive and shear stresses that are formed may be extremely high. This subject is well covered in the literature, starting with Heinrich Hertz's first analytical work over a century ago (Ref 1, 2, 3).

When dealing with contact stress fatigue, it is necessary to understand the difference between pure rolling and rolling plus sliding contact. Pure rolling of metals under compressive load is quite elusive and is difficult, if not impossible, to achieve because of the elasticity of the metals.

As pointed out in Chapter 2, "Mechanical Properties," in this book, all metals are elastic and deform elastically under load. In fact, the harder and stronger metals, such as those typically used in gears and rolling-contact

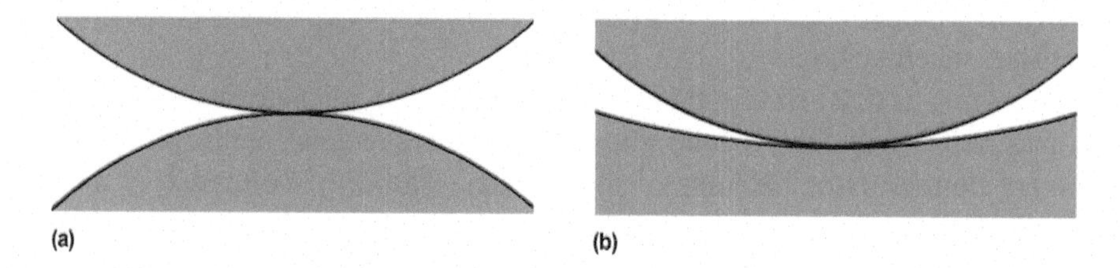

Fig. 1 (a) Sketch of counterformal, or convex, surfaces in contact; examples are gear teeth and roller or needle bearings rolling on a shaft, in an inner raceway, or on a flat surface. (b) Sketch of conformal surfaces, where a convex surface is in contact with a concave surface; examples are ball bearings in contact with an inner or outer raceway, roller or needle bearings in contact with an outer raceway, and a shaft in contact with a sliding bearing or on a flat surface.

bearings, can deform elastically to a greater degree than can relatively softer, lower-strength metals. This is shown graphically in Fig. 2 of Chapter 2 in this book, in which metals A and B, the hardest and strongest of those sketched, have the longest straight-line, or elastic, portions of the typical stress-strain curves.

Elasticity under heavy compressive loads results in lateral displacement of the material in the contact area, causing sliding and shearing forces at the interface between the contacting metals. This sliding under load generates heat that must be dissipated, usually by a lubricant. Elasticity under load also causes internal friction within the metals, generating additional heat. In fact, heat dissipation is one of the primary functions of a lubricant, in addition to reduction of friction at the interface.

Because of the many variations in rolling and sliding contact, as well as in metallurgical and geometrical variables, several types of fatigue failure can occur in rolling/sliding elements. In a real sense, there is a competition between the different modes of failure to determine which mode will dominate and cause failure of the element, the assembly, and sometimes the entire mechanism, depending on the degree of failure. It is, in fact, a race to failure, because all modes progress simultaneously, but usually only one mode causes total failure.

As pointed out in Chapter 3, "Stress versus Strength," in this book, elastic convex surfaces under heavy pressure deform elastically to form a bulge at the ends of the contact area. Also, surfaces in contact have the maximum shear stress a short distance below the surfaces when the members are either stationary or rolling with respect to each other. Since fatigue fractures are initiated by shear stresses, this subsurface location of maximum shear stress is of primary concern in rolling-element components.

Contact Stress Fatigue

All of the different types of fatigue discussed herein lead to the same final result: complete destruction of the original surface. However, the factors that cause this are very different, as are the various ways of trying to prevent this type of failure. The general term *contact stress fatigue* is where the fractures cause pieces of metal to separate from the surface, leaving cavities, or pits, in the surface. Contact stress fatigue is covered in this chapter as:

- Subsurface-origin fatigue
- Surface-origin fatigue
- Subcase-origin fatigue ("spalling" fatigue)
- Cavitation fatigue

Subsurface-Origin Fatigue. Pitting in hardened steel as a result of subsurface-origin fatigue occurs during essentially pure rolling motion of one element across or around another. This is most common in antifriction, or rolling-element, bearings, such as ball and roller bearings, needle bearings, and roller cams, but is also found on gear teeth. As pointed out above, "pure rolling" is really a misnomer, for there is always some degree of sliding—due to elastic deflection—of the parts under load.

Also, because the maximum shear stress is located a relatively short distance below the surface, this is the normal location for fatigue fracture to originate. Stress concentrations slightly below the surface are the usual sites of fracture origins. Since most geometrical stress concentrations are at the surface of metal parts, the most common geometrical stress concentrations within the metal are various types of inclusions inherent in the steel. There are many types of inclusions, but the most serious are the hard, brittle inclusions that are often angular in shape. Since inclusions are distributed at random within the steel, only those within the high shear stress region are likely to cause subsurface-origin pitting fatigue. Fracture will initiate where a larger inclusion coincides with a location of high shear stress.

Figure 2(a) shows two rolling elements under pressure, each of which has the same contact (compressive) stress at the surface. However, the depth of the maximum shear stress may differ, depending on the relative geometry of the parts involved. Any damaging inclusions within the high shear stress region can cause fatigue cracks to originate within the metal parallel to the surface. However, continued rolling across the damaged area will eventually cause cracks to reach the surface. The same procedure may be occurring at several locations within the component, and eventu-

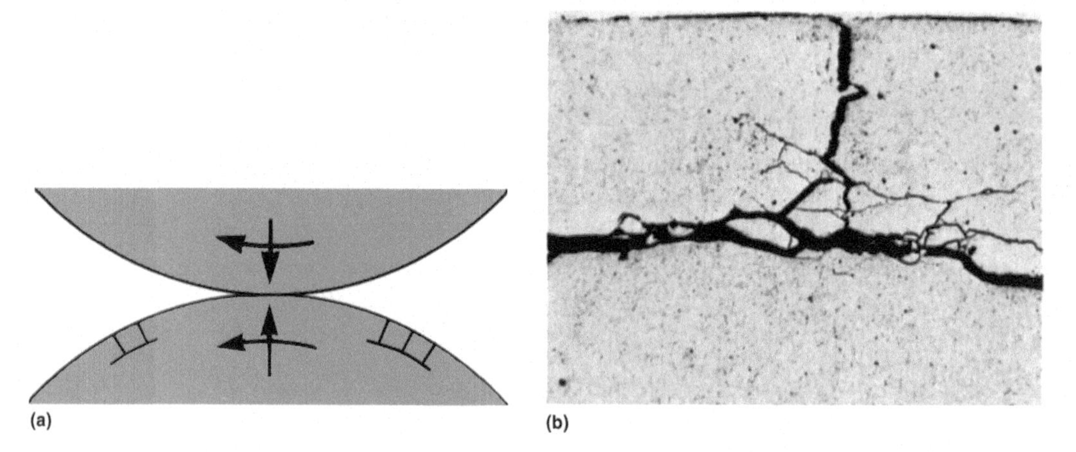

Fig. 2 Subsurface-origin pitting fatigue. (a) Sketch showing usual origin slightly below the surface where the shear stress is high. The fatigue cracks, which usually originate at stress concentrations such as hard, brittle inclusions, propagate parallel and perpendicular to the surface. When a small volume of metal is surrounded by cracks, it falls from the surface, leaving a cavity or pit with steep-sided walls and a flat bottom, until continued operation breaks down the sides of the walls. (b) Micrograph showing fatigue cracks parallel to the surface and perpendicular cracks going to the surface of the specimen. The total depth of this crack system was about 0.007 in., and the metal was case-hardened steel with a surface hardness near 60 HRC. The specimen was a 1-in.-diam roller, tested at a calculated compressive stress of 425,000 psi in pure rolling.

ally a volume of metal will be surrounded by cracks. When this happens, the particle of metal is lost from the surface, and an irregular-shaped cavity, or pit, is left in the surface. At first, the pit has sides perpendicular to the surface, but continued operation under the pressure of the mating roller each time it rolls across the pit rapidly causes the sides to fracture and break down. Thus, the pit does not stay in the original steep-sided shape very long, unless the rolling action is stopped.

Figure 2(b) shows an enlarged cross section through cracks in a rolling element before the surface material has spalled to form a pit. The horizontal crack was about 0.178 mm (0.007 in.) below the surface, and the vertical crack went up to the surface, but a volume of metal had not yet been surrounded by cracks sufficiently to release a particle. It is very difficult to locate the inclusions that cause this behavior; the chances of finding them in a random cut through the metal are very slight. Figure 3 shows such a pit, in which the test roller was stopped immediately after the pit was formed.

Since hard, brittle inclusions or other internal stress concentrations often contribute to this type of pitting, one may wonder why the antifriction bearing industry does not use cleaner steel to minimize subsurface pitting and permit higher service loads. The answer is that they do use cleaner steels! This problem has been recognized for many years, and the bearing industry is always alert to use high-quality, vacuum-melted steels to try to eliminate the problem. However, each time that an advance is made in eliminating inclusions and providing better-quality steel, the load and speed ratings are increased by each company that uses the improved metal. Customers are told that they can (a) use a higher load on the same

Fig. 3 A subsurface-origin pit in a carburized and hardened alloy steel test roller caused by fatigue in the manner shown in Fig. 2. When this specimen was tested in essentially pure rolling, a steep-sided, irregularly shaped pit was formed, and the test was stopped. The extremely high force needed to cause subsurface pitting is shown by the plastic deformation at the sides of the wear track formed by the mating roller, which was 5 in. in diameter

size bearing, (b) use a smaller bearing to do a certain job, (c) use higher speeds with the same size bearing, or (d) all of the above! This means that if the users do these things, the problem will always be with us, for it is impossible to eliminate all of the offending particles and other imperfections from the steel.

Investigation of the contribution of inclusions in the steel to contact fatigue failure should include examination of fracture surfaces by scanning electron microscopy and overall characterization of the steel cleanliness by metallographic examination. The fracture surfaces in open pits are generally smeared such that few fracture features remain, but frequently many cracks are a short distance below the surface that can be exposed for microscopic examination. It is also necessary to point out, as in Ref 1, that:

> a rolling element bearing does not have an unlimited life, a fundamental fact that does not appear to be generally appreciated. Even if a bearing is run under the recommended conditions of load, speed, and lubrication, and is protected against adverse external influences that otherwise would tend to reduce its life, failure will ultimately result by some process such as fatigue, wear, or corrosion.

Also, there are many other types of failure of antifriction bearings, so Ref 1, 2, or other references should be used in studying the failure of rolling-element parts.

Surface-Origin Fatigue. Surface-origin pitting of hardened steel surfaces also is destructive of the surface and, in the case of gear teeth, may also lead to bending fatigue fracture of the gear teeth. When sliding is added to the rolling action previously described, an entirely different and complicating set of circumstances arises. The maximum shear stress is no longer located below the surface of the steel but is brought up to the surface because of the influence of sliding friction and the associated traction forces.

In order to understand the state of stress at the surface of a rolling/sliding interface, it is necessary to analyze separately the relative motions of rolling and sliding. Rolling is shown schematically in Fig. 4, in which a small roller rotates counterclockwise against a larger roller rotating clockwise. There will be pure rolling (ignoring elastic deformation) if the surface velocities (not rpm) of the two rollers are identical. Points A, B, and C on each roller come into contact successively, as shown in Fig. 4(a). However, the same motion relative to the small roller can be achieved if it is held stationary and the large roller rolls around it with a clockwise planetary motion, as in Fig. 4(b). Again, points A, B, and C successively come into contact. Thus, the rolling direction—the direction of the contact point—is to the right, or clockwise, on the small roller in each example.

The above discussion assumes that there is only pure rolling between the two elements. The situation becomes more complicated if the rollers do *not* have the same surface velocity—that is, if there is also sliding at the interface. Assume that the larger roller is rotating with a higher surface velocity than is the smaller roller, as is shown in Fig. 5(a). The difference in the sliding tends to drag the surface of the small roller to the left, or counterclockwise, while the surface of the larger roller is dragged to the

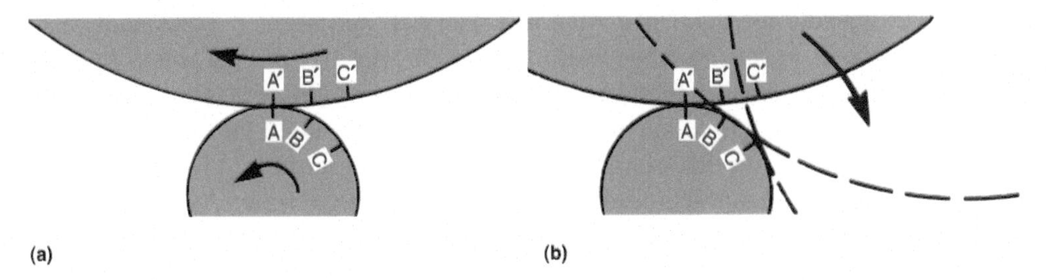

Fig. 4 Schematic of pure rolling (ignoring elastic deformation). (a) When two rollers of different sizes but the same surface velocity (not rpm) contact in pure rolling, point A will contact point A′, then point B will contact B′, point C will contact point C′, and other points around the periphery will contact in succession. However, note that the direction of rotation is toward the left and that the direction of rolling (or of the point of contact) is to the right, in the direction opposite to that of rotation. (b) This concept may be simplified by stopping one roller, such as the lower roller, as shown. If the mating roller rolls around the lower roller in a planetary motion, as shown, again points A, B, and C contact successively the corresponding points on the other roller. The effect on each roller is the same as before, for the relative motion is the same.

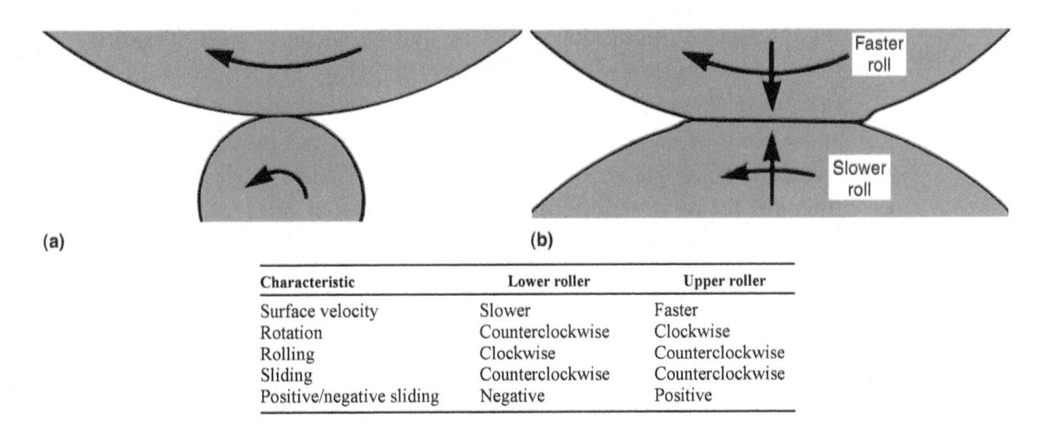

Characteristic	Lower roller	Upper roller
Surface velocity	Slower	Faster
Rotation	Counterclockwise	Clockwise
Rolling	Clockwise	Counterclockwise
Sliding	Counterclockwise	Counterclockwise
Positive/negative sliding	Negative	Positive

Fig. 5 Schematic of rolling/sliding contact. (a) The situation shown in Fig. 4 changes drastically if the rollers are externally driven and forced to rotate with different surface velocities. The upper roller is driven at a higher surface velocity than the lower roller, which introduces sliding into the interface. Now there is rolling-sliding contact, which greatly complicates the situation compared with the pure rolling shown in Fig. 4. (b) A closer look at the interface of (a) after heavy compression forces are applied to the two rollers shows that the faster upper roller tends to drag the surface of the slower lower roller to the left, increasing the bulge on that side, and reducing the bulge on the opposite side of the slower roller. However, the surface of the faster upper roller is itself held back by the action of the slower lower roller, increasing the bulge on the right side of the upper roller and reducing the bulge on the opposite side of the upper roller. The elastic deformation of each roller progresses like a wave around them, alternately applying shear stresses to and removing them from the surface of each. However, in the slower roller, the rolling and sliding are in opposite directions, creating the condition of negative sliding, which is most likely to cause surface pitting fatigue.

right, also counterclockwise on that roller. But note that the direction of sliding is opposite to the direction of rolling on the small roller, while the rolling and sliding directions are the same on the larger, faster roller.

The term *negative sliding* is used when rolling and sliding are in opposite directions, as on the small roller in this example; *positive sliding* occurs when the rolling and sliding directions are the same, as on the larger roller. Since the surface of the small, negative-sliding roller is, in effect, rolling in one direction and simultaneously being dragged in the opposite direction, the frictional, thermal, and shear stresses tend to be higher in this member than in the larger, positive-sliding roller. In addition, the smaller member will have a greater number rotations and more load applications on each surface point than does the larger member. Therefore, in this example, the small roller is far more likely to undergo destructive surface-origin pitting than is the larger roller. As is pointed out in Ref 1:

> In devices that undergo combined rolling and sliding, knowledge of the relative velocities and directions of rolling and sliding is necessary for definition of the wear mechanism. The direction of rolling is defined as the direction in which the point of contact moves; the direction of rolling is always opposite to the direction of rotation of a rolling element. On a given surface a condition of positive sliding exists if the direction of sliding is the same as the direction of rolling. Negative sliding occurs on the mating surface, where the directions of rolling and sliding are opposite to each other. Most surface fatigue failures originate in regions of negative sliding, because the shear stresses there are usually more severe than in regions of positive sliding. Negative sliding occurs on the dedenda of gear teeth, on the cam follower riding on a cam and in other devices, on the part that has the lower surface velocity in a rolling-sliding system.

The reasons for the severe stress conditions in negative sliding become obvious with study of Fig. 5(b). Note that the lower, slower roller is being dragged to the left, making a bulge on the exit side of the contact area. At the same time, the point of contact of the upper roller is moving to the right, trying to make a bulge on the entrance side of contact area at the right. The complex stress conditions that result from this negative sliding are the reasons that the maximum compressive stress that can be carried is less in a rolling/sliding situation than in a pure rolling situation. In other words, the addition of sliding reduces the load-carrying capacity. According to Ref 4, at about 65% rolling and 35% sliding, hardened steels commonly pit at 20 to 30 million stress cycles when the calculated contact stress is about 2,410 MPa (350,000 psi). When rolling alone is involved, the calculated contact stress must be 4,135 to 4,480 MPa (600,000 to 650,000 psi) to cause pitting in the same number of stress cycles.

Figure 6(a) shows a metallographic section through a nearly complete surface pit in a case-hardened steel roller test specimen. In this view, the direction of rotation and sliding is to the left, while the direction of rolling is to the right, the condition of negative sliding. Note that the surface has several small, diagonal surface-origin fatigue cracks slanting toward the lower right, for the surface was dragged to the left. The major crack system has progressed inward to a depth of about 0.255 mm (0.010 in.) and has started to come back toward the surface.

Meanwhile, the view from the original surface of the roller shows that the surface, which usually has a "frosted" appearance to the unaided eye, actually has many tiny V-shaped cracks when examined in a metallurgical microscope, as shown in Fig. 6(b). The V-shapes point in the direction of rotation and are actually the outer view of the small diagonal cracks shown on the cross section in Fig. 6(a). These are potential origins of continuously growing fatigue cracks that may form larger pits. Each time the mating roller rolls across from left to right, the lubricant is forced down into the crack and may act as a hydraulic wedge to help push the fatigue crack deeper, in addition to the cantilever beam effect that exists. As the crack expands in the V-shape on the surface and gradually becomes deeper, a V-shaped volume of metal falls out, leaving an arrowhead-shaped pit pointing in the direction of rotation, as shown in Fig. 7. With continued battering and high-stress contact from the mating surface, however, this V-shape rapidly breaks down; thus, these shapes are not often seen in actual operating parts that have run for some time after the original pit is formed.

Gear teeth are the principal hardened steel parts subject to surface-origin pitting. In essence, gear teeth are carefully shaped cantilever beams that have rolling and sliding forces, as shown in Fig. 8 and summarized in Table 1. It is significant that the dedenda (i.e., the regions below the pitch line) of all gear teeth are in negative sliding. This means that this is where surface-origin pitting is most likely to occur, other conditions being equal. Pure rolling occurs only on the pitch line of spur, bevel, and helical gears, but not at all on worms, spiral bevel, and hypoid gears and pinions. In fact, the direction of sliding undergoes a reversal at the pitch line, as shown in Table 1.

The driving gear is usually the smaller of the pair of gears and receives many more load applications per tooth, the difference depending on the size ratio between the gears. For this reason, the dedendum of the smaller gear is where pitting fatigue will originate if the parts have the same metallurgical properties. For this reason, the smaller gear often is made slightly harder than the larger gear to compensate for the difference in the number of load applications.

Since sliding is always present in the operation of gears, the lubricant is of extreme importance for the survival of any set of heavily loaded gears. In addition to the adhesive-wear problem discussed in Chapter 11 of this book, the reduction of surface friction is of critical importance in the effort

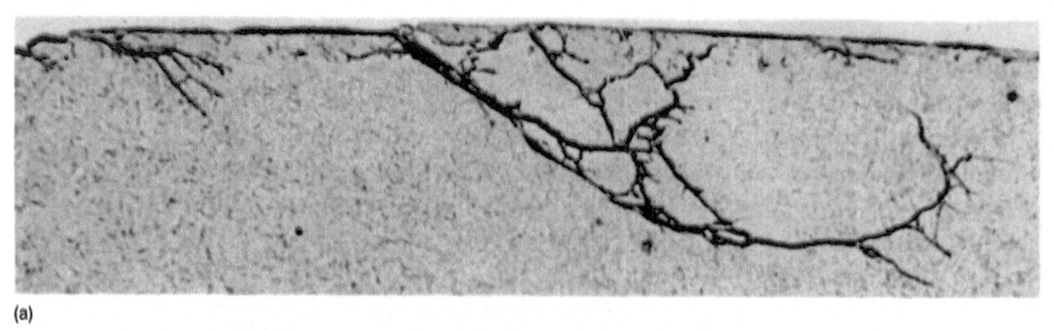

(a)

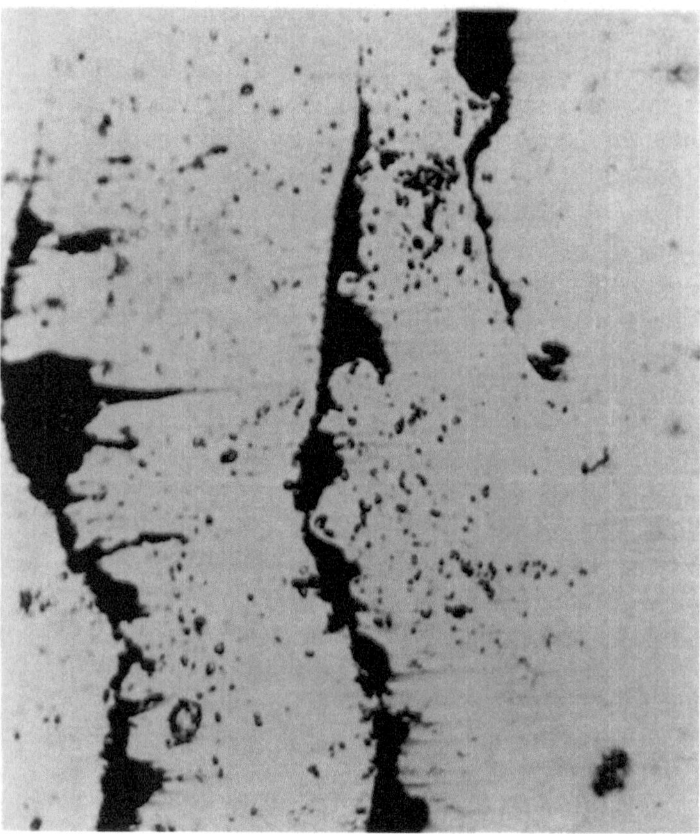

(b)

Fig. 6 (a) Cross section through a nearly complete surface pit (max depth about 0.010 in.) in a roller test specimen. The negative-sliding member (lower and slower roller in Fig. 5b) forms small diagonal cracks from the surface inward at an angle. Some of these cracks may proceed deeper into the hardened steel, as shown in this section. The orientation is the same as in Fig. 5(b), in which the sliding is to the left and the rolling is to the right, producing negative sliding. (b) Highly magnified view (1000×) of the surface of a negative-sliding roller. The many small, V-shaped cracks are the outer view of diagonal surface cracks like those shown in Fig. 6(a). As a crack progresses deeper into the steel, the V widens on the surface. Eventually a V-shaped volume of metal becomes totally surrounded by cracks, falls out, and leaves an arrowhead-shaped pit, which rapidly changes shape under continued loading. Sliding direction is to the left, rolling to the right.

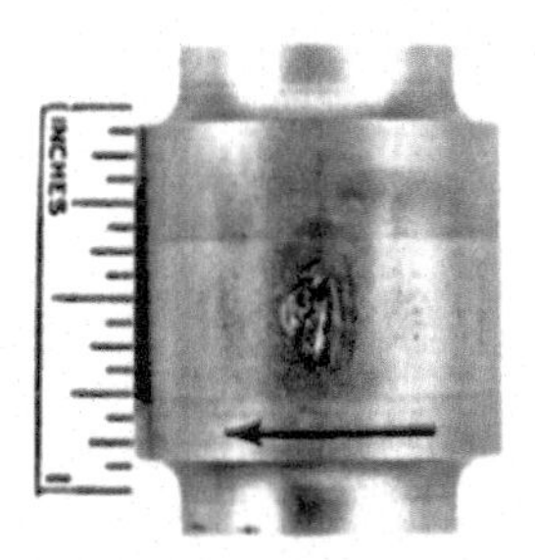

Fig. 7 Two views of a roller, showing a typical arrowhead-shaped surface pit. The arrows indicate the direction of rotation; the direction of rolling of the mating roller is to the right, while the direction of sliding is to the left. The upper view is a normal photograph of the pitted area; the lower view shows the entire surface "unwrapped" by a special photographic method. This specimen was tested at a calculated compressive stress of about 450,000 psi and ran for 14,229,000 revolutions prior to pitting failure. This was a carburized and hardened alloy steel, with a surface hardness near 60 HRC.

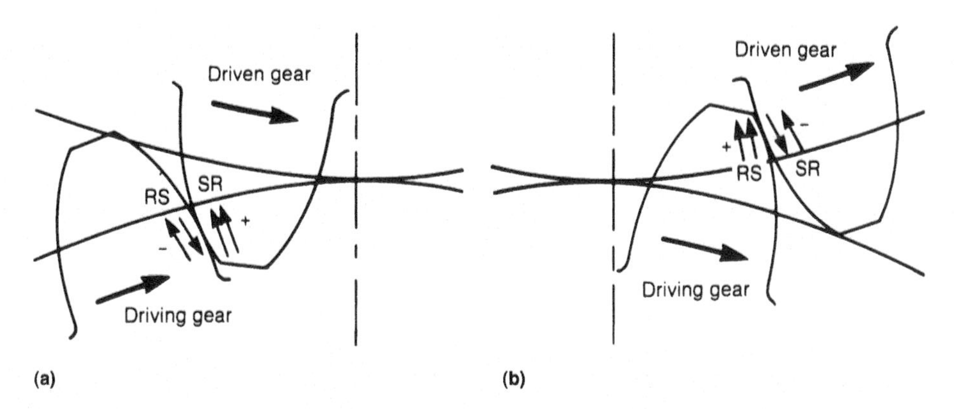

Fig. 8 Schematic of rolling-sliding action inherent in gear teeth: (a) beginning of contact, (b) end of contact. As gear teeth contact, rolling (R) and sliding (S) stresses are formed by the relative movement. Pure rolling occurs only at the pitch line, and on each gear the direction of sliding reverses at the pitch line. Analysis of the relative motions of gears reveals that on both the driving and driven gears there is negative sliding below the pitch line and positive sliding above the pitch line. Since negative sliding is more likely to cause surface-origin pitting, damage is most likely to occur below the pitch line on the smaller gear, which is usually the driving gear.

Table 1 Movements in mating gears

Gear	Rolling direction(a)	Sliding direction(a)		
		On dedendum (below pitch line)(b)	On addendum (above pitch line)(c)	With resect to pitch line
Driving	Up	Down	Up	Away
Driven	Down	Up	Down	Toward

(a) Up, toward top; down, toward root. (b) Negative sliding: rolling and sliding are in opposite directions. (c) Positive sliding: rolling and sliding are in the same direction.

to resist pitting fatigue. Because of the difficulty in increasing the life and/or load-carrying capacity of gear systems, the development of pits usually represents the limit of gear service, in the absence of gross metallurgical or operational anomalies.

Extensive research with roller test specimens (Ref 5–7) has indicated that optimum conditions for long-term resistance to pitting fatigue on gears are provided with a surface hardness near 60 HRC, smooth surface finishes, and at least 10 to 20% retained austenite on the surface of case-hardened steel. The function of the retained austenite apparently is to permit some degree of plastic deformation by increasing the contact area in order to reduce the actual compressive stress. Bending fatigue strength, however, is reduced by excessive retained austenite.

Figure 9 shows the damage that surface-origin pitting fatigue can cause to operating gear teeth.

Another type of surface-origin pitting fatigue occurs on conformal surfaces, such as a shaft rotating within a sliding bearing. Fatigue failure of sliding bearings is usually the result of long, hard service under severe repetitive, compressive forces, such as occur on the upper half of engine connecting rod bearings (which transmit the explosive forces to the crankshaft) or the lower half of the main crankshaft bearings (which resist bending of the crankshaft due to the explosive forces). Locations of highest stress are about 35° on each side of the top center of the upper halves, and about 35° on each side of the bottom center of the lower halves (Ref 8). From these origins, the fatigue pits usually spread out to wider locations on the bearings until the bearings are completely destroyed. In addition to the explosive forces imposed on the bearings, there are also centrifugal and inertial forces that together cause the total load spectrum on the individual bearings.

The surface materials for sliding bearings are typically soft, nonferrous bearing metals, such as tin-base or lead-base Babbitt metals, copper-lead alloys, bronzes, aluminum alloys, certain powder-metal alloys, and trimetal bearings (consisting of three layers of metals). Fatigue fracture in sliding bearings normally originates at the surface because that is where the shearing stresses are highest due to the sliding action on the surface. The lubricant system and geometrical characteristics of the bearing and of the mating surface are critical to maintain a proper lubricant film thick-

(a)

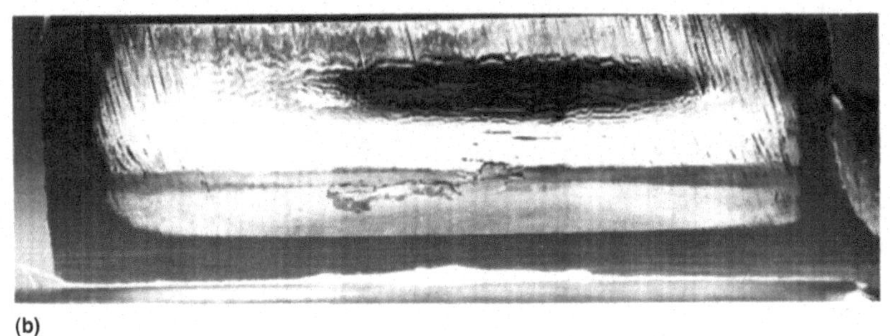

(b)

(c)

Fig. 9 Surface-origin pitting fatigue. (a) Typical surface deterioration due to pitting fatigue on gear teeth. In a standard gear system, the pitch line is near the center of the height of the teeth. Pitting fatigue usually starts slightly below the pitch line and then rapidly spreads to adjacent areas, causing complete surface deterioration. This heavily loaded gear was made of case-hardened alloy steel, with a surface hardness near 60 HRC. (b) Heavily loaded final drive pinion with surface pitting near the pitch line, which was low on this gear tooth. The rippling shown is plastic deformation caused by the sliding action under heavy contact pressure and is a warning that the metal is close to failure. The parallel diagonal marks are tool marks resulting from the shaving operation and are not involved with the service in any way. (c) Fracture of gear tooth at pitch line. Another consequence of pitting fatigue on the active profile of gear teeth is the fact that the pit itself is a stress concentration that can cause abnormal bending fatigue fracture of gear teeth, which are really carefully shaped cantilever beams. This gear tooth pitted just below the pitch line and then fractured as a result of bending fatigue. The normal location for bending fatigue fractures of gear teeth is at the root fillet, not midway up the tooth, as in this example. This gear was made from a medium-carbon steel, induction hardened to about 55 HRC.

ness. Obviously, if an oil film can prevent the surfaces from contact, neither fatigue failure nor other problems associated with sliding contact can occur. References 1 and 2 give guidance in identifying and solving failures encountered by sliding bearings.

Subcase-origin fatigue damage to case-hardened rolling/sliding surfaces, such as gear teeth and certain roller mechanisms, also can completely destroy the contacting surfaces. In this type of failure, very large pieces are suddenly lost from the surface, and extensive damage may result. However, this type of failure is relatively easy to prevent, once the mechanism is identified.

Attention is called to Chapter 3, "Stress versus Strength," in this book, particularly Fig. 9(d) which shows the principle behind subsurface-origin failures in bending fatigue. Fatigue failure at the base of the hardened case of any part is possible if the stress exceeds the fatigue strength at that location. The same principle holds for contact stress fatigue.

Subcase-origin fatigue also is known as "spalling" fatigue or "case crushing" (Ref 9). However, the term *subcase fatigue* is more descriptive of the mechanism involved and is preferred. "Case crushing" is particularly misleading in that it implies static fracture, which may be accurate in instances of severe overloading, but this is not a fatigue mechanism.

As shown in Fig. 10, subcase fatigue is somewhat similar to subsurface fatigue, discussed earlier in this chapter. However, the difference is in the scale of the damage to the contact surface. Subsurface fatigue results in pits originating a few thousandths of an inch from the surface, whereas subcase fatigue typically originates much deeper—usually slightly below the case depth, which may be 1 mm (0.040 in.) or more from the surface, depending on the heat treatment of the particular part. At any rate, fatigue can originate deep within the part as a result of contact stress fatigue. The fatigue cracks propagate parallel and perpendicular to the surface. These cracks tend to be very long, as shown on the hypoid rear axle pinions in Fig. 10(b) and (c). Surface cracking is the first visible indication of the subsurface fatigue, although extensive cracking has already occurred below the surface. These hidden cracks could be detected by ultrasonic instrumentation if suspected. Continuing service rapidly leads to the severe destruction shown in Fig. 10(c), in which the long, undermined pieces come out as large fragments. This leaves large, long, gouged-out cavities that completely destroy the surfaces involved.

Correction of this type of problem is relatively simple, because this type of failure indicates that the shear strength is inadequate below the case. The strength can be raised and the failures prevented by changing the heat treatment to increase the case depth and/or increasing the core hardness (and strength) by using a steel with higher carbon or alloy content. However, any metallurgical changes must be made judiciously and carefully, because increasing the case depth or core hardness too much may through harden the component, which could lead to brittle fracture

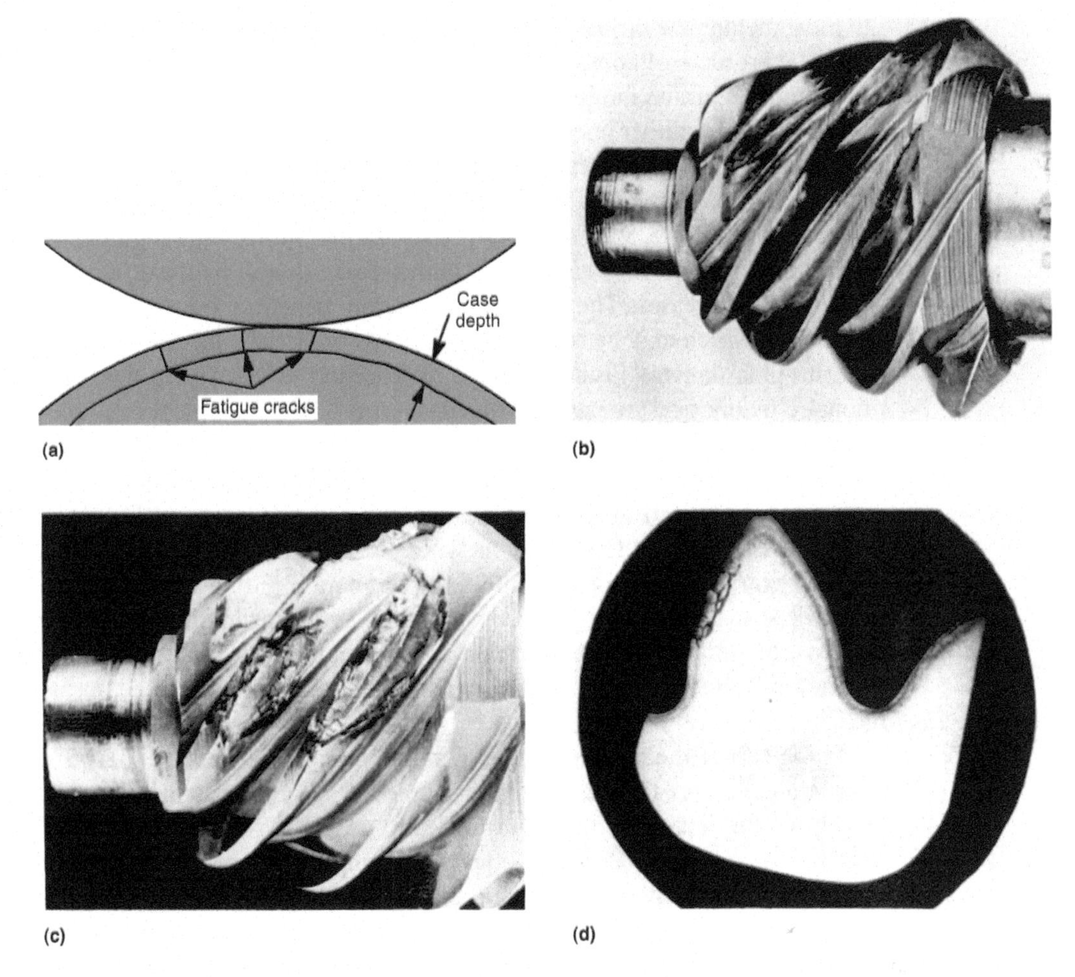

Fig. 10 Subcase-origin fatigue. (a) As the name implies, subcase fatigue cracks originate deep within the steel at the region below the case, where the core metal is comparatively soft in relation to the case itself. The fatigue cracks spread laterally, parallel to the surface, and then join and cause cracks that come out to the surface. By the time surface cracks are observed, however, the fatigue is well developed within the gear teeth. Large chunks can come from the surface at one time; deterioration is not gradual as in surface-origin fatigue. (b) Hypoid pinion with many long cracks on the surface and tips of the teeth. Although the teeth in this photograph appear to be relatively intact, they are actually undermined with fatigue cracks. The cracks themselves are several inches long. (c) Hypoid pinion similar to that shown in (b). Long, finger-length pieces of metal have come out of the surface, leaving a longitudinally ridged and gouged appearance. The ridges are the locations where two adjacent fatigue cracks joined and sent a crack through the case to the surface. This and the gear in (b) were made of low-carbon alloy steel, case carburized to a surface hardness near 60 HRC. (d) Cross section through a gear tooth showing subcase-origin fatigue cracks similar to those sketched in (a). The original fatigue cracks are those parallel to the surface below the dark-etched case. When these cracks join and run to the surface, complete removal of the large fragments of metal occurs. Core hardness was low on this gear, contributing to this type of failure.

depending on the component shape. Also, the residual stress pattern would be changed drastically, potentially causing other problems.

Cavitation Fatigue. Since the general title of this section is "Contact Stress Fatigue," it seems appropriate to include a type of pitting fatigue that is caused by vibration and movement in various liquids, of which water is the most common. Since many liquids are corrosive in some ways to most metals, the problem of environmental reaction becomes entwined with the problem of contact stress fatigue.

Cavitation pitting fatigue can be a serious problem in marine propellers, diesel-engine cylinder liners, pump impellers, hydraulic pumps and equipment, turbines, torque converters, piping components, and miscellaneous other parts that contact liquids in environments with vibrations or high flow rates (Ref 1, 2, 9). The pits from cavitation pitting can range in size from very small to very large, from pinhead size to golf ball size, or even larger in some cases. The pits can completely penetrate the thickness of the metal, which may be several inches thick; obviously, this can result in catastrophic damage to the structure, in addition to destroying the functional efficiency of the parts involved. Figure 11 shows typical examples of cavitation pitting fatigue.

Cavitation pitting is characterized by mechanical-induced damage from cyclical stresses (fatigue), sometimes aggravated by corrosion. The damage occurs in low-pressure regions where a rapid pressure drop creates a rapidly vibrating liquid-metal interface. Cavitation pitting fatigue can be visualized by studying Fig. 12, considering that the events shown occur in microseconds. The frequency for a specific application depends on the vibration frequency of the parts involved or flow rate of the fluid across the surface.

In Fig. 12(a), the metal wall is moving to the right, against the inertia of the liquid. In Fig. 12(b) the metal has reached the end of its travel to the right, but the liquid is still being pushed to the right. As the metal moves back to the left (Fig. 12c), the liquid is still moving to the right because of its inertia. Small cavities, or negative-pressure "bubbles," form in the liquid at the interface with the metal as the two materials momentarily move away from each other. In Fig. 12(d), the metal reaches the end of its travel to the left, pulling the cavities along with it. In Fig. 12(e), the metal is again moving to the right but collides with the liquid moving to the left. The cavities collapse violently, or "implode," because of the inertia both of the metal and the liquid as they move toward each other and increase the effective fluid pressure.

The collapsing cavities implode on the metal with compressive stresses estimated at several hundred atmospheres, equivalent to many thousands of pounds per square inch. Repeated implosions in the same area lead to the formation of fatigue cracks and pits. Since the geometry of the vibrating system and the properties of the liquid are relatively constant, the cavities form in clusters at certain preferred locations. With constant repetition of the pounding, the fatigue mechanism progresses until pits form in the metal surface at these locations. These pits often penetrate deeply into the metal surface as the pressure-change events occur within the pits, and deep, irregularly shaped pits develop.

Corrosion may enter the picture if surface films are formed on the clean, unprotected, virgin metal in the pit when the system is at rest. Then, when the motion resumes, any surface film is rapidly destroyed by the very high

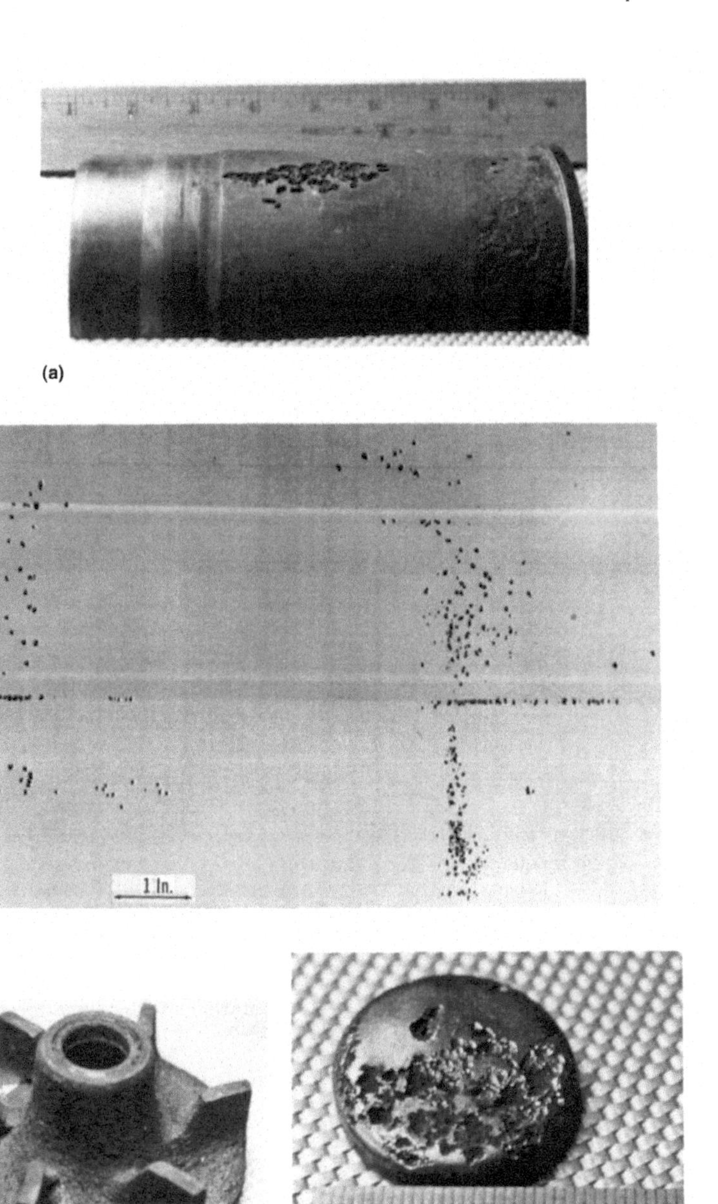

(c) (d)

Fig. 11 Cavitation pitting fatigue. (a) Cavitation pitting on a gray cast iron diesel-engine cylinder sleeve. The pitted area is several inches long, and the pits nearly penetrated the thickness of the sleeve. Note the clustered appearance of the pits at preferred locations. (b) Cavitation pitting on another gray cast iron diesel-engine cylinder sleeve unwrapped by a special photographic process. Again, note the clustered locations, with the most severe pitting on the thrust side of the sleeve, against which the piston slides on the power stroke of the combustion cycle. The lighter pitting at left is on the opposite, or antithrust, side of the sleeve. (c) Cavitation pitting at preferred locations on the vanes of a gray cast iron water-pump impeller. This impeller rotated in a clockwise direction; the arrows show some of the pits that were formed in the metal on the suction side of the vanes. (d) Cavitation pitting that perforated this steel freeze plug from a gasoline engine, causing leakage of coolant that could have damaged the engine. Vibration of the wall of the engine block at this location caused this type of damage on the coolant side.

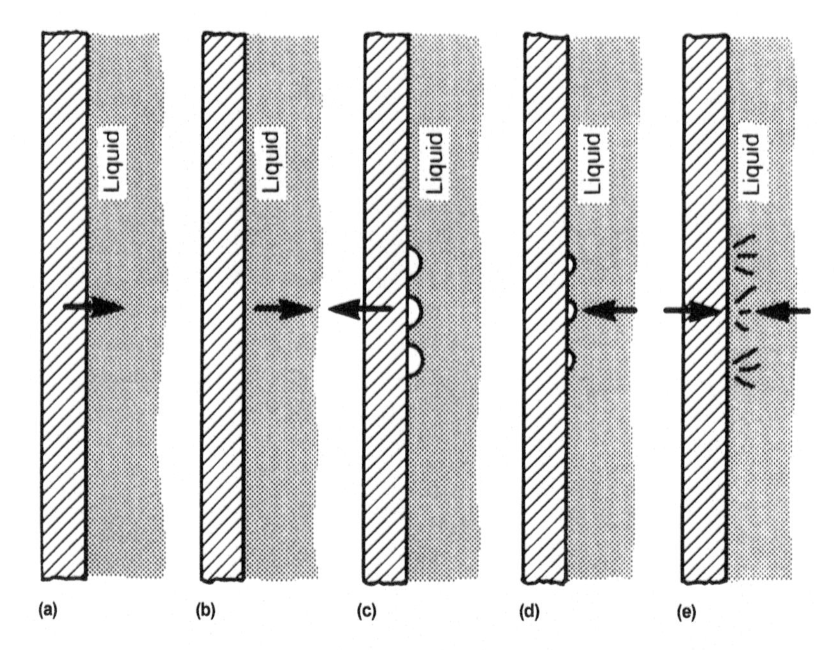

Fig. 12 Mechanism of cavitation pitting fatigue. Serial sketches show a metal wall vibrating to the right and left against a liquid, which in all cases is to the right of the wall. The events shown can occur in a very short time, on the order of microseconds. (a) The metal moves to the right against the stationary liquid, which resists movement because of its inertia. (b) The metal reaches the end of its travel, but the inertia of the liquid causes it to continue to move to the right. (c) The metal starts moving toward the left, away from the liquid, which cannot catch up to the wall because of its inertia. Consequently, cavities or voids are formed in the liquid on the surface of the metal. These are essentially negative-pressure bubbles. (d) The metal reaches the end of its travel to the left and the liquid tries to catch up to the metal and collapse the cavities. (e) The metal starts moving to the right as the cavities collapse violently, or implode, against the metal. As the vibration of the metal continues, the formation and collapse of these cavities in the liquid can cause pitting fatigue on the surface of the metal. If the pounding continues at the same locations on the metal, pits will eventually lead to complete perforation of the metal wall.

compressive forces encountered, whereby corrosion then becomes a factor in the pitting process. However, even glass and ceramics can have cavitation pitting in inert liquids. Apparently, corrosion is not necessary for cavitation pitting fatigue.

It is not easy to prevent cavitation fatigue in a liquid-metal system. The reason is that cavitation pitting fatigue is a function of metal properties, design of the part, vibrational characteristics of the entire mechanism, and the pressure and flow characteristics of the liquid.

Some possible ways to try to reduce the cavitation fatigue problem include:

- *Increase the stiffness of the part:* This should reduce its amplitude of vibration and increase the natural vibration frequency of the part. It may be possible to increase wall thickness or to add stiffening ribs to change its vibration characteristics.

- *Increase the smoothness of the surface:* Because the cavities tend to cluster in certain low-pressure areas, the effect may be dispersed if there are no surface peaks and valleys.
- *Increase the hardness and strength of the metal:* Because cavitation pitting is essentially a fatigue phenomenon, the metal properties are important. However, an increase in hardness and strength may only delay the problem, not prevent it. Gray cast iron is a frequent victim of cavitation pitting because of its low strength. It has poor resistance to pitting but is often used in parts subject to this type of failure due to its low cost.
- *Streamline the fluid flow:* It may be possible to eliminate corners on trailing edges or in low-pressure regions that could cause cavitation pitting. This has obvious relevance to the design of marine propellers, which may pit in the low-pressure regions, usually near the trailing edge on the forward side. In a closed system like a diesel engine or a piping system, for example, it may be possible to streamline the fluid flow by smoothing geometric features to disperse the cavities so that they collapse on different locations of the surface.
- *Increase the nominal pressure on the liquid or use inhibitors to increase the vapor pressure on the liquid:* If the cavities (or "bubbles") cannot form in the liquid, they cannot cause damage to the metal.
- *Live with the problem:* If none of the above possible solutions is practical for a given application, it may be necessary to simply live with the problem and replace pitted parts during regularly scheduled maintenance. This is the way that we live with other inevitable fatigue and wear problems, such as those occurring in bearings, tires, shoes, and the like.

Summary

The wear caused by contact stress fatigue is the result of a wide variety of mechanical forces and environments. Each of the types of contact stress fatigue discussed in this chapter is unique and must be properly identified before corrective action can be taken. Table 2 summarizes the characteristics of surface, subsurface, and subcase fatigue on mating metal surfaces. Features and corrective actions for contact stress fatigue are:

- *Subsurface-origin fatigue* causes relatively small pits in the surfaces of contacting metal parts under essentially pure rolling conditions, which can be minimized if metal with fewer hard, brittle inclusions is used.
- *Surface-origin fatigue* causes relatively small pits, initially arrowhead shaped, in the surfaces of contacting metal parts under combined rolling and sliding conditions, particularly with negative sliding. Usually

Table 2 Characteristics of contact stress fatigue

Characteristic	Surface pitting	Subsurface pitting	Subcase fatigue
Location of origin	Surface, often at "micropits"	Short distance below surface, usually at a nonmetallic inclusion	Near case-core boundary in case-hardened parts
Initial size	Small	Small	Large
Initial area depth ratio	Small	Small	Large
Initial shape	Arrowhead, then irregular	Irregular	Gouged and ridged
Crack angle with respect to surface	Acute	Roughly parallel at bottom, perpendicular at sides	Roughly parallel at bottom, perpendicular at sides
Apparent occurrence	Gradual	Sudden	Sudden

only minimal improvement is possible in life or load-carrying capacity if mechanical, metallurgical, and lubrication conditions are optimum. This type of fatigue usually marks the limit of load-carrying capacity of gears or bearings.

- *Subcase-origin fatigue* causes removal of very large fragments from the surface of contacting case-hardened metal parts and relatively easy to prevent by judiciously increasing case depth, core hardness, or both.
- *Cavitation fatigue* causes formation of clusters of many pits in metal surfaces vibrating or moving rapidly relative to a liquid, which may be very difficult to prevent, depending upon design, material, liquid, and vibrational characteristics.

REFERENCES

1. *Failure Analysis and Prevention,* Vol 11, *ASM Handbook,* 9th ed., ASM International, 1986, p 133–134, 159, 163–171, 483–513, 592–594
2. *Failure Analysis and Prevention,* Vol 11, *ASM Handbook,* 10th ed., ASM International, 2002, p 722–726, 792–793, 999–1000, 1002–1014
3. *Fatigue and Fracture,* Vol 19, *ASM Handbook,* ASM International, 1996, p 355–362, 345–354
4. *Failure Analysis and Prevention,* Vol 10, *Metals Handbook,* 8th ed., American Society for Metals, 1975, p 152
5. J.P. Sheehan and M.A.H. Howes, "The Role of Surface Finish in Pitting Fatigue of Carburized Steel," Paper 730580, Society of Automotive Engineers, 1973
6. J.P. Sheehan and M.A.H. Howes, "The Effect of Case Carbon Content and Heat Treatment on the Pitting Fatigue of 8620 Steel," Paper 720268, Society of Automotive Engineers, 1972
7. M.A.H. Howes and J.P. Sheehan, "The Effect of Composition and Microstructure on the Pitting Fatigue of Carburized Steel Cases," Paper 740222, Society of Automotive Engineers, 1974
8. *Engine Bearing Service Manual,* 7th ed., Federal Mogul Bearings, Inc., 1956, p 93

9. R. Pedersen and S.L. Rice, Case Crushing of Carburized and Hardened Gears, *SAE Trans.*, Vol 69, 1961, p 370–380

SELECTED REFERENCES

- L.E. Alban, *Systematic Analysis of Gear Failures,* ASM International, 1985
- *Bearing Steels: The Rating of Nonmetallic Inclusions,* STP 575, ASTM, 1975
- J.B. Bidwell, Ed., *Rolling Contact Phenomena,* Elsevier, 1962
- R.E. Denning and S.L. Rice, "Surface Fatigue Research with the Geared Roller Test Machine," Paper 620B, Society of Automotive Engineers, 1963
- *Friction, Lubrication, and Wear Technology,* Vol 18, *ASM Handbook,* ASM International, 1992
- W.E. Littmann and C.A. Moyer, "Competitive Modes of Failure in Rolling Contact Fatigue," Preprint 620A, Society of Automotive Engineers, 1963
- G.J. Moyer and J.D. Morrow, "Surface Failure of Bearings and Other Rolling Elements," Engineering Experiment Station Bulletin 468, University of Illinois, 1964
- J.O. Smith and C.K. Liu, Stresses Due to Tangential and Normal Loads on Elastic Solids with Application to Some Contact Stress Problems, *J. Appl. Mech. (Trans. ASME),* Vol 20, 1953, p 157–166.

Understanding How Components Fail, Third Edition
Donald J. Wulpi
Edited by Brett Miller

CHAPTER 13

Aqueous Corrosion Failures*

CORROSION, defined as the deterioration of a metal due to chemical or electrochemical reactions with its environment (Ref 1), is of enormous economic cost to society. The most recent U.S. cost study was conducted in 1998 and estimated the direct costs of corrosion at 3.1 percent of the gross domestic product (GDP) (Ref 2). The study concludes that the indirect costs, including outages, delays, and litigation costs, are on par with the direct costs, indicating that the total cost of corrosion is approximately 6.2 percent of the GDP. Applying this percentage to the 2012 GDP (about $15.5 trillion), corrosion and corrosion prevention during 2012 will cost approximately $961 billion in the United States alone. Although these numbers may now be somewhat inaccurate, the staggering figures show that corrosion is enormously costly to society. In some cases, corrosion can result in catastrophic failures that lead to loss of life, environmental impacts, or both.

This chapter outlines the major types of corrosion, their interactions, their complicating effects on fracture and wear, and some possible prevention methods. There is a massive body of literature on the subject; selected references are given at the end of the chapter.

Anyone who has worked with corrosion and corrosion testing will agree that slight changes in metals, their design, or their environment can significantly affect the corrosive behavior. Sometimes baffling corrosion problems are encountered in which the actual conditions and interactions are not known or understood. For this reason, it is extremely important to try to obtain as much firsthand information as possible about the circum-

*_Understanding How Components Fail, Second Edition_, by Donald J. Wulpi, Chapter 13, Corrosion Failures, ASM International, 1999, reviewed and revised by David M. Norfleet and John A. Beavers, Det Norske Veritas (USA), Inc.

stances of a corrosion problem or of the corrosive effect complicating a fracture or wear problem, because the various modes of failure are frequently combined. The failure analyst must be very careful that the corrective measures taken are truly corrective; in some cases, they may make the problem worse rather than better. As in all failure analysis work, it is good to "make haste slowly" in order to properly understand and identify the problem before means of prevention can be put into practice.

Life Cycle of a Metal

Corrosion is a natural process in which the chemical action of the refining process is reversed. In their natural chemically stable state, metals are found primarily either as oxides or as sulfides in the ores. The addition of large amounts of energy during the refining process to strip away the oxides or sulfides produces relatively pure metals in a less chemically stable state (metastable state). These refined metals can be used for various purposes either alone or by alloying with other metals. However, the process of reversion to the natural, chemically stable condition progresses inexorably unless prevented by deliberate actions. To illustrate this process for a common metal, iron is found as iron oxide (rust) in the ore, refined to iron (or steel) to be used for some purpose, but eventually will revert to iron oxide (rust). To paraphrase a well-known saying, "ashes to ashes, rust to rust."

Figure 1 is a schematic showing the life cycle of a typical metal; the vertical scale is the energy level, while the horizontal scale is time. The ore at the lower left is in the natural condition, at a low energy level. The thermal or electrical energy added during the refining process strips away the oxygen and/or sulfur from the metal (step a), changing the metal into an unnatural, somewhat chemically unstable condition at a higher energy level (step b). The refined metal may be remelted and cast, hot or cold

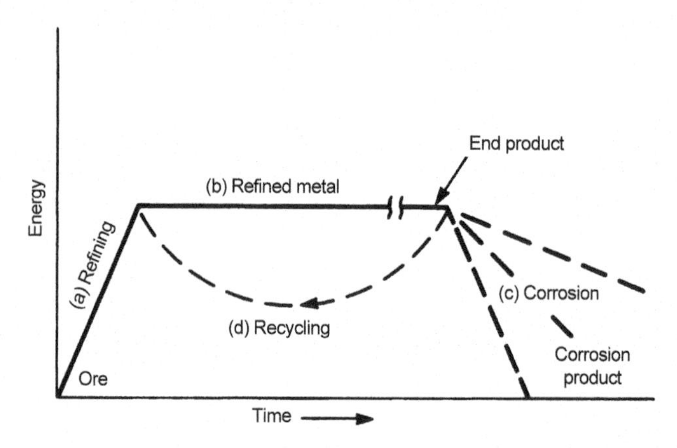

Fig. 1 Life cycle of a typical metal

formed, or machined into useful shapes to serve some purpose in the manufacture of an end product. However, the refined metal will tend to deteriorate at some rate (step c) and revert to the lower energy, original chemically stable, orelike condition, thereby releasing energy. We sometimes take advantage of the energy released by corrosion to convert it into electrical energy, such as in an ordinary dry-cell battery. However, most energy that is generated by corrosion is dissipated and wasted without being harnessed for useful purposes. Of course, recycling or remelting is an efficient way of bypassing deterioration due to corrosion and reusing the metal (step d).

From a philosophical standpoint, understanding this inexorable, inevitable process of corrosion makes it easier to understand the problems in preventing corrosion and to be more aware of the tricky nature of the problem. Corrosion will win—eventually—unless all possible means are taken to prevent its occurrence. Attempts to prevent corrosion may be compared to sticking a finger into a hole in a dike to prevent a leak; it may be possible to stop that leak, but inevitably there will be other leaks.

Basic Nature of Corrosion

The definition of corrosion in the first sentence of this chapter refers to "chemical or electrochemical" reactions with the environment. The chemical reference is to a relatively simple chemical dissolution of a metal, such as when hydrochloric acid dissolves iron. However, the purist will say that even this is an electrochemical reaction of a metal with a highly aggressive environment. Thus, it is generally agreed that all aqueous forms of corrosion are electrochemical in nature and that an understanding of the electrochemical reactions involved is necessary to understand the basic nature of corrosion. There are nine commonly referred to forms of aqueous corrosion (Ref 3):

- Galvanic corrosion
- Uniform corrosion
- Crevice corrosion
- Pitting corrosion
- Environmentally induced cracking
- Hydrogen damage
- Intergranular corrosion
- Dealloying/dezincification
- Erosion corrosion

In addition, there are corrosion mechanisms that form subsets within each of the nine listed above. For example, microbiologically influenced corrosion typically manifests itself as pitting corrosion; while stress corrosion cracking is a subset of environmentally induce cracking. One of the complicating factors in studying corrosion is that there are many types of corrosion, and in some cases, more than one type may be occurring simultaneously. For this reason, it is vital to recognize each of the types of corrosion and to understand how to control them. The extent of corrosion related failures is much too large to consolidate all of the various mechanisms and modes into one chapter. Therefore, several of the more common failure modes with accompanying case studies are presented in this chapter, including *galvanic corrosion*, *uniform corrosion*, *localized corrosion (pitting and crevice corrosion)*, *microbiological influenced corrosion*, *stress corrosion cracking*, and *corrosion fatigue*.

Galvanic Corrosion

Galvanic corrosion, caused by differences in contacting metals and/or their environment, is one of the most serious and technically challenging forms of corrosion.

The basic principles of the electrochemical reactions that cause galvanic corrosion are identical to those of a simple battery, as shown in Fig. 2. Three conditions are necessary:

- two dissimilar metals (or one may be a metal, the other graphite);
- an electrical connection between the two metals (typically physical contact between the two) providing an electron pathway;
- an electrolyte, which is an ionically conductive medium, contacting the two metals providing a pathway for ion exchange.

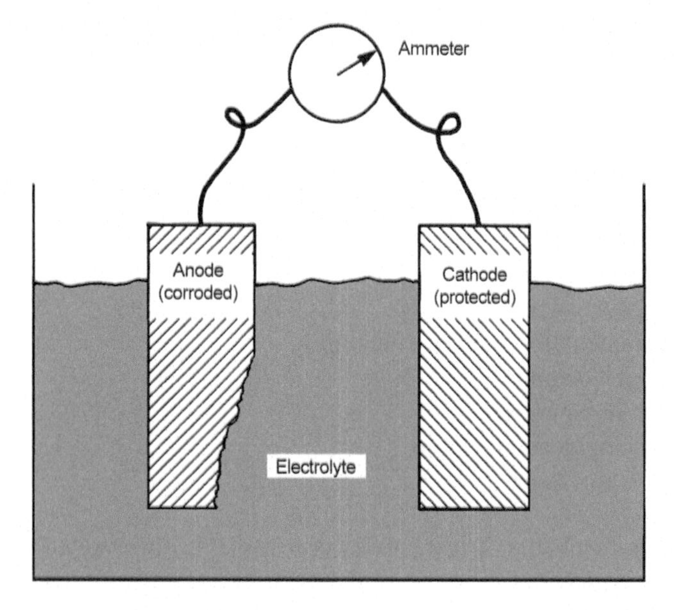

Fig. 2 Galvanic cell showing the basic principles of the electrochemical nature of corrosion

If these conditions are met, an electric current is generated, and one of the two materials will exhibit an increase in its corrosion rate (anode), while the other material will exhibit a decreased corrosion rate (cathode) and is protected. The reaction kinetics will dictate the speed at which the corrosion takes place.

The various metals (and graphite, a nonmetal) are usually listed in a sequence that has the most anodic (easily corroded) metals at one end and the most cathodic (easily protected) metals at the other end. For practical purposes, the electrolyte for this sequence is seawater, or approximately 3 to 5% sodium chloride and other salts dissolved in water. This sequence is called the galvanic series and is shown in Table 1 for many pure metals and select alloys.

The position of a given metal, or of a pair of dissimilar metals, in the galvanic series is of great importance in determining its corrosion properties. The usual series has the most anodic metals at the lower end. These are the most active or most readily corrodible metals. The least active or least corrodible metals are at the upper end. Because these include gold, platinum, silver, and other precious metals, they are also known as the "noble" metals, which strongly resist corrosion.

When two metals make electrical contact in an electrolyte, the farther apart they are in the galvanic series, the more likelihood there is that the more anodic metal will corrode. For example, aluminum in contact with gold in seawater will be rapidly corroded, while aluminum in contact with iron (steel) will corrode less rapidly. An ordinary dry-cell battery consists of a zinc anode separated by an electrolytic paste from a manganese dioxide cathode. When the circuit is closed, the electric current that is generated corrodes the zinc anode. When the zinc is depleted, the electric current ceases and the battery is dead.

The occurrence and severity of galvanic corrosion can be affected by several factors:

- A given metal may be either high or low in the series depending on the nature of the surface film. For example, many stainless steels may be either active (anodic) or passive (cathodic) depending on whether a passive film of chromium oxide is present on the metal surface. This film can be an air-formed film, a film formed in the operating environment, or a film formed by a nitric acid "passivating" treatment, which also removes foreign metal particles that could impair corrosion resistance.
- The severity of galvanic corrosion can be affected by the relative area of the anode and the cathode. For example, if a large "anode" metal is connected to a small "cathode" metal in an electrolyte, there may be little or no corrosion of the anode because the electrical effect is spread out, or dissipated, over a large area. However, if a small anode metal contacts a large cathode metal in an electrolyte, the reverse is true: the

Table 1 Galvanic series in seawater at 25 °C (77 °F)

Cathodic end (least easily corroded)
Platinum
Gold
Graphite
Titanium
Silver
Chlorimet 3
Hastelloy C
Inconel alloy 625
Incoloy alloy 825
Type 316 stainless steel (passive)
Type 304 stainless steel (passive)
Type 410 stainless steel (passive)
Monel alloy 400
Inconel alloy 600 (passive)
Nickel 200 (passive)
Copper alloy 922, cast (leaded tin bronze M)
Copper alloy 923, cast (leaded tin bronze G)
Copper alloy 715 (copper nickel, 30%)
Copper alloys 651, 655 (silicon bronze)
Copper 110 (ETP copper)
Copper alloy 230 (red brass, 85%)
Copper alloys 608, 614 (aluminum bronze)
Copper alloys 443, 444, 445 (admiralty brass)
Copper alloy 270 (yellow brass, 65%)
Chlorimet 2
Hastelloy B
Inconel alloy 600 (active)
Nickel 200 (active)
Copper alloys 464, 465, 466, 467 (naval brass)
Copper alloy 675 (manganese bronze A)
Copper alloy 280 (Muntz metal, 60%)
Tin
Lead
Type 316 stainless steel (active)
Type 304 stainless steel (active)
50-50 lead-tin solder
Type 410 stainless steel (active)
Ni-Resist (high-nickel cast iron)
Cast iron
Wrought iron
Low-carbon steel
2117, 2017, 2024, in this order
Aluminum alloys
Cadmium
5052, 3004, 3003, 1100, 6053, in this order
Aluminum alloys
Galvanized steel or galvanized wrought iron
Zinc
Magnesium alloys
Magnesium
Anodic end (most easily corroded)

Source: Ref 1

anode metal may corrode rapidly, for it essentially becomes an electrical stress concentration.

- The galvanic series should be used as a starting point in a particular corrosion study, but it should not be considered highly accurate because of variations in the metals, in the electrolyte, and in their relative size.

Galvanic corrosion may be prevented, or minimized, by any or all of the following measures:

- Prevent electrical (electron) current flow by physical separation or by electrically insulating the dissimilar metals from each other by using nonconductive materials such as plastics, waxy coatings, certain heavy greases, paint, and the like. The current flow will take the path of least resistance, and that path may not always be obvious, such as structure grounds and the like.
- Try to eliminate the electrolyte; if it is not present, there can be no galvanic corrosion.
- If different metals must be used together, choose those that are close together in the galvanic series. For example, contact between aluminum and steel in an electrolyte may cause gradual pitting and deterioration of the aluminum. However, if brass or copper replaces the steel, deterioration of the aluminum will be much more rapid in the same environment.
- Use a large anode metal and a small cathode metal to take advantage of the area effect. For example, plain steel rivets through aluminum sheet or plate may be satisfactory, but aluminum rivets through steel will corrode rapidly in an electrolyte. Similarly, copper rivets in steel plate may be satisfactory, but steel rivets through copper plate will corrode rapidly in an electrolyte.
- In a closed system, it may be possible to use corrosion inhibitors in the electrolyte. This is the principle used in automotive engine antifreezes, for the cooling system of an automotive engine may contain such dissimilar metals as gray cast iron, aluminum, copper, brass, tin-lead solder, and steel. The antifreezes periodically should be drained and replaced, as the inhibitors tend to become less effective with time. Plain water lacks the necessary corrosion inhibitors and therefore should never be used as an engine coolant except in an emergency, and then only for as short a time as possible.
- The principle of galvanic corrosion can be used to protect a structural metal by contact with a sacrificial (expendable), more anodic metal. The most common example is the use of zinc to protect iron or steel. The steel may be dipped into molten zinc (galvanized), electroplated with zinc, or coated with a zinc-rich primer or zinc-rich polymeric coating. In any case, the zinc is gradually corroded, or sacrificed, in order to protect the steel. Other active metals, such as aluminum and cadmium, also tend to protect iron or steel from corrosion but are used less frequently for various reasons. Coatings of aluminum and aluminum-zinc alloys on steel, however, are widely used for resistance to both high temperatures and aqueous corrosion, as are required in automotive exhaust systems, for example. The same cathodic protection

principle is used with magnesium anodes buried in the ground to protect steel pipelines from corrosion. The anodes must be replaced periodically, since they are sacrificed to protect the steel. Magnesium anodes are also used within glass-lined water heaters in order to help retard corrosion of the steel shell if the glass lining is cracked. Replaceable zinc anodes are attached to the hulls of steel ships and the legs of offshore oil rigs for protection of the steel in the same way.

Uniform Corrosion

The most common type of corrosion is ordinary uniform corrosion, such as rust on iron or steel. Other metals also corrode uniformly. Aluminum, copper, brass, and many other metals may form either protective or semi-protective films on the surfaces, depending on the nature of the metal and of the environment.

The term *uniform corrosion* is really a misnomer, however, for the corrosion actually occurs as a result of microscopic corrosion reactions (reduction and oxidation) occurring on the surface of the metal. The rates at which these reactions occur depend on the local chemistry and impurity content of the metal. Thus, the corrosion is "uniform" only on a macroscale, not on a microscale.

Since uniform corrosion is the most common form of corrosion, it is of the most economic significance and damages the greatest tonnage of metal. From a technical standpoint, however, uniform corrosion is fairly predictable and is relatively easy to live with, provided other types of corrosion are not present. Controls for uniform corrosion are relatively simple measures:

- Use a more suitable material, such as a more noble metal or stainless steel. However, other considerations, such as mechanical or physical properties of the metal, as well as economics and availability, are usually the controlling factors in the choice of material.
- Use coatings of various types to protect the metal. These may be:

  a. Paint or another type of coating to prevent contact of the environment with the metal surface. Painting is a time-honored method of corrosion protection and is remarkably effective as long as the paint film is intact. However, if corrosion originates on the opposite, unprotected side of thin, painted sheet steel, repainting the painted side will have no long-term benefit.
  b. Various oxide coatings are frequently used for improved corrosion resistance. An example is anodized aluminum, which is essentially aluminum with a relatively thick aluminum oxide film formed on the surface. A thin oxide film naturally forms on aluminum when exposed to air but is easily removed and destroyed, although it will

reform. Various types of oxide coatings are frequently applied to steel for decorative and corrosion-resistance purposes. However, like certain other coatings, oxides are ineffective if damaged and are subject to pitting by certain environments, particularly chlorides and other halides.

c. Plating with a more active sacrificial metal, as previously discussed, is usually quite effective for relatively long periods. Plating with a less active (more cathodic) metal can be very effective if the base metal is completely and uniformly coated without pinholes, cracks, scratches, abrasions, and the like. For example, plating steel with lead or tin is very effective in corrosion resistance, unless the coating has some locations that are open to the base metal. If an electrolyte can reach the steel through a small opening, the steel will corrode rapidly because the steel now is the anode and the small anodic area is exposed to the electrolyte.
d. Cladding of one metal with a different alloy is another method of protection that is very effective in certain cases. Cladding may consist of two or three metals flat rolled together, as in coins that consist of two nickel-alloy outer layers on each side of a copper alloy, in a "sandwich" construction. Two aluminum alloys of different properties are frequently rolled together; a layer with better corrosion resistance may be joined to another layer, which may have better mechanical properties.

- Another way to cope with the relatively predictable nature of uniform corrosion is to let the part corrode. This may be the most economical way of handling the problem. Railroad rails, for example, are not protected from corrosion by painting, plating, or the like, but may become covered with an oily or greasy coating that tends to resist rusting on the sides and bottom. However, the upper contact surface remains completely unprotected because of the heavy contact stress of the steel railway car wheels on the steel rails.

Localized Corrosion (Pitting and Crevice Corrosion)

Localized corrosion can be defined as corrosion confined primarily to small areas on a metal surface that corroded at a higher rate in comparison with surrounding areas. Two examples include pitting corrosion and crevice corrosion. *Pitting corrosion* is a form of localized corrosion that occurs on a boldly exposed surface. *Crevice corrosion* is a form of localized corrosion that occurs at or near an occluded region between the metal surface and another material. Crevice corrosion is similar to galvanic corrosion, but in this case, the metal is the same but the environment is different within the crevice than on the boldly exposed surface. In the case of stainless steels, the passive film that develops on the surface protects the underlying metal from chemical attack. The local environment within a crevice can produce low pH conditions that can break-down the passive film and produce crevice corrosion.

Crevice corrosion is very difficult to combat without careful control of design, materials, engineering, and quality. *Crevice corrosion* is the commonly used term for differential oxygen-concentration cell corrosion, a more descriptive (but unwieldy) term for this corrosion mechanism. One form of crevice corrosion is called *poultice corrosion*, referring to accumulations of moist or wet particles, such as dirt, sand, and the like, on the metal surface.

Since most corrosion is caused by oxidation of reactive metals, areas of high oxygen concentration would be expected to corrode more readily than those of lower oxygen concentration. However, this is not the case. A crevice, or joint, between two surfaces or the metal under a poultice of moist dirt or debris is more likely to corrode than the more exposed metal outside the joint or pile of dirt. The contact region, where there is little oxygen, becomes the anode in the corrosion cell and will corrode, while the region exposed to the higher oxygen content becomes the cathode and is protected. Concealed metal at the edge of the joint or under debris tends to pit and eventually perforate the metal thickness. Figure 3 shows a very large hole in ¼-in.-thick steel plate caused by crevice corrosion under a pile of debris.

The practical problems caused by crevice corrosion are enormous. This is the major cause of auto-body corrosion, which originates within the concealed joints and panels of the steel body. When a brown, rusty stain appears on a rocker panel, door, hood, fender, or trunk lid, it is already too late. Corrosion has already occurred in crevices, joints, or under dirt and debris in the presence of moisture—frequently laden with salt from deicing or from sea air. When this corrosion reaches the exterior surface, it first stains the paint and eventually perforates the metal, leaving holes with which we are all too familiar. Painting the outside stain does not stop the underlying corrosion. In addition, the corrosion greatly weakens thin steel structural members such as box frames, cross members, and rocker panels, which are intended to provide strength and stiffness to the vehicle.

Crevice corrosion may also occur under fasteners, such as bolted or riveted joints, if moisture can penetrate and remain. This can occur even if both metals are the same but is aggravated with dissimilar metals in contact because galvanic corrosion may also become a factor, particularly with a small anode and large cathode to bring the area effect into play. Aluminum and steel are frequently joined without consideration of the crevice, galvanic, and area-effect corrosion problems that may be caused. Figure 4 shows an aluminum plate that was riveted to a steel plate. In the presence of an electrolyte, an appreciable amount of white (aluminum)

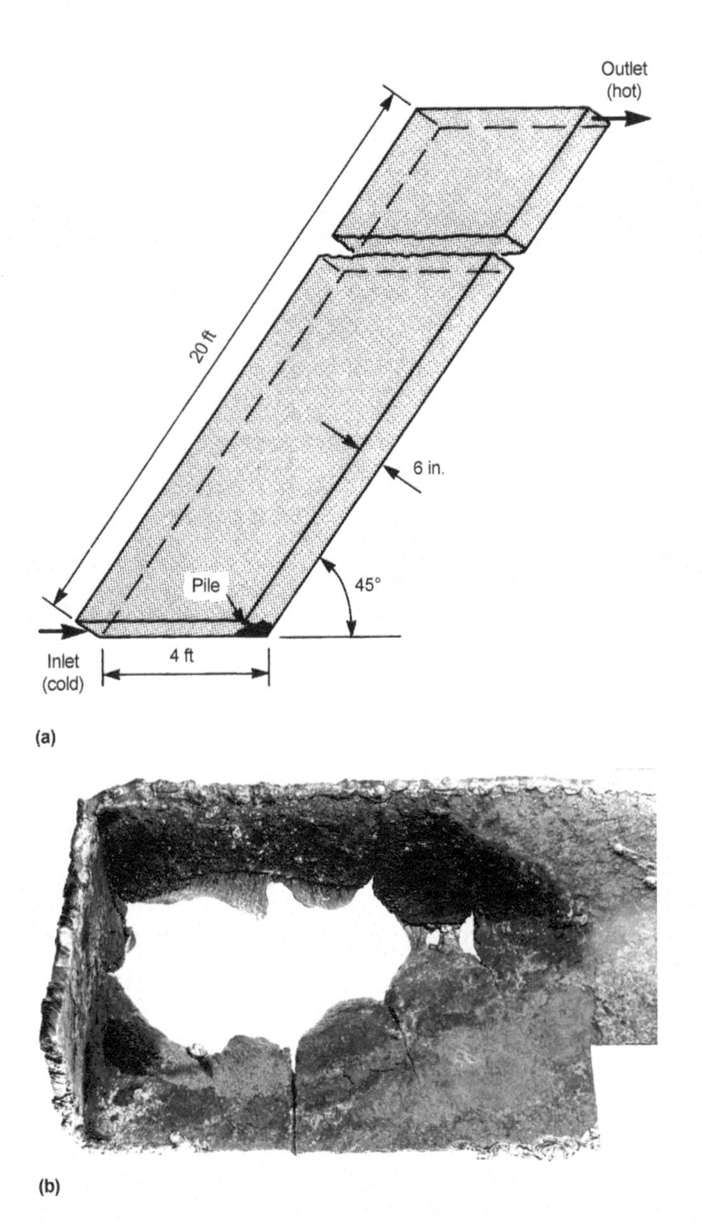

Fig. 3 A hole 4-¼ in. long caused by crevice corrosion in ¼-in.-thick steel plate. The photograph (b; at ⅓×) shows the inside of the lower right corner of a large steel box (a) that acted as a heat exchanger or panel to cool extremely hot exhaust gases in a steel mill. Cold water entered at the lower left corner, and the heated water left at the upper right corner of the panel, tilted about 45°. The recirculating mill water contained many particles of rust, dirt, etc., which gradually accumulated in the dead corner at the lower right, opposite the inlet. The pile of particles apparently caked together to form a large cementlike plug (dark area around hole in b) over the gradually corroding steel underneath. The corroded hole, with very thin edges, grew to the very large size shown without leaking. Eventually, however, the plug could not support the water over the hole; it then burst, releasing a large volume of water.

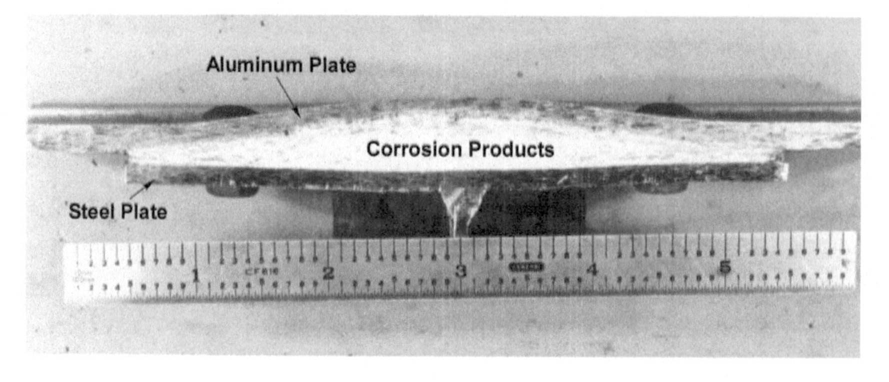

Fig. 4 An aluminum plate riveted to a steel plate resulted in a galvanic couple and crevice corrosion that produced a significant amount of corrosion products between the two plates. The stresses generated as a result of the volume change of the corrosion products were sufficient to plastically deform the aluminum plate.

corrosion products accumulated between the two plates as a result of combined effects of a galvanic couple and crevice corrosion. The volume change associated with the formation of the corrosion products caused a stress large enough to plastically deform the aluminum plate.

The Statue of Liberty in New York Harbor had been in bad condition because the copper exterior (cathode) was held in place with a steel framework and bolts (anode), which had corroded seriously at the joints after almost 100 years of exposure to sea air. After several years of restoration, the steel framework has been replaced with stainless steel parts, along with other controls to keep this magnificent symbol alive for many years to come.

Once the insidious nature of crevice corrosion is understood, the prevention methods become rather obvious, even though they may not be easy to put into practice. Consider these possible solutions:

- Avoid bolted or riveted joints unless the metals are coated, ideally both before and after joining. However, if the alternative is some form of welding, the heat of welding may destroy the coated surfaces. In either case, it may be necessary to spray or dip the completed assembly with paint or a waxy or greasy corrosion-resistant material that has the ability to flow and cover any uncoated areas. A sacrificial anodic metal may also be used to protect the structural metal.
- Close or seal existing crevices, because if moisture or another electrolyte cannot reach the crevice, there can be no corrosion. In many cases, however, it is impossible to keep moisture from entering the crevices. Then it is necessary to provide drain holes, tubes, and such, to continuously or periodically drain the moisture.
- Inspect and remove deposits frequently; if foreign particles such as dirt, rust, sand, and the like, are not present, they cannot cause crevice

corrosion. Another possible solution is to use filters, traps, or settling tanks to remove particles from the system. However, all of these require periodic maintenance to remove the debris that accumulates.
- Use solid nonabsorbent gaskets or seals, such as those made from solid rubber or plastic. These tend to seal a joint and keep the electrolyte out. Obviously, the surfaces must be smooth to promote sealing, and the clamping force must be adequate for the application.

As is apparent from these possible solutions, there is no single way in which crevice corrosion can be eliminated; prevention differs with the individual case and the conditions.

Microbiologically Influenced Corrosion

Microbiologically influenced corrosion (MIC) occurs in the presence of bacteria or microorganisms, whereby the by-products produced by the microorganisms are harmful to the material and, in turn, lead to corrosion. Microorganisms can be present in solutions or accumulated solids or can form biofilms on a material's surface. The high concentration of bacteria that occurs in the presence of accumulated solids or biofilms can lead to concentrated corrosion cells that promote localized corrosion attack. Susceptible materials include carbon steels, copper alloys, aluminum alloys, nickel-based alloys, and stainless steels.

In the pipeline industry, MIC has been a concern since the 1920s and 1930s (Ref 4) due to the susceptibility of carbon steel, the proximity of the pipelines to organic-material-laden soil, and the organic products that flow through them. Five common forms of bacteria are associated with MIC of carbon steels: aerobic (require oxygen), anaerobic (do not require oxygen), sulfate reducing (SRB), acid producing (APB), and iron-related bacteria (IRB). Different tests have been developed to determine the presence and quantity of bacteria in a system. The tests provide a snapshot of the bacteria concentrations at the time of sampling and may not be representative of the bacteria concentrations at the time of active corrosion. For this reason, there are no definitive tests for MIC (Ref 5). The failure analyst must piece together all data, including visual inspection, metallography, morphology of the pits, and the like, as well as eliminating other possible forms of corrosion, to reach the conclusion that MIC contributed to the failure.

Figure 5(a) is a photograph of the inner surface of a 12-in.-diameter crude line that experienced an in-service leak. The leak occurred at the 6:00 orientation, in an area that had accumulated water and deposits. Samples collected from the deposits within the pit and surrounding area tested positive for high concentrations of bacteria. In cross section (Fig. 5b), the failure pit shows evidence of undercutting, a characteristic often associated with anaerobic MIC.

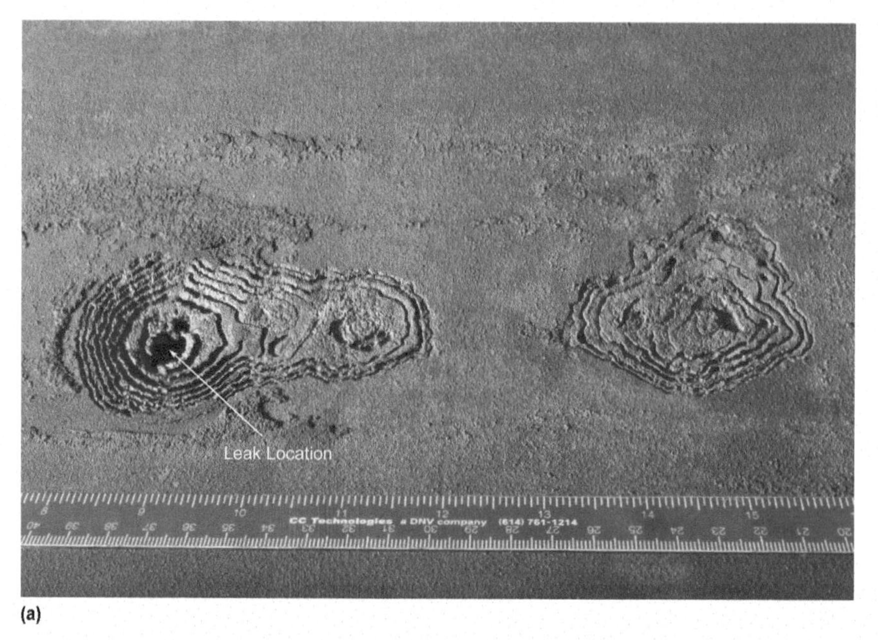

(a)

(b)

Fig. 5 (a) Inner surface of a 12-in.-diameter crude oil pipeline that experienced an in-service leak. (b) Metallographic cross section of the leak location showing undercutting within the pit, a pitting morphology that is typically associated with MIC.

Stress-Corrosion Cracking

The phenomenon of stress-corrosion cracking (SCC; a subset of environmentally induced cracking) is a major problem in many industries because it may result in brittle fracture of a normally ductile metal. It is defined as cracking under the combined action of a specific corrosive environment and tensile stresses; the stresses may be either applied (external) or residual (internal). The cracks themselves may be either transgranular or intergranular, depending on the metal and the corroding agent. As is normal in all brittle fractures, the cracks are typically perpendicular to the tensile stress. Usually there is little or no obvious visual evidence of corrosion.

The classic example of SCC is the so-called season cracking of brass cartridge cases, as shown in Fig. 6. The term arose during the Indian campaigns of the British army during the 1800s, when serious problems resulted from cracking of the thin-wall necks of cartridge cases (which were stored in ammunition dumps near the horse corrals!) during the monsoon seasons. High temperature and humidity, plus traces of ammonia in the air, caused SCC in the severely deformed thin sections, locations that were subjected to high tensile hoop stress when the bullets were inserted. We now know that most zinc-containing copper alloys, such as the 70% copper, 30% zinc alloy called cartridge brass, are susceptible to SCC when the surface is tensile stressed and is in the presence of certain chemicals, such as moist ammonia, mercurous nitrate (see Ref 6), and amines.

SCC is a slow, progressive type of fracture, somewhat similar to fatigue. The crack or cracks grow gradually over a period of time until a critical size is reached; the stress concentration may then cause a sudden fracture of the remaining metal. In other instances, as in a cartridge case, the crack will grow away from the high-stressed origin and then stop when it is no longer highly stressed in tension. Although there is no one generally accepted theory of its mechanism (Ref 7), SCC has several unique characteristics:

- For a given metal or alloy, only certain specific environments cause this type of failure.
- Pure metals are much less susceptible to SCC than are impure metals, but pure binary alloys, such as copper-zinc, copper-gold, and magnesium-aluminum alloys, generally are susceptible.

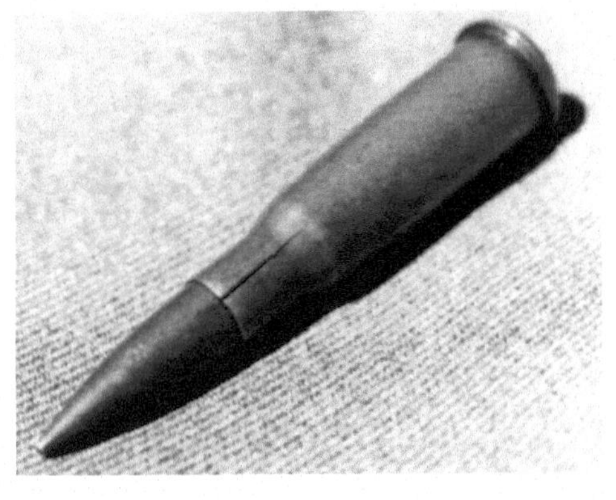

Fig. 6 Stress-corrosion crack in thin neck of a cartridge case

- The cracks tend to occur in clusters or colonies where multiple cracks initiate on the metal surface.
- The cracks typically branch as they propagate through the metal.
- Certain aspects of the metallurgical structure of an alloy (e.g., grain size, crystal structure, and number of phases) influence the susceptibility of the alloy to SCC in a given environment.

Nearly all metals are susceptible to SCC in the presence of tensile stresses in specific environments. Ordinary carbon and alloy steels are subject to SCC in sodium hydroxide, nitrates, carbonate-bicarbonate, and even alcohols. Austenitic stainless steels, such as those of the 200 and 300 series, are subject to SCC from chlorides (and other halides). A more complete list of environments that may cause SCC in many common metals is shown in Table 2.

The tensile stress required to initiate SCC can emerge from residual or applied stresses. It is usually felt that tensile residual stresses (including assembly stresses) are more often the cause of SCC than are tensile applied stresses. (Ref 7, 8). Residual stresses (see Chapter 4, "Residual Stresses," in this book) are frequently the result of welding.

Table 2 Some environments that can cause stress-corrosion cracking of metals and alloys under certain conditions

Material	Environment
Aluminum alloys	$NaCl-H_2O_2$ solutions
	NaCl solutions
	Seawater
	Air, water vapor
Copper alloys	Ammonia vapors and solutions
	Amines
	Water, water vapor
Gold alloys	$FeCl_3$ solutions
	Acetic acid-salt solutions
Inconel	Caustic soda solutions
Lead	Lead acetate solutions
Magnesium alloys	$NaCl-K_2CrO_4$ solutions
	Rural and coastal atmospheres
	Distilled water
Monel	Fused caustic soda
	Hydrofluoric acid
	Hydrofluorosilicic acid
Nickel	Fused caustic soda
Carbon and alloy steels	NaOH solutions
	$NaOH-Na_2SiO_2$ solutions
	Calcium, ammonium, and sodium nitride solutions
	Mixed acids (H_2SO_4-HNO_3)
	HCN solutions
	Acidic H_2S solutions
	Moist H_2S gas
	Seawater
Stainless steels	Acid chloride solutions such as $MgCl_2$ and $BaCl_2$
	$NaCl-H_2O_2$ solutions
	Seawater
	H_2S
	$NaOH-H_2S$ soutions
Titanium	Red fuming nitric acid

Source: Ref 8

Tensile stresses are also generated in other ways, such as in shrink fits, bending or torsion during assembly, crimping, and the like. The only requirement for SCC is that there is a tensile stress on the surface of a susceptible metal in a critical environment. The stress need not exceed the yield strength of the metal, but the higher the stress, the less critical the environment, and vice versa.

In certain metals, particularly many austenitic stainless steels, the heat of welding causes sensitization, or depletion of chromium due to formation of complex chromium carbides in the grain boundaries. Because chromium is the major element that makes "stainless steels" corrosion resistant, SCC can occur alongside the carbides in the grain boundaries, where there is little or no chromium. In order to solve this problem, it is necessary to do one of two things: use a stainless steel of very low carbon content (0.03% or less), so that there is little or no carbon to deplete the chromium from the grain boundaries, or use a stainless steel containing an element that forms carbides even more readily than does chromium. Those commonly used are titanium, or niobium (columbium) plus tantalum, which are the bases for the type 321 and type 347 variations, respectively, of the standard type 304 stainless steel.

Preventing SCC is simple in principle: change the alloy or remove either the tensile stress or the corrosive environment, since both are required for the problem to occur. In the real world, of course, this is easier said than done. One way that has been found to be very effective is to form compressive residual stresses on the surface of the part by mechanical methods, such as shot peening, grit blasting, surface rolling, and the like. This will increase resistance to crack initiation and ultimate fracture. If a tensile residual stress is high, it may be possible to stress relieve the part or assembly by a thermal treatment. In fact, warming to subcritical temperatures, depending on the metal involved, usually is all that is necessary.

Identification of SCC is not always easy, for it may be confused with another type of fracture. For example, Fig. 7 shows a stress-corrosion crack in a high-strength steel part that has a fracture pattern that could be mistaken for fatigue fracture, because it appears to have fatigue beach marks. However, because the part had not been cyclically stressed, it could not be fatigue (Ref 9). The beach mark-like pattern observed is the result of differences in the rate of penetration of corrosion on the surface as the crack advanced. As frequently happens, the crack progressed relatively slowly until it reached the critical size and then fractured suddenly in a brittle manner. The texture of the fracture surface is sometimes helpful in differentiating a stress-corrosion fracture from a fatigue fracture, because the surface texture in a fatigue fracture usually is quite smooth in the origin region, gradually becoming rougher toward the final rupture. This change is usually not seen in a stress-corrosion fracture.

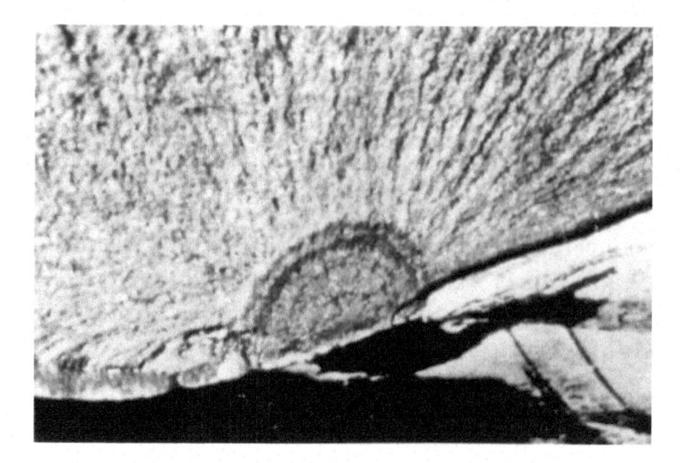

Fig. 7 Stress-corrosion crack in a high-strength steel part (4×). The fracture surface appears to have the characteristic beach mark pattern of a fatigue fracture. However, this was a stress-corrosion fracture in which the pattern was caused by differences in the rate of corrosion penetration. Final fracture was brittle. Source: Ref 9

Metallurgical cross sections are helpful in differentiating SCC from other forms of cracking. Figure 8(a) shows cracking that developed on the external surface of a Type 316 stainless steel pipe section exposed to a high-chloride environment. The presence of multiple cracks on the surface of the pipe is an indicator of SCC. Figure 8(b) is a photomicrograph of a metallographic cross section removed from an area of cracking. The primary crack originated at the external surface and propagated through the cross section in a heavily branching manner, consistent with SCC.

SCC is sometimes confused with hydrogen-embrittlement cracking. In fact, Ref 10 states:

> It is difficult and sometimes impossible to distinguish with certainty between hydrogen cracking and stress-corrosion cracking failures that have occurred in service by exposure to hydrogen gas, hydrogen sulfide, and water and dilute aqueous solutions. Several basic characteristics to be observed in investigating failures of these types are (a) history of the metal or part, (b) cracking origin, (c) crack pattern (d) evidence of little or no corrosion on the fracture surfaces, and (e) microscopic features.

The complexity of SCC and of the many factors such as alloy, heat treatment, microstructure, stress system, part geometry, time, environmental conditions, and temperature makes it obvious that suspected instances of environment-related fractures must be carefully studied and analyzed, taking into consideration all available information before deciding on the dominant failure mechanism.

(a)

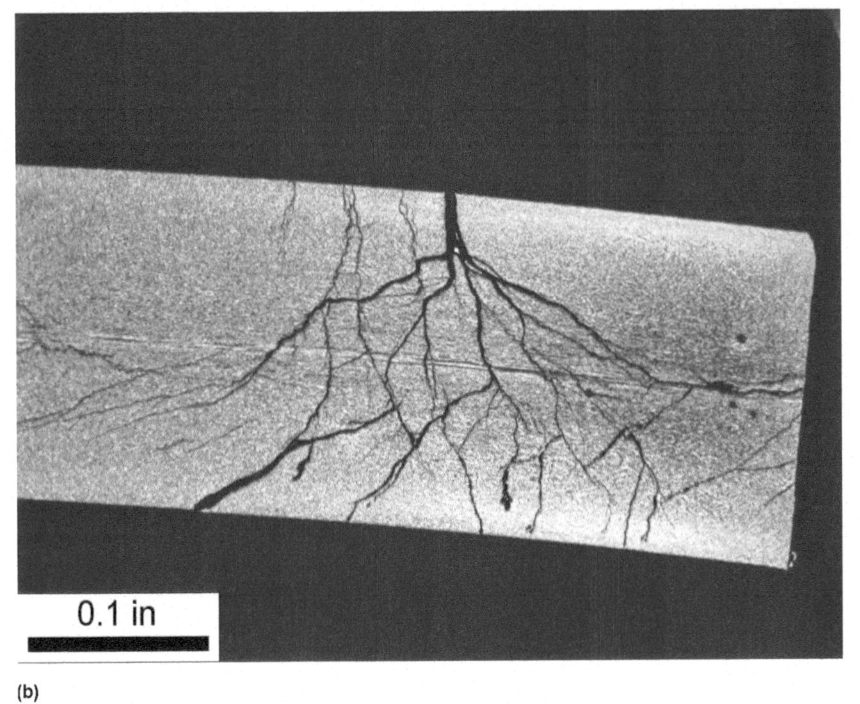

(b)

Fig. 8 (a) A type 316 stainless steel pipe section exposed to a high-chloride environment, resulted in stress-corrosion cracking on the external surface. (b) A photomicrograph of a metallographic cross section removed from a location of cracking in (a). There is a distinct branching morphology to the cracking that is characteristic of stress-corrosion cracking.

Corrosion Fatigue

While SCC is the result of a static, or steady stress, corrosion fatigue (a subset of environmentally induced cracking) is essentially fatigue fracture aggravated by the effects of the environment. Reference 1 defines corrosion fatigue as:

> Effect of the application of repeated or fluctuating stresses in a corrosive environment characterized by shorter life than would be encountered as a result of either the repeated or fluctuating stresses alone or the corrosive environment alone.

In general, we think of the corrosion that occurs during cyclic stressing as reducing fatigue life, as pointed out above. Even air can affect fatigue behavior in certain alloys compared with their behavior in a vacuum. Figure 9 shows the tendency of 2024-T3 aluminum to form normal fatigue striations (incremental crack growth front following a single cycle) when tested in air but to have a relatively flat and featureless fracture region when tested in vacuum (Ref 9). This implies that air was necessary for the formation of the fatigue striations, at least under the conditions of the test.

In real life, the type of cyclic stressing strongly influences the life under aggressive corrosion conditions (Ref 1). The longer and more frequently a fatigue crack is opened to the corrosive environment, the more severe will be the effect of the environment on shortening the fatigue life. In many

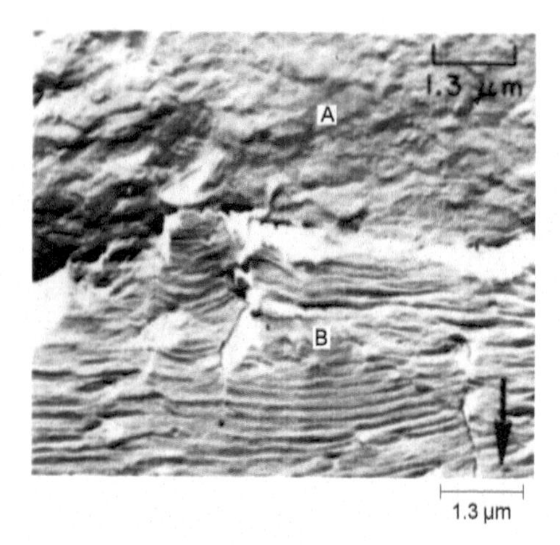

Fig. 9 Fatigue fracture in aluminum alloy 2024-T3 tested first in vacuum (region A) and then in air (region B) (7500×). The arrow at the lower right indicates the direction of crack propagation. Note the flat, featureless fracture surface formed while testing in vacuum (region A), with no correlation between fracture appearance and the cyclic nature of the imposed loading. In contrast, the regular fatigue striations formed while testing in air (region B) correlate with the crack advance of each loading cycle. Source: Ref 9

cases, fatigue is initiated from small pits on the corroded surface, which act as stress concentrations (Fig. 10, 11). In other cases, it appears that the fatigue crack initiates first and then is made to grow more rapidly by the moisture or other corrodent that enters the crack by capillary action. Probably a combination of the two mechanisms is most common.

Identification of corrosion-fatigue fractures uses the same techniques and thought processes as in all failure analyses. However, the effect of the environment on the fracture complicates the analysis. The origin area of a fatigue fracture is most likely to be the most severely corroded because it has been exposed to the environment the longest time. Thus, discoloration

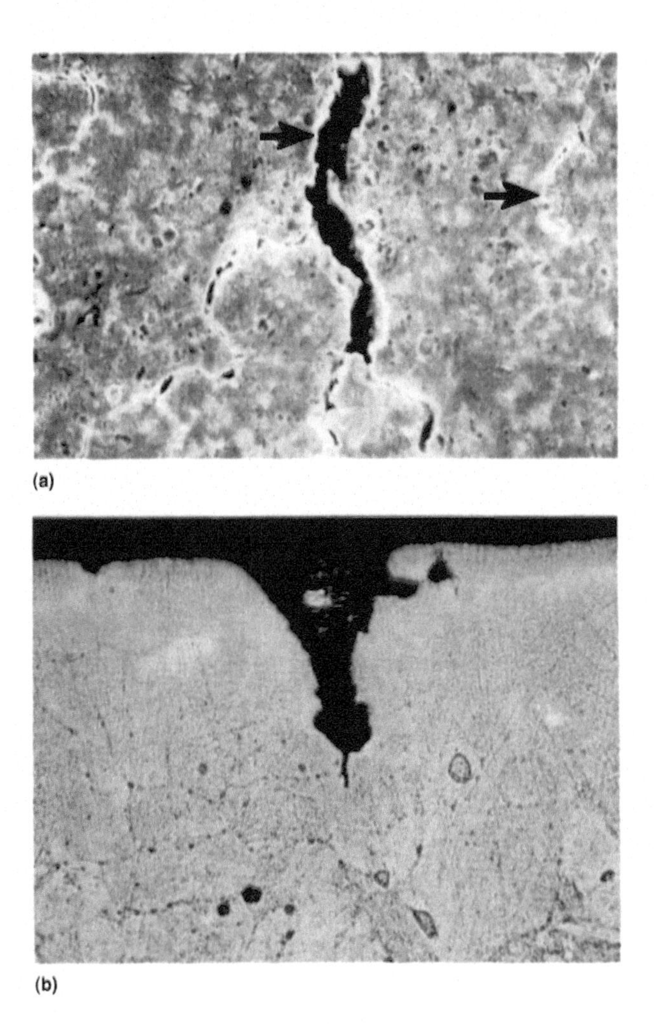

Fig. 10 (a) Scanning electron photograph of corrosion pits on the surface of a gas turbine airfoil showing both large and small pits (arrows) that led to fatigue fractures (300×; shown at 75%). The material is the precipitation-hardening stainless steel 17-4PH. (b) Photomicrograph of a polished and etched section through a pit like those shown in (a) (500×; shown at 75%). Note the start of a fatigue crack growing from the bottom of the pit. Pitting was caused by continued operation in a seawater environment.

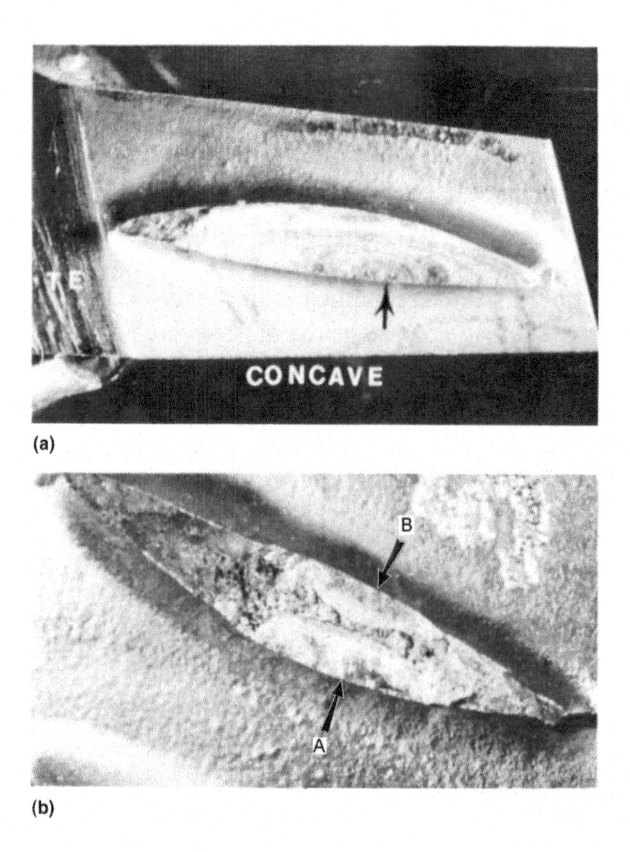

(a)

(b)

Fig. 11 First-stage compressor blades that fractured due to corrosion fatigue originating in corrosion pits like those shown in Fig. 10. Note that (a) had one fatigue origin (arrow) on the mid-pressure side (5×; shown at 70%). Arrows in (b) show fatigue origins on both the suction (A) and the pressure (B) sides (6.5×; shown at 70%). These fatigue fractures in cast 17-4PH material are ambient-temperature fractures. Pitting was caused by seawater corrosion.

from severe corrosion frequently makes the location of the origin easier to identify; at the same time, however, this corrosive film can obscure the origin and make detailed study much more difficult. Of course, as noted above, there may be many origins, rather than just one. A complicating factor, on this or any other type of fracture surface, is that corrosion will occur over the entire fracture surface if the broken part is not removed from the corrosive environment, immediately cleaned, and protected from further corrosive and mechanical damage. Comparison of the two mating fracture surfaces, if both are available, may give some insight into when the corrosion occurred.

Prevention of corrosion fatigue is rather straightforward in theory but frequently much more difficult to put into practice. It is usually impractical simply to use a higher-strength metal without addressing the corrosion issue. The corrosion will probably occur just as rapidly, and fracture actually may be more rapid because of the increased sensitivity of higher-strength metals to stress concentrations. More effective measures include:

- Reduce or eliminate corrosion by any of the conventional means, such as painting or plating, if these are practical for the application. It may also be possible to reduce the aggressiveness of the environment by adding inhibitors or by changing the concentrations of the solution in a closed system.
- Change the material to one more resistant to the environment, such as a stainless steel, or a different nonferrous alloy system. However, this should be done only as a last resort, for many other potential problems arise, such as economics, availability, general engineering suitability, and manufacturing difficulties.
- Reduce the resultant tensile stress that is causing the fatigue problem. In some cases, it may be possible to reduce the applied stress by decreasing the load applied to the part. In other cases, it may be necessary to redesign the part to increase the section size, but again, this is a major decision.
- Reduce the resultant tensile stress by increasing the compressive residual stresses on the critical surface(s). As discussed in Chapter 4, "Residual Stresses," in this book, this may be accomplished in many ways. Some of the simplest ways involve using mechanical prestressing, such as shot peening or surface rolling. Shot peening puts into compression a relatively shallow layer below the surface (usually several thousandths of an inch), which may be inadequate if the corrosion pits penetrate this layer. Deeper penetration of the compressive stress may be accomplished by surface rolling or by pressing of grooves in critical areas. However, both of these are limited by geometry. Surface rolling, for example, is usually done on propeller shafts of ships at the location where the hub for the propeller fits. This is a critical location because the rotating propeller hub must be separated from the stationary rear support or bearing by a sealing system, which often does not work properly. As a result, water, often seawater, corrodes the steel shaft, sometimes resulting in corrosion fatigue that could mean disastrous loss of the propeller at sea.

As with most types of failures, corrosion-fatigue problems may be solved by a variety of approaches. The particular solutions chosen depend upon the individual situation, the critical nature of the part, current and proposed metals, availability, engineering and manufacturing problems, and economics.

Summary

The subject of corrosion failure is extremely complex and cannot be covered easily by one chapter. However, corrosion must be considered in every failure analysis, for it is not always obvious. In fact, corrosion has a subtle influence on many types of failure, particularly those involving fracture and/or wear.

The types of corrosion considered here are galvanic, uniform, pitting, crevice, microbiologically influenced corrosion, stress-corrosion cracking, and corrosion fatigue. There are other categories and subcategories that also reflect the environmental influence on metal failures. The list of references for further study should be consulted for more complex problems.

REFERENCES

1. *Corrosion,* Vol 13, *ASM Handbook,* ASM International, 1987, p 4, 83–87, 291–302
2. G.H. Koch, M.P. Brongers, and N.G. Thompson, *Direct Costs of Corrosion in the United States,* Vol 13A, *ASM Handbook,* ASM International, 2003, p 959–968
3. D. Jones, *Principle and Prevention of Corrosion,* 2nd ed., Prentice-Hall, Inc., 1996
4. A. Tiller, A Review of the European Research Effort on Microbial Corrosion between 1950 and 1984, *Biologically Induced Corrosion,* Vol. 8, NACE, 1986.
5. A. Wagner and F. Mansfield, *Microbiologically Influenced Corrosion-Corrosion Testing Made Easy,* NACE, 1997
6. "Standard Test Method for Mercurous Nitrate Test for Copper Alloys," B154, *Annual Book of ASTM Standards,* ASTM, 2012.
7. *Failure Analysis and Prevention,* Vol 11, *ASM Handbook,* ASM International, 1986, p 203–224
8. M. Fontana, Stress Corrosion, lesson 5, *Corrosion,* Metals Engineering Institute course, American Society for Metals, 1968
9. *Fractography and Atlas of Fractographs,* Vol 9, *Metals Handbook,* 8th ed., American Society for Metals, 1974, p 31, 91
10. *Failure Analysis and Prevention,* Vol 10, *Metals Handbook,* 8th ed., American Society for Metals, 1975, p 238

SELECTED REFERENCES

- T. Burstein, L.L. Shreir, and R.A. Jarman, *Corrosion,* 3rd ed., Butterworth-Heinemann, 1994
- S.L. Chawla and R.K. Gupta, *Materials Selection for Corrosion Control,* ASM International, 1993
- S.K. Coburn, Ed., *Corrosion Source Book,* American Society for Metals, 1984
- B. Craig and D. Anderson, Ed., *Handbook of Corrosion Data,* 2nd ed., ASM International, 1995
- E.D.D. During, Ed., *Corrosion Atlas,* 3rd ed., Elsevier Science, 1997
- U.R. Evans, *An Introduction to Metallic Corrosion,* 3rd ed., Edward Arnold and American Society for Metals, 1981
- M.G. Fontana, *Corrosion Engineering,* 3rd ed., McGraw-Hill, 1986

- R.H. Jones, Ed., *Stress-Corrosion Cracking: Materials Performance and Evaluation,* ASM International, 1992
- J.H. Payer, *Corrosion,* course 14, Materials Engineering Institute, ASM International, 1994
- R.W. Revie and H.H. Uhlig, *Corrosion and Corrosion Control: An Introduction to Corrosion Science and Engineering,* Wiley, 1985
- P.A. Schweitzer, Ed., *Corrosion and Corrosion Protection Handbook,* 2nd ed., Marcel Dekker, 1989
- G.M. Ugiansky and J.H. Payer, Ed., *Stress Corrosion Cracking,* ASTM, STP 665, 1979
- L. S. Van Delinder, *Corrosion Basics: An Introduction,* National Association of Corrosion Engineers, 1984

Understanding How Components Fail, Third Edition
Donald J. Wulpi
Edited by Brett Miller

CHAPTER 14

Elevated-Temperature Failures*

ELEVATED-TEMPERATURE FAILURES are perhaps the most complex type of failure because all of the modes of failures discussed previously can occur at elevated temperatures (with the obvious exception of low-temperature brittle fracture). Elevated temperatures greatly complicate both the analyses of the failures that occur and the possible solutions. Therefore, these failures must be examined and considered very carefully because many of the different failure modes tend to interact. As a rule, the type of failure is established by examination of the fracture surfaces and comparison of operating conditions with available data on elevated-temperature fractures. In addition to fractographic evaluation, optical microscopy analysis in elevated-temperature failures is required when conditions are such that a change in metallurgical structure of the failed component occurs (Ref 1).

The term *elevated temperature* needs definition. Normally the useful static strength of a metal is limited by its yield strength, as discussed in Chapter 2, "Mechanical Properties," in this book. However, as the temperature increases, the useful static strength of a metal is limited by the time-dependent factor of creep, which is time-dependent strain occurring under stress (Ref, 2, 3). Each metal or alloy must be considered individually because of differences in their properties. Approximate values for the lower limit of elevated-temperature behavior for several metals and alloy systems are shown in Table 1.

The proper material must be selected for the intended design and usage conditions. Each material or alloy has both limits and advantages. In ser-

**Understanding How Components Fail,* Second Edition, by Donald J. Wulpi, Chapter 14, Elevated-Temperature Failures, ASM International, 1999, reviewed and revised by Daniel J. Benac, Baker Engineering and Risk Consultants, Inc.

Table 1 Approximate values for the lower limit of elevated-temperature behavior for several metal and alloy systems

Metal	Temperature	
	°F	°C
Aluminum alloys	300	150
Titanium alloys	600	315
Carbon and low-alloy steels	700	370
Austenitic, iron-base high-temperature alloys	1000	540
Nickel-base and cobalt-base high-temperature alloys	1200	650
Refractory metals and alloys	1800–2800	980–1540

Source: Ref 1

vice at elevated temperature, the life of a metal component is predictably limited, whether subject to static or dynamic loads. In contrast, at lower temperatures, and in the absence of a corrosive environment, the life of a component in static service is unlimited, if the operational loads do not exceed the yield strength of the metal.

The principal types of elevated-temperature failure mechanisms are creep, stress rupture, low-cycle or high-cycle fatigue, thermal fatigue, overload, hot corrosion, embrittlement, sulfidation, carburization, oxidation, distortion, and combinations of these as modified by environment. Elevated-temperature problems are real concerns in industrial applications such as boilers, steam and gas turbine parts, containers in contact with molten metal, heat-treating furnaces and equipment, cement mills, incinerators, exhaust valves and manifolds in internal combustion engines, power plant equipment, petroleum refinery furnaces, aerospace structures, and steam-reformer tubes.

Generally, the type of failure in these applications is established by examination of fracture surfaces and comparison of operating conditions in which the failure occurred with available data on creep, stress rupture, tension, elevated-temperature fatigue, and thermal-fatigue properties. This analysis is usually sufficient for most failure investigations, but more thorough analysis may be required when stress, time, temperature, and environment have acted to change the metallurgical microstructure of the component.

Creep

By definition, creep is time-dependent strain, or gradual change of shape, of a part that is under stress. It usually is considered the result of tensile stress, but creep can and does occur under all types of stress. Typically, creep begins at approximately one-third to one-half of the alloy melting point (~0.3 to 0.5 T_M). Factors such as time-temperature exposure, geometry, and grain structure affect creep behavior. Creep rate will actually decrease with increased grain size. Some components such as turbine blades are processed to form a single-grain (crystal) structure, which has good creep resistance. Creep is not limited to metals. Three analogies may illustrate the point: old, tightly stretched rubber bands can creep over a relatively long period of time to the extent that they are virtually useless. Also, creep is the reason that old shoes are more comfortable than new shoes, for the materials have had an opportunity to adjust over time to the shape of the feet. Wooden parts, such as some beams and fence rails under bending loads, will gradually sag over relatively long periods of time. Gradual change of shape under compressive, torsional, bending, and internal-pressure stresses may or may not lead to fracture. In the following discussion, creep is assumed to be caused by tensile stress.

Creep usually is considered to occur in three stages, as shown schematically in Fig. 1, which plots strain, or elongation, due to tensile stress against time at fixed values of temperature and stress. Following initial elastic strain resulting from the immediate effects of the applied load, the metal undergoes increasing plastic strain at a decreasing strain rate. This is the primary, or first, stage of creep, occurring within the metal during the first few moments after the load is applied. However, the creep rate usually slows as crystallographic imperfections within the metal undergo realignment, leading to secondary creep.

Stage 2, or secondary creep, is essentially an equilibrium condition between the mechanisms of work hardening and recovery. The metal is still stretching under tension but not as rapidly as in the first stage. The duration of secondary creep depends upon the temperature and stress level on the metal, as shown in Fig. 2. Here a steel alloy was tested at a specific temperature under four different stress levels, which cause different types of behavior. The lowest stress level causes little strain, or change in shape, while successively higher stress levels rapidly lead to fracture.

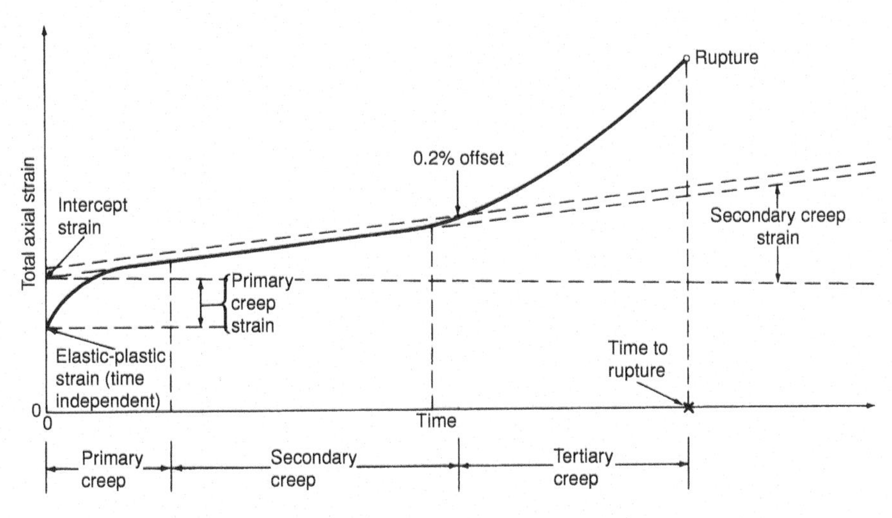

Fig. 1 Schematic tension-creep curve, showing the three stages of creep. Source: Ref 4

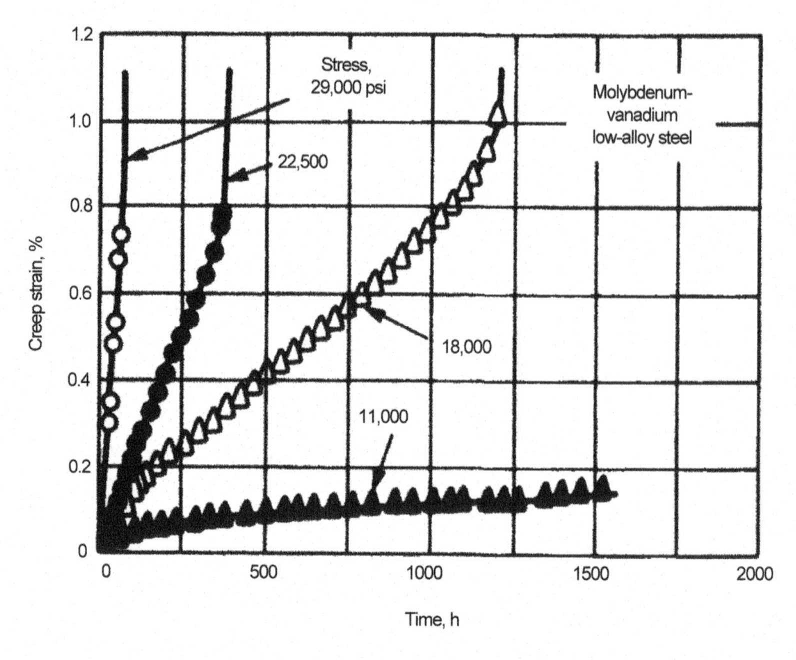

Fig. 2 Creep curves for a molybdenum-vanadium low-alloy steel under tension at four stress levels at 600 °C (1112 °F). Source: Ref 2

Stage 3, or tertiary creep, is the gradual increase in strain prior to fracture. It may result from metallurgical changes within the metal that permit rapid increases in deformation, accompanied by work hardening that is insufficient to retard the increased flow rate of metal. In tension, tertiary creep may be accelerated by a reduction in cross-sectional area resulting from cracking or localized necking. Environmental effects, such as oxidation or corrosion, also may increase the tertiary creep rate.

Under certain conditions, some metals may not exhibit all three stages of plastic deformation. For example, at high stresses or temperatures, the absence of primary creep is not uncommon, with secondary creep or, in extreme cases, tertiary creep following immediately upon loading. At the other extreme, notably in cast alloys, no tertiary creep may be observed, and fracture may occur with only minimum extension. Both of these phenomena are shown in Fig. 3.

Identification of failures caused by creep is usually relatively easy, primarily because of the deformation frequently involved. However, depending upon the alloy, creep fracture may be macroscopically either brittle or ductile. Brittle fracture is intergranular and occurs with little or no elongation or necking. Intergranular fractures typically start at grain-boundary triple points under high stress and low creep temperatures (Fig. 4a) or at creep voids (cavitation) under high temperatures and low stress (Fig. 4b). Ductile fracture is transgranular and typically is accompanied by discernible elongation and necking. The type of fracture depends not only on

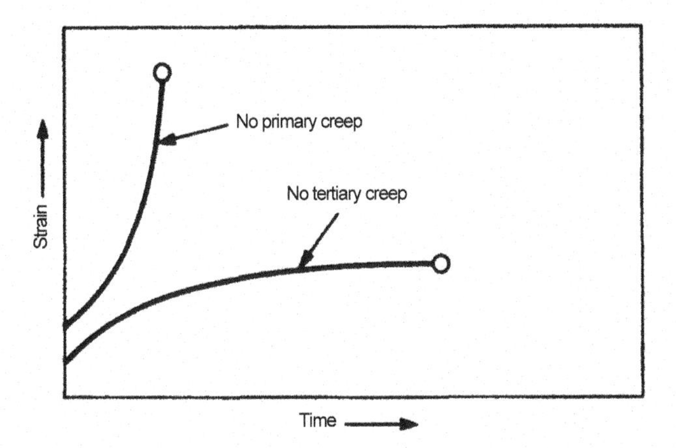

Fig. 3 Creep curves showing no primary creep and no tertiary creep. Source: Ref 2

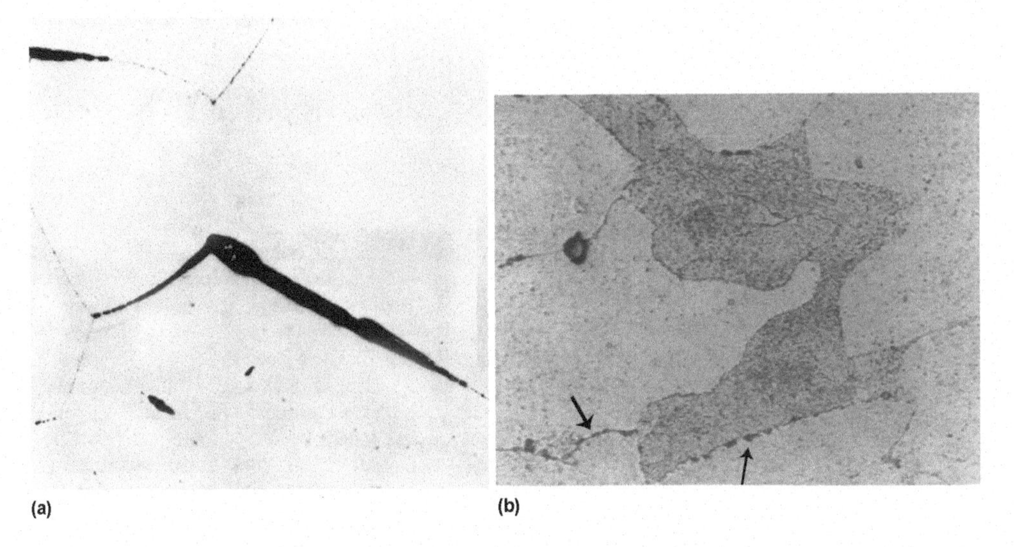

Fig. 4 Comparison of wedge-shaped cracks and creep voids: (a) triple-point stress rupture (60×); (b) creep cavitation damage (arrows) in a desuperheater inlet header (1000×)

temperature but also on strain rate. At constant temperature, the occurrence of either transgranular or intergranular fracture depends on strain rate. Conversely, at constant strain rate, the type of fracture depends on temperature. In general, lower creep rates, longer rupture times, or higher temperatures promote intergranular fracture. Sometimes components subjected to elevated temperatures do not fail or crack but distort.

The classic failure of a tube or pipe subjected to internal pressure and exposed to temperatures that exceed the creep or rupture strength of the material is a fish-mouth rupture. This rupture occurs because the internal stresses are highest in the hoop or the circumferential direction, causing the wall to yield, thin, and form a longitudinal crack. This type of failure can occur during short time exposures such as a fire, from overheating in equipment operations, or under long time conditions under creep conditions. Examples of high-temperature fractures of tubing are shown in Fig. 5 to 7.

Stress Rupture

The terms *stress rupture* and *creep fracture* are often used interchangeably. They are similar, except that in a stress rupture failure, usually the stress is higher and the time to fracture is less than in a creep condition. For example, creep and stress rupture tests are often performed to evaluate a material's life expectancy (Ref 5). During a stress rupture test, usually only the nominal stress, temperature, time to fracture, and total strain are recorded. There are insufficient data to plot the complete creep curve.

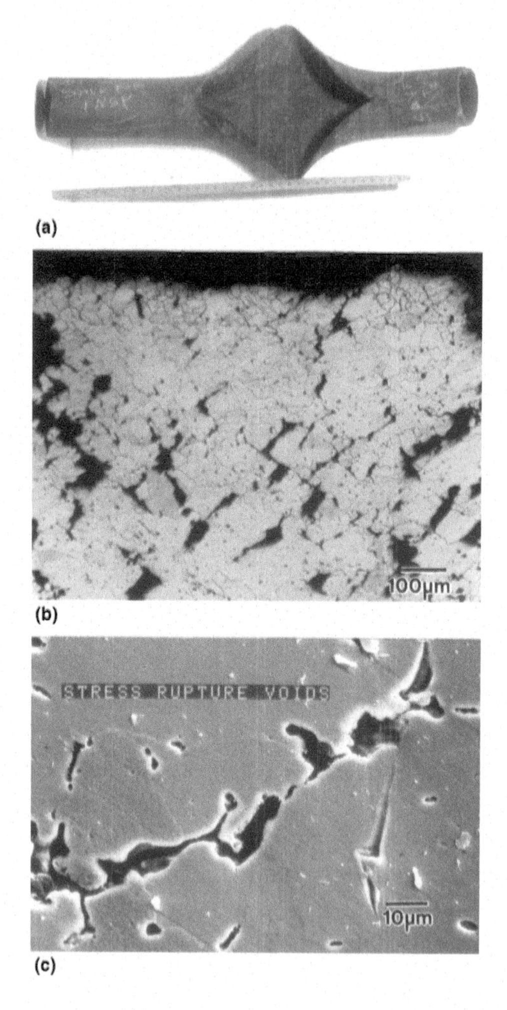

Fig. 5 Stress rupture of heater tube: (a) heater tube that failed due to stress rupture; (b) and (c) stress rupture voids near the fracture. Source: Ref 3

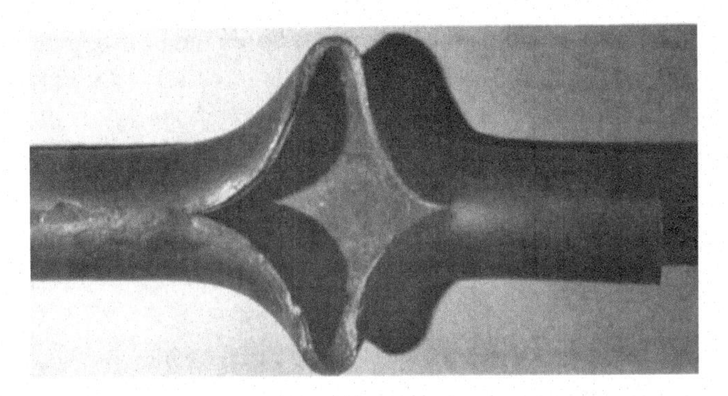

Fig. 6 Thick-lip "fishmouth" failure of a 2-in.-diam superheater tube. The tube bent away from the fracture due to the reaction force of the escaping steam. The material was ASME SA-213 T22 (0.15 maximum C, 1.90–2.60 Cr, 0.87–1.13 Mo). Hardness was 96–98 HRB. Scale about 0.012 in. thick is present on the inside, tending to prevent heat transfer and causing overheating to about 1525–1575 °F.

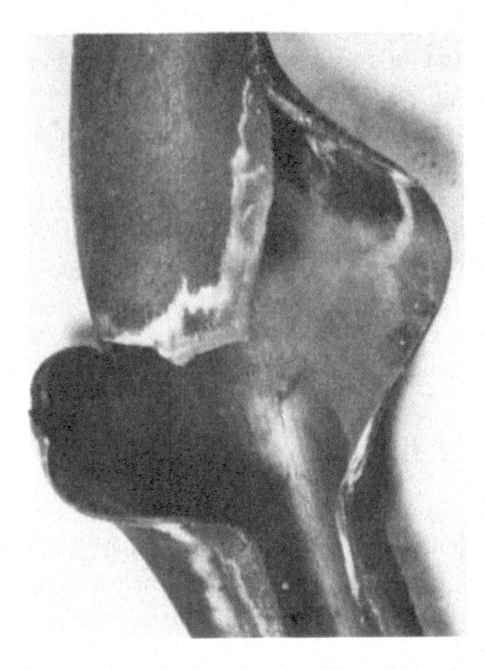

Fig. 7 Thin-lip rupture in a boiler tube caused by rapid overheating. This rupture exhibits a "cobra" appearance as a result of lateral bending under the reaction force imposed by escaping steam. The tube was a 2-½ in. outside diameter, 0.250 in. wall boiler tube made of 1.25Cr-0.5Mo steel (ASME SA-213). Source: Ref 2

Overheating Failure

The term *overheating failure* is often misused but generally means a failure resulting from operation of a component at a temperature higher than expected in design. Time at temperature is an important factor, and these types of failures are often called *short-term* and *long-term* overheating failures. A short-term overheating failure is one in which a single incident or a small number of incidents exposes the component to an excessively high temperature (hundreds of degrees above normal), to the point where deformation or yielding occurs. Overheating results from abnormal conditions such as loss of coolant flow and excessive temperature excursion.

Long-term overheating occurs over a period of months or years. There may be evidence of heavy oxidation, minimum distortion, and secondary cracking.

Elevated-Temperature Fatigue

The effect of temperature on fatigue strength is marked: fatigue strength decreases with increasing temperature. However, the precise relationship between temperature and fatigue strength varies widely, depending on the alloy and the temperature to which it is subjected.

In some cases the part may operate at elevated temperature with alternate steady-state and fatigue-type (cyclic) stressing. In these components, the combined creep-fatigue loads result in substantially decreased life at elevated temperatures compared with that anticipated in simple creep loading, and this effect must be considered in failure analysis. If oxidation on the fracture appears to be maximum near the surface of the part, the fracture is primarily due to fatigue; however, if oxidation appears to be relatively uniform on the fracture surface, steady-state or static loads may have been more significant. However, other failure analysis observations must also be made, such as deformation or its absence and the patterns on the fracture surface.

Thermal Fatigue

Fatigue may be caused either by cyclic mechanical stressing or by cyclic thermal stressing. Thermal-fatigue cracks are the result of repeated heating and cooling cycles, producing alternate expansion and contraction. When the metal cools, it contracts, causing residual tensile stresses if it is restrained from moving freely. If this alternate expansion and contraction continues, fatigue cracks will form and will propagate each time the metal is cooled, as in the engine exhaust valve in Fig. 8. See Chapter 4, "Residual Stresses," in this book for more information on thermal residual stresses.

Thermal cycles may be caused by friction, as in brake drums and clutch plates, where the surface is frequently heated and expanded by the friction but is prevented from expanding freely by the colder, stronger metal below the surface. Compressive yielding occurs in the hot surface layer, causing tensile residual stresses when the metal contracts during cooling. This condition frequently causes thermal-fatigue cracks, usually called *heat checking*. This network of multiple cracks on the friction surface may be

Fig. 8 Thermal-fatigue crack in the hardfacing alloy on an exhaust valve from a heavy-duty gasoline engine (~2.5×). Advanced burning originated from the large crack. Additional thermal-fatigue cracks are also present on the valve face. Engine efficiency rapidly deteriorates from increasing loss of compression when very hot exhaust gases blast through the open passage (also called *blowby*).

harmless unless the cracks wear the mating surface or unless one or more cracks progress to cause complete fracture.

Engine exhaust manifolds also are subject to thermal fatigue, particularly on heavy-duty engines. They may become literally red-hot under certain conditions and then cool down when the engine is stopped. If the manifold is not permitted to "float" or move freely in an axial direction, tensile residual stress may be generated when it cools, eventually causing fatigue fracture if repeated often enough.

Thermal fatigue may be prevented in many parts by designing curves rather than straight lines into the system. When this is done, heating and cooling cycles simply cause distortion of the curves, rather than forming tensile residual stresses upon cooling. Expansion loops and bellows in elevated-temperature piping and tubing systems operate on this principle.

Thermal-fatigue cracks usually initiate at the surface and then propagate oriented perpendicular to the surface and the plane of maximum stress. They may occur singularly but often occur in multiples. Because the crack initiates externally, the amount of corrosion or oxidation along the surface of a thermal-fatigue crack is inversely proportional to the depth of the crack. Thermal-fatigue cracks usually progress transgranularly. However, in the presence of intergranular oxidation, which may help to initiate the cracks, thermal-fatigue cracks tend to propagate along the grain boundaries.

Metallurgical Instabilities

Stress, time, temperature, and environment may act to change the metallurgical structure during testing or service and thereby contribute to failure by reducing strength. These microstructural changes are referred to as *metallurgical instabilities*. Sources of instabilities include transgranular-intergranular fracture transition, recrystallization, aging or overaging, intermetallic-phase precipitation, delayed transformation to equilibrium phases, order-disorder transition, general oxidation, intergranular corrosion, stress-corrosion cracking, slag-enhanced corrosion, and contamination by trace elements.

Life assessment methods for elevated-temperature failure mechanisms and metallurgical instabilities that reduce life or cause loss of function or operating time of high-temperature components are described in further detail in Ref. 4 and 6.

Environmentally Induced Failure

A critical factor in the performance of metals in elevated-temperature service is the environment and resulting surface-environment interactions. In fact, the most important source of elevated-temperature failure requiring premature replacement of a component is environmental degradation of material. Control of environment or protection of materials (by coatings or self-protective oxides) is essential to most elevated-temperature applications.

General oxidation can lead to premature failure; grain-boundary oxidation may produce a notch effect that also can limit life. Some environments may be more harmful than others. Attack of fireside surfaces of steam-boiler tubes by ash from vanadium-bearing fuel oils can be quite severe; external tube metal loss occurs, leading to thinning and increased tube strain. Also, vanadium-ash attack and hot corrosion in general are equally harmful in gas turbines. The high-temperature attack occurs because of the formation of low-melting-point vanadium- or sulfur-bearing deposits that attack the coatings and the metallic component. Figure 9 shows the hot-corrosion attack of a nickel-alloy turbine blade.

In all elevated-temperature failures, the characteristics of the environment must be carefully considered. These include not only the temperature itself but also whether the elevated temperature is steady or fluctuating, the rate of temperature change (which will affect differential expansion and contraction), the thermal conductivity of the metals involved, the characteristics of the fluids (both liquids and gases) that are in contact with the surfaces, and the way in which the fluids contact the metal surfaces. The nature of fluid contact is most important in certain types of parts that have high gas or liquid flow rates at elevated temperature, which can cause erosion problems.

Corrosion and Corrosion-Erosion. Certain types of parts function in environments where high rates of fluid flow at high temperature are normal. Typical of those in gaseous environments include engine exhaust valves (Fig. 10), blades and vanes in the hot sections of gas and steam

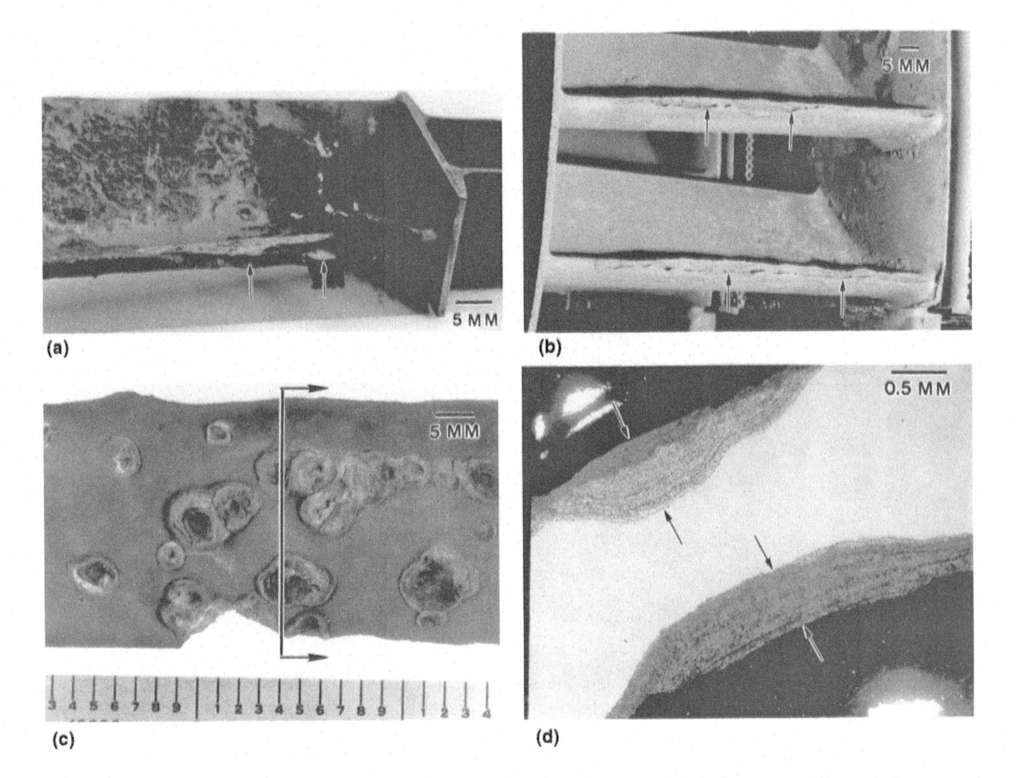

Fig. 9 Hot-corrosion attack of René 77 nickel-base alloy turbine blades. (a) A land-based, first-stage turbine blade. Notice the deposit buildup and flaking and splitting of the leading edge. (b) Stationary vanes. (c) A land-based, first-stage gas turbine blade that had type 2 hot-corrosion attack. (d) A metallurgical cross section showing the layered corrosion attack of the blade shown in (c). Source: Ref 4

turbine engines and generators (Fig. 11), certain locations (particularly inlets and outlets) in various types of furnaces, and many ducts and pipes that conduct hot gases. Typical parts in high-temperature liquid environments are certain piping systems, pumps, rotors, propellers, nozzles, and the like, that experience high rates of fluid flow.

The problem with these parts is that the combination of high temperature and high-velocity fluid flow often results in erosive wear at critical locations that frequently destroys the efficiency of the parts and their assemblies. Chapter 11, "Wear Failures—Abrasive and Adhesive," in the book points out that erosive wear is caused by high-speed, low-stress particles that tend to cut, or erode, any materials that they slide, or impinge, against. In general, elevated temperatures reduce metal strength and hardness. Thus, any part that changes the direction of high-temperature, high-velocity fluids is subject both to increased erosion from mechanical action and to increased corrosion from the chemical action of the fluid on the part.

Elevated-temperature corrosion and corrosion-erosion problems are usually difficult to prevent. Possible solutions include development and selection of special alloys that can resist elevated-temperature corrosion-

Fig. 10 Severe localized erosion-corrosion of two gasoline-fueled engine exhaust valves made from a nickel-base superalloy operating between 1400 and 1500 °F. The exhaust gas damage in the underhead radius and stem was identified as lead oxide corrosion, aggravated by bromine from the gasoline.

Fig. 11 High-temperature erosion-corrosion in a turboprop engine blade. Most of the 1-⅓-in.-long blade had been damaged by sulfidation, or hot corrosion caused by excessive sulfur in the fuel. This uncoated INCO 713 turbine blade was one of many such rotary blades subjected to this type of high-temperature erosion-corrosion. Nickel aluminide coatings are sometimes applied to the nickel-base alloy to improve the life of the blades.

erosion, use of protective coatings, and use of cooling techniques to lower the temperature of operating parts. Frequently, protective coatings of various types may be very useful, particularly if they diffuse into the base metal and form corrosion- and wear-resistant compounds. Various types of aluminum-rich coatings are frequently used, as are alloys with cobalt, nickel, or chromium. It is vital to have testing equipment suitable to reproduce the types of failures encountered in service.

General Oxidation. Oxidation has long been recognized as a severe limitation to the utilization of metals at high temperatures. In certain applications, the major elevated-temperature problem is general oxidation, or scaling. This is particularly true when the metal is subjected to repetitive heating and cooling cycles, in which scale, a layer of oxidation products, forms because of exposure at high temperature to an oxidizing atmosphere, usually air. The scale flakes off when the metal cools because of differences in thermal expansion and contraction characteristics of the scale and the base metal. Since the scale is not a metal but a metal-oxide compound, the properties may be completely different.

As a group, the ferritic stainless steels alloyed with chromium are usually considered to be superior in oxidation resistance to many other iron-base alloys. In fact, Ref 7 states that:

> The main advantage of ferritic stainless steels alloyed with chromium for high-temperature use is their good oxidation resistance, which is comparable to that of austenitic grades. In view of their lower alloy content and lower cost, ferritic steels should be used in preference to austenitic steels, stress conditions permitting. Oxidation resistance of stainless steel is affected by many factors, including temperature, time, type of service (cyclic or continuous), and atmosphere. For this reason, selection of a material for a specific application should be based on tests that duplicate anticipated conditions as closely as possible.

Because of the need for good oxidation resistance in automotive exhaust systems and in their catalytic converters, as well as in many other applications, the standard ferritic stainless steels—particularly type 409—are widely used, along with some other steels developed specifically for their superior oxidation resistance. Under favorable conditions, these steels form a tightly adhering oxide scale that expands and contracts with the base metal, thus proving suitable for the application, provided that there is no need for high strength at the elevated temperatures. In the latter case, it is necessary to consider highly specialized alloys with the necessary properties.

Metal Dusting. Metal dusting, similar to oxidation attack, has occurred in petrochemical plant equipment. Metal dusting can be distinguished from conventional oxidation attack of carburized metal since it occurs randomly in localized areas and progresses more rapidly. Metal dusting generally takes place at temperatures from 480 to 815 °C (900 to 1500 °F), although it has occurred at temperatures as high as 1095 °C (2000 °F) in strongly reducing atmospheres such as those containing large amounts of hydrocarbon gases.

Carburization. The problem of carburization of steel—particularly carburization of stainless steels used in elevated-temperature furnace environments—is common to many industrial applications. Typical is the carburization, with or without oxidation, of stainless steel resistance-heating elements and various components and fixtures of heat-treating furnaces. Simultaneous carburization and oxidation of stainless steel heating elements result in a form of attack sometimes referred to as "green rot." This form of attack is common to nickel-chromium and nickel-chromium-iron alloys.

The basic problem with carburization of elevated-temperature materials is that the added carbon rapidly combines with chromium to form various types of chromium carbides. Thus, carburization depletes chromium—a major high-temperature and corrosion-resisting element—from the grain boundaries, sometimes resulting in intergranular fracture. In addition, the carbon changes the density and thermal expansion characteristics of the metal, tending to cause residual stresses.

Carburization of austenitic stainless steels and elevated-temperature alloys may be easily detected by changes in their magnetic properties. Austenitic stainless steels are normally nonmagnetic; however, when carbon is added by diffusion at elevated temperature, the alloy becomes magnetic as the chromium in the matrix of the base metal is depleted by combination with carbon to form the chromium carbides previously mentioned.

Sulfidation. For carbon steels, sulfidation becomes a problem at temperatures exceeding 232 °C (450 °F) in process equipment containing hydrocarbons bearing sulfur compounds. General metal wastage occurs.

High-Temperature Hydrogen Attack (HTHA). HTHA can occur in carbon and low-alloy steels when they are used for piping and pressure vessels that are exposed to elevated temperatures and high-pressure hydrogen. The result can be sudden and catastrophic brittle failure. Atomic hydrogen permeates the steel and reduces iron carbide (Fe_3C) to form methane (CH_4). The methane does not diffuse from the metal, and its pressure may exceed the cohesive strength of the metal and cause fissuring between grains. Figure 12 shows the effects of HTHA.

Liquid-Metal Contact and Embrittlement. Failure as a result of liquid-metal contact is not limited to temperatures that we normally consider to be high. Mercury is a liquid metal that can cause severe stress-corrosion cracking problems by room-temperature contact with many high-strength steels and with copper, aluminum, nickel, and titanium and their alloys.

The one type of elevated-temperature failure is liquid metal embrittlement (LME), which is the brittle failure of a metal that is coated with a film of liquid metal and subjected to tensile stresses.

Many high-temperature alloys frequently cannot be used in contact with liquid (or molten) metals used in coating and in other industrial processes. Molten lead, however, can be contained in some high-temperature alloy vessels provided that it is covered with a protective layer of pow-

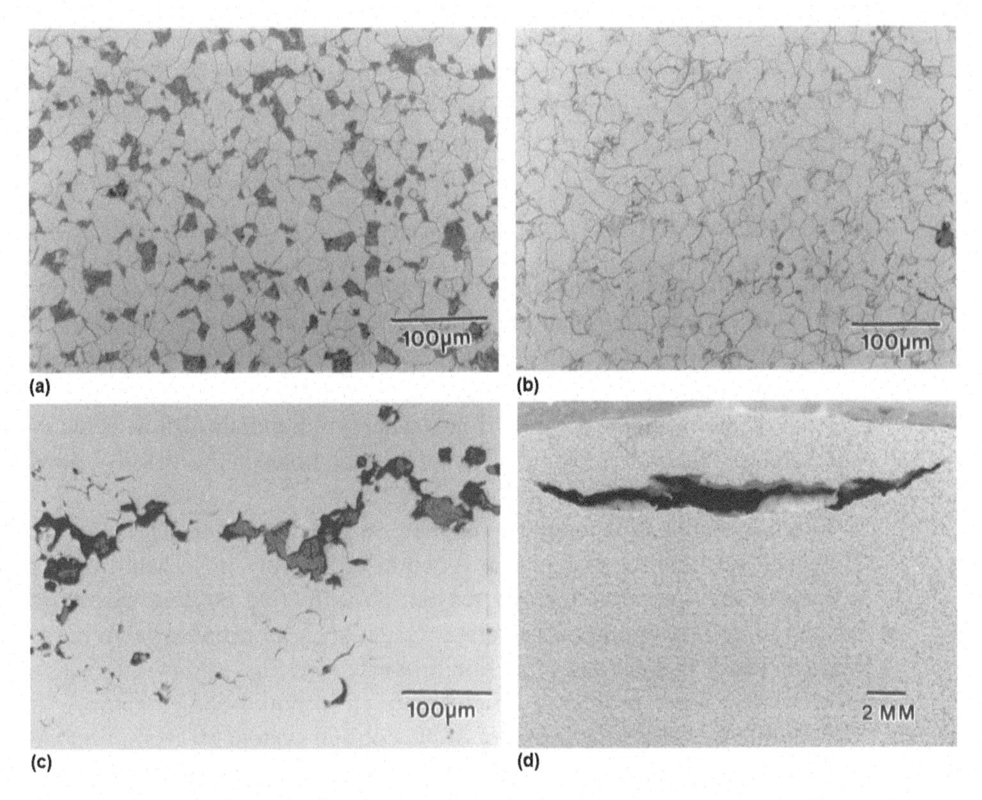

Fig. 12 Hydrogen-damaged refinery platformer line (carbon steel, 0.5% Mo). (a) Undamaged microstructure. (b) Decarburization region caused by hydrogen depleting the iron carbides. (c) Microfissuring at inclusions. (d) Hydrogen blister caused by methane gas formation. (a) and (b), nital etch; (c) and (d), unetched. Source: Ref 4

dered charcoal; this prevents formation of lead oxide, which is highly corrosive to most high-temperature alloys (Ref 2).

Molten zinc, used in hot-dip galvanizing of fabricated parts and structural I-beams, is commonly contained in tanks or vats made from plain carbon boilerplate steel. The principal requirement of a galvanizing-tank material, aside from strength, is the ability to resist the corrosive attack of molten zinc.

The reason that high-temperature alloys are frequently undesirable for use in contact with liquid metals is that the temperature causes precipitation of chromium carbides in the grain boundaries, or sensitization. As mentioned previously, this depletes the surrounding areas of chromium; grain-boundary corrosion, cracking, and fracture then can occur in the presence of the liquid metal.

A common problem with molds used to die cast zinc, aluminum, magnesium, and copper is heat checking, or thermal-fatigue cracking of the surfaces in contact with the liquid metal being cast. This condition results in raised ridges in the casting because the liquid metal flows into the cracks in the molds. It is necessary to keep the temperature of the die itself quite high so that there is little differential expansion and contraction that can cause tensile residual stresses and ultimate cracking on the surfaces. The dies need to be preheated when starting up; then the temperature is controlled by water flowing through cooling passages in the dies.

Cooling Methods

The cooling techniques mentioned earlier are potential means of prevention of elevated-temperature failures. In gaseous-flow mechanisms, it is often desirable and possible to have cool air or other gases flow through or past the parts involved to reduce the temperatures. This is commonly done in the hot sections of gas turbines where tremendous airflow is available; some of the incoming air is routed through holes in the airfoil-shaped blades and vanes, as shown in Fig. 13.

Internal combustion engines may be cooled by either liquid or air. However, no cooling system can function effectively if its heat-transfer properties are impaired for any reason. An effective cooling system is critical to engine operation; severe damage, such as exhaust valve burning, can occur if it does not function properly (see Fig. 14).

If exhaust-valve problems are encountered in liquid-cooled engines, it is desirable to inspect the integrity of the cooling system by sectioning the cylinder head to determine if the liquid-coolant passages are blacked or dammed. This can occur if the sand cores that form the coolant passages were cracked prior to casting. Molten metal can flow into the cracks, effectively damming the passages. Also, the casting should be cut along the section lines shown on the engineering drawing to determine if the foundry

Fig. 13 Hollow, air-cooled stationary vane in an aircraft turboprop engine. Note that the trailing edge is also air cooled by the air that blows between the two thin metal surfaces. The metal is the nickel-base casting alloy INCO 713. Damage on the trailing edge is secondary, caused when a foreign object was sucked into the engine during service.

(a)

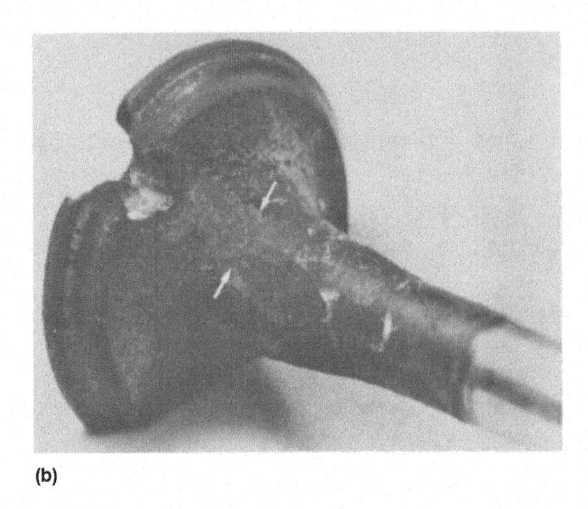

(b)

Fig. 14 (a) Incipient burning on the valve face of an automotive engine exhaust valve caused by microporosity in the hardfacing alloy (~2.5×). (b) Severe, destructive burning (guttering) in an automotive-engine exhaust valve (~2×). The very hot exhaust gases (blowby) rushing through the cavity removed the underhead deposits adjacent to the burned area (arrows). Continued operation of the valve in (a) would ultimately lead to this type of damage.

is making the cores (and cooling passages) too small and the metal walls too thick. Foundries may tend to do this in order to reduce foundry problems and also, incidentally, to increase the tonnage of metal produced and shipped.

Similarly, air-cooled engines cannot function properly if the cooling system does not operate properly. Air-cooled engines may overheat if the cooling fins on the outside of the engine are clogged with dirt, leaves, or other debris. It does not take much dirt or insulating deposit on cooling surfaces to greatly reduce the thermal conductivity and heat-transfer capabilities of the metal.

Certain mechanisms that operate in very high temperatures can survive only with the aid of their cooling systems. Two spectacular examples are the oxygen lances used in basic-oxygen steelmaking converters and water-cooled cupolas used in making various types of cast iron. The oxygen lances are essentially double-walled tubes that are inserted directly above the liquid metal in the converter. The oxygen blast into the molten metal causes extremely high temperatures at the lance. Only the recirculating cooling water in the tube prevents the lance itself from melting. Similarly, plate steel is sometimes used to make cupolas for melting cast iron. Again, the only thing that keeps the molten iron from damaging the steel shell is the external water-spray cooling system. Obviously, it is critical that the cooling system operates properly to extract heat from the steel shell, which is otherwise unprotected from the molten metal within it.

Summary

Identification and prevention of elevated-temperature failures can be extremely complex and require extensive evaluation. To summarize the principal types of elevated-temperature failure:

- *Creep:* gradual change of shape while under stress
- *Elevated-temperature fatigue:* reduced fatigue life at elevated temperature
- *Thermal fatigue:* repetitive changes in temperature on the surface when there is not necessarily any applied external force
- *Metallurgical instabilities:* changes in the microstructure of the metal during elevated-temperature service
- *Environmentally induced failure:* reactions of the environment with the metal components
- *Corrosion and corrosion-erosion:* combinations of environmental effects, which are frequently critical
- *General oxidation:* scaling, which may be a major problem under certain conditions
- *Metal dusting:* a form of carburization that occurs rapidly in localized areas due to carburizing gases and/or process streams containing carbon and hydrogen
- *Carburization:* diffusion of carbon, causing metallurgical changes in the metal that can lead to failure by erosion or fracture
- *Sulfidation:* corrosion of a metal or alloy resulting from a reaction with sulfur compounds in high-temperature environments
- *High-temperature hydrogen attack:* reaction of absorbed hydrogen with carbides in a steel to form methane causing decarburization and loss of strength at elevated temperatures and pressures
- *Liquid-metal contact:* thermal and metallurgical changes caused by contact with liquid, or molten, metal, common in certain industries

ACKNOWLEDGMENT

Much of the text in this chapter was taken from Ref 2.

REFERENCES

1. Elevated-Temperature Failures, lesson 7, *ASM Practical Failure Analysis Course,* ASM International, 2002
2. *Failure Analysis and Prevention,* Vol 11, *ASM Handbook,* ASM International, 1986, p 225–238, 263–303, 602–627
3. Creep and Stress Rupture Failure, *Failure Analysis and Prevention,* Vol 11, *ASM Handbook,* ASM International, 2002, p 728–737
4. Elevated Temperature Life Assessment for Turbine Components, Piping and Tubing, *Failure Analysis and Prevention,* Vol 11, *ASM Handbook,* ASM International, 2002, p 289–311
5. Failure Avoidance Brief: Estimating Heater Tube Life, *Practical Failure Analysis,* Vol 9 (No. 1), February 2009, p 5–7
6. Failure Analysis and Life Assessment of Structural Components, *Failure Analysis and Prevention,* Vol 11, *ASM Handbook,* ASM International, 2002, p 228–242
7. *Properties and Selection: Stainless Steels, Tool Materials, and Special-Purpose Metals,* Vol 3, *Metals Handbook,* 9th ed., American Society for Metals, 1980, p 195

SELECTED REFERENCES

- G. Bernasconi and G. Piatti, Ed., *Creep of Engineering Materials and Structures,* Applied Science Publishers, 1978
- H.E. Boyer, Ed., *Atlas of Creep and Stress Rupture Curves,* ASM International, 1988
- J. Bressers, Ed., *Creep and Fatigue in High Temperature Alloys,* Applied Science Publishers, 1981
- J.R. Davis, Ed., *ASM Specialty Handbook: Heat-Resistant Materials,* ASM International, 1997
- M.J. Donachie, Ed., *Superalloys: Source Book,* American Society for Metals, 1983
- *Fatigue and Fracture,* Vol 19, *ASM Handbook,* ASM International, 1996
- A.K. Khare, Ed., *Ferritic Steels for High-Temperature Applications,* American Society for Metals, 1983
- G.Y. Lai, *High Temperature Corrosion of Engineering Alloys,* ASM International, 1990
- R. Viswanathan, *Damage Mechanics and Life Assessment of High-Temperature Components,* ASM International, 1989

Understanding How Components Fail, Third Edition
Donald J. Wulpi
Edited by Brett Miller

CHAPTER 15

Fracture Mechanics*

FRACTURE MECHANICS can provide helpful quantitative information on the circumstances that led to a failure, and it can be used to prescribe preventive measures to avoid the recurrence of failures in similar components. A fracture mechanics approach is needed when a material or structure contains a cracklike defect. Conventional approaches based only on a critical stress or strain are not adequate in these cases.

A full fracture mechanics analysis may not be required in cases where the costs are relatively low, people have not been injured, and even when litigation is involved. Many failures can be prevented by specifying proper design and appropriate material properties, along with careful attention to manufacturing and quality control procedures and with other factors. However, a fracture mechanics approach can often be used to determine the key factors leading to the failure. Fracture mechanics can be used to determine whether:

- Design deficiencies led to the failure.
- The material was not adequate.
- Limitations on operating conditions, such as maximum load, or maximum temperature, were not followed.

Fracture mechanics is a well-developed quantitative approach to the study of failures. However, because it is often difficult to obtain exact values for many of the inputs of the analysis when a structural failure occurs, fracture mechanics is not always used. Fracture mechanics requires an understanding of three components of the analysis: stresses, defect sizes, and material properties. The stress analysis requires knowledge of the applied

*_Understanding How Components Fail, Second Edition,_ by Donald J. Wulpi, Chapter 15, Fracture Mechanics, originally by E.J. Kubel, Jr., reviewed and revised by John D. Landes, University of Tennessee, Knoxville.

loads and the geometry of the component. Often, loading information may not be known or may be falsely reported. Defect size characterization may require a nondestructive evaluation inspection or can come from a post-failure examination of the fracture surfaces. Material properties required are tensile properties and fracture toughness information; sometimes fatigue information is also required.

When fracture mechanics analysis indicates that failure should not have occurred based on the information given for the above, one or more of these items may be incorrectly assumed, and the conditions of the failure must be investigated more thoroughly. An application of the fracture mechanics techniques may be able to determine which input was incorrectly assumed.

Fracture Toughness and Fracture Mechanics

Fracture toughness is the resistance of a material to the extension of a cracklike defect. Fracture toughness is often characterized in terms of a material being ductile or brittle. A ductile material may have higher toughness compared with a brittle material. Ductile fracture occurs by the initiation, growth, and coalescence of voids in the material. This usually results in a fracture behavior that is slow and stable for much of the fracture process. Ductile fracture often occurs with the material undergoing yielding and plastic deformation.

Brittle fracture occurs as an unstable event where the crack propagates rapidly and failure is sudden. Brittle fracture occurs when a stress condition at the crack tip is achieved and often occurs under nominally linear-elastic deformation in the component. Ductile fracture is strain controlled and often occurs when the component has nominal yielding. As a result, the level of fracture toughness measured is generally higher for ductile fracture behavior than it is for brittle fracture conditions.

Fracture toughness testing may require more material or expense than is warranted. Hence, some more conventional tests may give information about relative fracture toughness levels in a material. One relative measure of toughness is the area under a standard tension stress-strain curve taken to specimen fracture; this is a measure of the energy absorbed by the material during the tensile test. For example, the toughness of steels shown in Fig. 1 increases progressively from very brittle steel A to very soft and ductile steel E. Fracture occurs in the brittle steel with little or no elongation or plastic strain. In contrast, the ductile steel sustains a lower load with high strain before breaking.

The tension test gives only a relative estimate of toughness and is best used for comparing the relative toughness of two or more materials. Sometimes toughness under high-strain-rate loading conditions is required. Tests such as the Charpy impact test use a standard notched-bar specimen, which is broken at a known temperature in a single-blow pendulum-type impact machine. Figure 2 illustrates a Charpy testing machine and a standard ASTM type A Charpy impact test specimen. Notch-toughness results are reported in joules or foot pounds of energy absorbed by the test specimen. However, notch-toughness test results do not give quantitative toughness values and cannot be used directly in engineering design calculations.

The inadequacy of tests such as the Charpy impact test to characterize fracture properties in design-significant terms became apparent around the

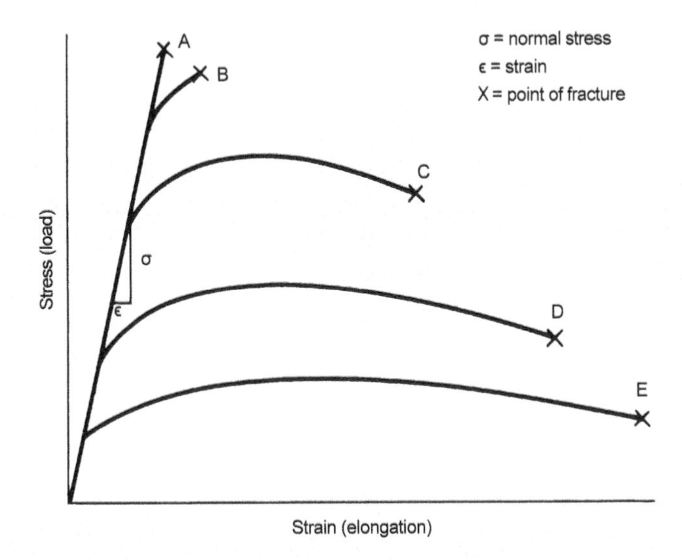

Fig. 1 The area under a stress-strain curve taken to specimen fracture gives a rough estimate of the toughness of steels. In general, toughness varies inversely with strength.

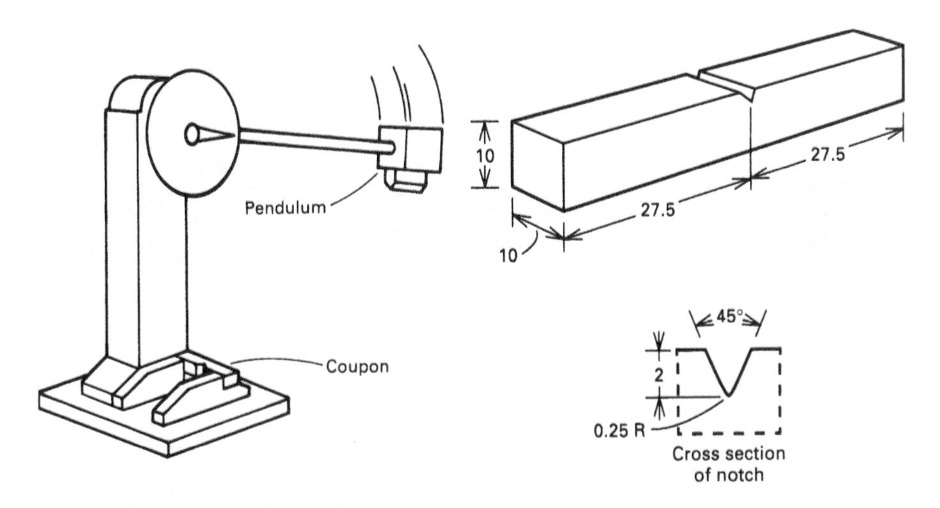

Fig. 2 Typical Charpy testing machine and Charpy V-notch specimen used to determine notch toughness in accordance with ASTM E 23 (dimensions in millimeters). The test coupon is chilled to the desired temperature and then quickly placed into the anvil to be broken.

time of World War II. Notch toughness became the focus of attention when a number of catastrophic failures occurred as a result of brittle fracture of plates in some 250 welded transport ships, 19 of which broke in two. (See example in Fig. 1 of Chapter 8, "Brittle Fracture," in this book.) The investigation of these brittle fractures led to new methods to evaluate notch toughness. What engineers needed was a toughness measure that could be determined using a simple laboratory test and could be used in a quantitative way to predict the flaw size at which fracture would occur in components containing defects. Conversely, for a given flaw size, it would be possible to predict the maximum safe operating stress.

Through the concepts and application of fracture mechanics, a new parameter called *fracture toughness* was developed. This was based on the parameter, K, the crack-tip stress intensity factor. Fracture toughness is a material property, as is yield strength. Fracture toughness is defined by the American Society for Testing and Materials (ASTM) as "a generic term for measures of resistance to crack extension," (Ref 1). A general definition of fracture toughness is the ability of a material to withstand fracture in the presence of cracks. A fracture toughness value based on K was developed as a quantitative measure of a material's resistance to crack advance and hence gave a parameter that could be used for design against failure or for the selection of more fracture-resistant materials.

Linear-Elastic Fracture Mechanics

Two categories of fracture mechanics are linear-elastic fracture mechanics (LEFM) and elastic-plastic fracture mechanics (EPFM). LEFM is used if there is only a small amount of plastic deformation at or near the crack tip. This requires a high level of tensile strength and a low level of fracture toughness. Also, the dimensions of the specimen or structure play a role. Larger dimensions contribute to more linear elastic behavior. Some materials that are designed using LEFM concepts are high-strength steels, titanium, and aluminum alloys. EPFM is used when there is extensive crack-tip plasticity. EPFM is generally used for the lower-strength, higher-toughness materials.

The LEFM approach to fracture analysis assumes that a part or specimen contains a cracklike defect that is in a linear-elastic stress field. A critical value of K is used as a fracture toughness parameter. Most structures contain defects of various types, as shown in Table 1. Cracks and cracklike flaws often are present in sizes below the limit of sensitivity of nondestructive inspection tests. It is not practical or economical to fabricate defect-free structures, so the only thing that realistically can be done is to limit the defect size by quality control in fabrication and by appropriate inspection. Development of fatigue cracks can be controlled using proper fatigue design practices and in-service inspection.

A crack in a loaded part or specimen generates its own stress field ahead of a sharp crack, which can be characterized by a single parameter, crack-tip stress intensity factor, K. K represents a single parameter that includes both the effect of the stress applied to a sample and the defect size. It can have a simple relation to applied stress and crack length, or the relation can involve complex geometry factors for complex loading, various configurations of real structural components, and variations in crack shapes. The fracture of a cracked body is controlled by K, sometimes labeled K_c.

Modes of Loading

Figure 3 defines three modes of loading: mode I, opening mode, has the stresses perpendicular to the crack plane; mode II, sliding, or shear, mode, has the stresses parallel to the crack plane but in the plane of the plate; and mode III, tearing mode, has the stresses parallel to the crack plane but out of the plane of the plate. The great majority of all actual cracking and fracture cases are mode I problems. A crack in the very early stage of development will turn into a direction in which it experiences only mode I loading, unless it is prevented from doing so by geometrical confinement. For this reason, fracture mechanics is generally confined to mode I. Combined modes may be important for cracks that are not aligned perpendicular to

Table 1 Typical defects in alloys

Metallurgical microstructure	Processing defects	Operations defects
Inclusions (oxides, sulfides, constituents)	Casting pores	Corrosion pits
Large precipitates (carbides, intermetallics, dispersoids)	Powder contaminants (hairs, ceramics, other metals)	Wear/fretting damage
Clusters or bands (inclusions, precipitates)	Weld defects	Surface oxidation and carburization
Brittle surface coatings	Forging/extrusion "cracking" and geometry discontinuities	Hydrogen attack
Local "soft spots" (precipitate-free zones)	Tool marks and dents	Creep voiding/cracking
Local "hard spots" (solute-enriched phases)	...	...

Note: In this table, the term *defect* is defined as "a feature, sized between 1 μm and 1 mm, that impairs the mechanical integrity and performance of a component."

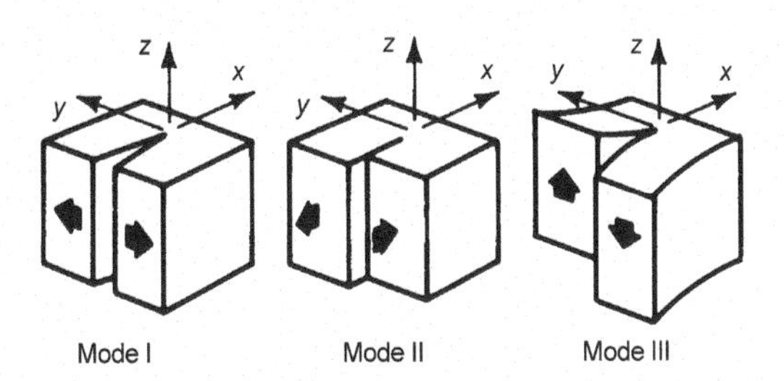

Fig. 3 Modes of loading. Mode I (opening mode): tension stress in the *y* direction, or perpendicular to crack surfaces. Mode II (edge-sliding mode): shear stress in the *x* direction, or perpendicular to the crack tip. Mode III (tearing mode): shear stress in the *z* direction, or parallel to the crack tip.

the crack plane but are at a slanted angle. In mode I loading, the critical stress intensity factor is labeled K_{Ic} if the fracture occurs under linear elastic and plane strain conditions. The K_{Ic} fracture toughness value must be measured by a standard test method; in the United States ASTM test standards are used, and ASTM E 399 (Ref 2) is used to determine K_{Ic}.

Plane Strain and Plane Stress

In practice, a two-dimensional state of stress is assumed in a bulk material when one of the dimensions of the body is small relative to the others. A two-dimensional state of stress called *plane strain* develops when plastic deformation at the crack tip is severely limited. This is promoted by thick sections, high strength, and limited ductility. In contrast, a two-dimensional state of stress called *plane stress* develops when much more plastic deformation occurs around the crack tip. This occurs when the section is relatively thin compared with the planar dimensions. The difference between plane strain and plane stress is based on the presence or absence, respectively, of transverse constraint in material deformation in the vicinity of the crack tip. Real structures do not contain either completely plane-strain or plane-stress conditions, but this assumption makes it possible to perform a two-dimensional stress-field analysis on a three-dimensional problem. Material thickness is linked with plane-stress and plane-strain behavior due to its dependence on the stress state at the crack tip.

In general, increasing specimen thickness decreases the fracture toughness to a limiting (critical) value. Figure 4 shows that at some specified

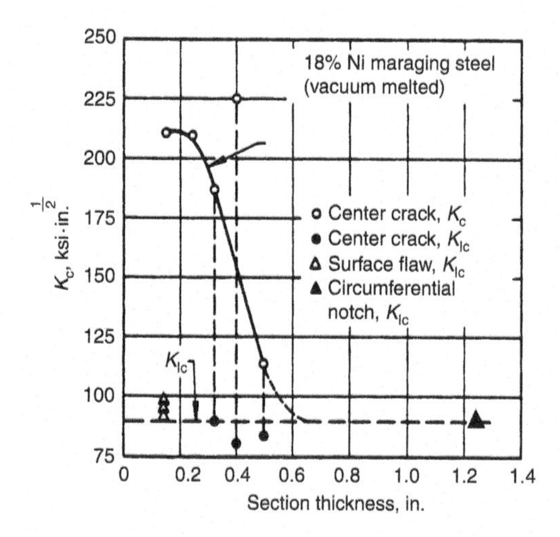

Fig. 4 Effect of section thickness of 18% maraging steel (2 GPa, or 300 ksi minimum yield strength) on measured values of the critical stress-intensity factor, K_{Ic}. At thicknesses greater than about 15 mm (0.6 in.), the critical stress-intensity factor, K_c (the value of K at which crack propagation becomes rapid), drops to the critical plane-strain stress-intensity factor.

thickness (which is different for each material and heat-treated condition), crack propagation is governed by plane-strain conditions, and the critical stress-intensity factor reaches a minimum value designated as the critical plane-strain stress-intensity factor (K_{Ic}). This factor is extremely important in materials evaluation because, unlike other parameters, it is essentially independent of specimen dimensions, provided that plane-strain conditions are met. Therefore, K_{Ic} provides a lower bound value of toughness and can be used as a conservative design parameter.

Plane-strain conditions also govern initiation of slow crack growth, so K_{Ic} both characterizes total fracture in thick sections and indicates the stress at which cracks propagate in thinner sections. Using K_{Ic} in design and materials selection improves the likelihood of selecting the right material, properly evaluating the potential danger from the presence of a flaw, and preventing catastrophic failure.

Factors Affecting Fracture Toughness

In general, fracture toughness decreases with increasing strength. One possible way to increase toughness while maintaining acceptable strength levels is to manipulate microstructure and/or chemistry. Microstructural factors that improve fracture properties include fine grain size, low volume fraction of widely spaced inclusions, and special features, such as transformation-induced plasticity (TRIP) in steels and crack-deflection mechanisms in aluminum-lithium alloys.

Figure 5 shows the inverse relation between strength and fracture toughness of several different types of steels. Development of a good combination of strength and toughness in these steels is linked to size and

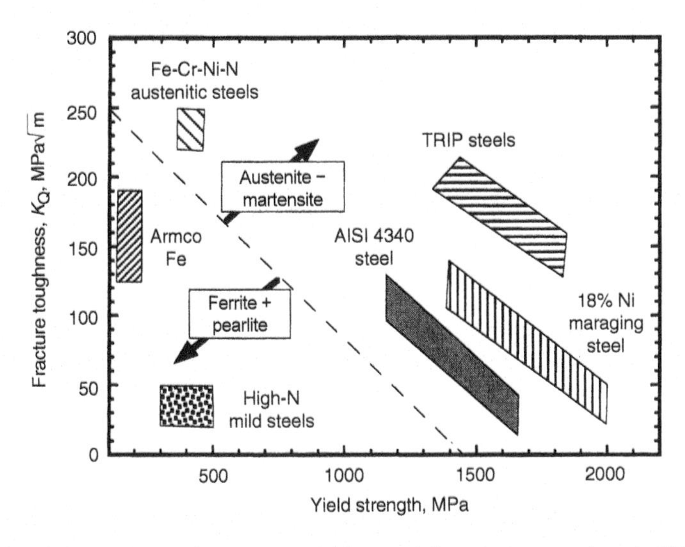

Fig. 5 Fracture toughness versus yield strength for some structural steels. TRIP, transformation-induced plasticity.

distribution of carbide and nitride precipitates, relative amounts of martensite and austenite phases, and grain size. Table 2 summarizes the effects of microstructure on toughness.

Even if the microstructure is exactly as desired, processing and fabrication methods affect fracture toughness, can degrade toughness properties, and are even more important for fatigue properties. Processing frequently involves hot mechanical working, such as forging, extrusion, and hot pressing, which can produce highly directional grain structures and texturing. Fracture toughness is affected by both of these types of directionality, and fracture-toughness values usually are listed in terms of the orientation relative to the principle direction of working. Six standardized (by ASTM) orientations are: L-S, L-T, S-L, S-T, T-L, and T-S, where L is length, or longitudinal rolling direction, extrusion direction, or axis of forging; S is thickness, or short transverse; and T is width, or long transverse. As shown in Fig. 6, the first letter denotes the direction of the applied load in mode I, and the second letter denotes the direction of crack growth.

Specimen-loading rate and test temperature also affect fracture toughness. ASTM E399 (Ref 2) for plane-strain fracture toughness testing recommends a specific range of loading rate because many materials, such as

Table 2 Effects of microstructural variables on fracture toughness of steels

Microstructural parameter	Effect on toughness
Grain size	Increase in grain size increases K_{Ic} in austenitic and ferritic steels
Unalloyed retained austenite	Marginal increase in K_{Ic} by crack burning
Alloyed retained austenite	Significant increase in K_{Ic} by transformation-induced toughening
Interlath and intralath carbides	Decrease K_{Ic} by increasing the tendency to cleave
Impurities (P, S, As, Sn)	Decrease K_{Ic} by temper embrittlement
Sulfide inclusions and coarse carbides	Decrease K_{Ic} by promoting crack or void nucleation
High carbon content (>0.25%)	Decrease K_{Ic} by easily nucleating cleavage
Twinned martensite	Decrease K_{Ic} due to brittleness
Martensite content in quenched steels	Increase K_{Ic}
Ferrite and pearlite in quenched steels	Decrease K_{Ic} of martensitic steels

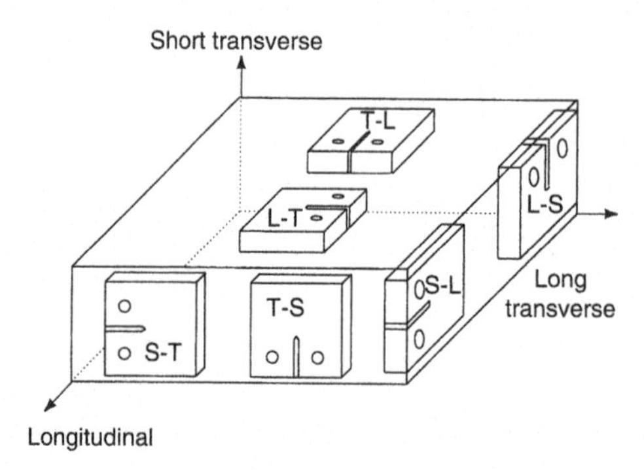

Fig. 6 Specimen orientation scheme showing longitudinal, long transverse, and short transverse directionality. Six possible specimen designations are: L-S, L-T, S-L, S-T, T-L, and T-S. The first letter denotes the direction of the applied load; the second letter denotes the direction of crack growth.

steels, produce different fracture-toughness values at different loading rates. In general, at higher temperatures, the material yield stress decreases, and there is a higher fracture toughness. Conversely, at low temperatures, the yield stress is high, and the fracture toughness is lower.

Crack Growth

Some consider the single most prevalent initiator of brittle fracture to be the fatigue crack, which conservatively accounts for at least 50% of all brittle fractures in manufactured products.

Conventional fatigue data generally are obtained with small laboratory specimens and are plotted as S-N (total-life fatigue analysis) curves, but the absolute value of the stress required to produce fracture at a given number of cycles depends on specimen configuration. (Refer to Chapter 10, "Fatigue Fracture," in this book for an explanation of S-N testing.)

Use of S-N and ε-N techniques (crack-initiation fatigue analysis) is satisfactory in situations where a component or structure can be considered a continuum (no cracks present). However, they are not valid if cracklike discontinuities exist. When a cracklike defect is present, a fracture mechanics approach is used where the rate of crack growth, labeled da/dN, is measured as a function of the K range during loading, ΔK.

Fracture mechanics can be used to aid in the design and predict service life of pressure vessels and other engineering structures in which subcritical flaw growth or time-dependent fractures, such as those stemming from stress-corrosion cracking or fatigue, are important. An adequate fracture mechanics analysis of a structure subjected to cyclic loading, a hostile environment, or both requires data on the rate of growth at which a subcritical-size flaw will grow to critical size under prescribed loading and environmental conditions.

There are three distinct stages (as discussed in Chapter 10, "Fatigue Fracture," in this book) that occur during the fatigue life of a structure: crack initiation, crack propagation, and final rupture. The presence of a preexisting flaw (crack) reduces or eliminates the initiation stage. For many design considerations, the second stage, fatigue-crack growth, is of primary importance. If fatigue data are expressed in terms of da/dN, crack growth per cycle, versus ΔK, the range of crack-tip stress intensity factor, the basic curve obtained is independent of specimen configuration and is a material property.

The advantage of the fracture mechanics approach to crack-growth studies is that the variables—applied load and crack length—can be incorporated into the parameter AK, which describes the stress-intensity conditions corresponding to a given crack growth rate under cyclic loading. This allows application of crack growth-rate data to a wide variety of initial crack lengths, applied loads, specimen shapes, and stress patterns. The cyclic life in terms of cycles to failure can be determined from this.

Crack growth can be caused by cyclic stresses in a benign environment (fatigue), sustained loading in an aggressive environment (stress corrosion), or the combined effects of cyclic stresses and an aggressive environment (corrosion fatigue). Environmentally induced acceleration of fatigue-crack growth increases with increasing yield strength and is related to the environmentally enhanced crack growth threshold, K_{IEAC}, of the environment material system under consideration. This emphasizes the need to conduct fracture-toughness and fatigue-crack growth-rate tests in environments that may be encountered in actual service.

Because fracture mechanics deals only with crack growth, it does not consider the initiation phase of stress-corrosion cracking. Therefore, fracture mechanics is most useful to evaluate resistance to the growth of stress-corrosion cracks in materials that contain preexisting flaws or for research on crack-growth kinetics.

Case History: Hydrotest Failure of a Carbon Steel Pressure Vessel

The following case history illustrates the use of fracture mechanics in failure analysis.

A carbon steel pressure vessel failed while being hydrotested in the fabricating shop. The origin of the failure was determined to be a small surface flaw at the toe of a nozzle weld. The preexisting thumbnail crack was about 1.5 mm (0.06 in.) deep and 15.2 mm (0.60 in.) long (see Fig. 7). Upon reaching a pressure of 14 MPa (2000 psi), the vessel fractured, ejecting two large sections of the wall. The temperature of the water being used for the hydrotest was 15 °C (60 °F), and the ambient temperature was about 10 °C (50 °F).

The pressure vessel, about 760 mm (30 in.) in diameter and 4.6 m (15 ft) long with a 33.3 mm (1–5/16 in.) wall thickness, was intended for use at an oil-production facility. It was fabricated from ASTM A 515 grade 70 steel plate, in accordance with the ASME Boiler and Pressure Vessel Code, Section VIII, Division 1. The steel was in the as-rolled condition, and the vessel was not required to be stress relieved after welding. A 515 pressure vessel plate is carbon steel made to a coarse-grain practice (silicon killed) and is intended for elevated-temperature use. However, the code permitted the alloy to be used to –30 °C (–20 °F), the minimum design metal temperature of the vessel.

Failure analysis included visual examination, macrofractography, microstructural examination, mechanical testing (hardness, tensile, and Charpy V-notch), and stress analysis.

Fracture toughness was estimated from the Charpy toughness; the critical stress-intensity factor is roughly 15.5 times the square root of the Charpy value on the lower shelf. The value for K_{Ic} at the failure temperature, obtained in this manner was 37.4 MPa$\sqrt{m}$ (34 ksi$\sqrt{in.}$).

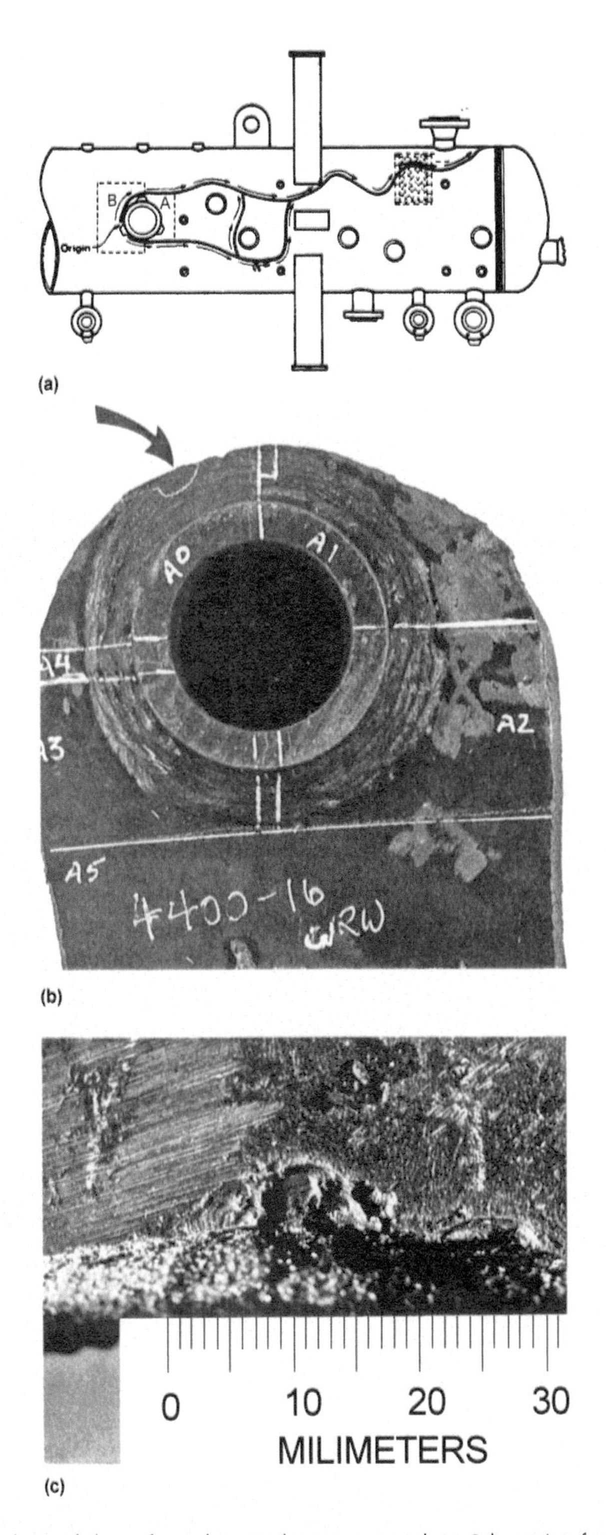

Fig. 7 Hydrotest failure of a carbon steel pressure vessel. (a) Schematic of pressure vessel that failed during hydrotesting showing the location of the origin of the failure and the path of the propagating fracture. A and B indicate sections of the vessel selected for examination. (b) Inside surface of specimen A showing the fracture origin at the toe of a nozzle weld (arrow). (c) Macrograph of the fracture origin showing the thumbnail crack on the fracture surface.

The nominal stress roughly perpendicular to the fracture surface at the origin in the vessel wall corresponding to the internal pressure at the time of failure was calculated to be 79 MPa (11.5 ksi). However, several factors existed that contributed to the presence of a much higher stress at the origin:

- No postweld heat treatment or stress relief heat treatment, resulting in residual stresses
- Steep weld-toe angle causing a stress concentration factor of 4 to 5
- The possibility of the multiaxial stress state resulting from overall and local geometries, allowing stress at the failure origin to approach uniaxial yield strength without being relieved by plastic flow

Considering these factors, it was assumed that a local stress level of 415 MPa (60 ksi) existed at the origin at the time of failure.

A stress-analysis calculation was made using the estimates of the stress and the measured preexisting crack size at the arc strike. A Q value (geometry factor) of 0.75 was used for the 10:1 ratio of crack length to depth. A value of the critical stress intensity (Kc) at failure of 33 MPa · m (30 ksi · in.) was calculated.

The chemical analysis and tensile tests on the steel used to make the pressure vessel met the requirements of A 515 grade 70. The fabrication and testing of the vessel were in accordance with the ASME Boiler and Pressure Vessel Code in effect at the time. However, the toughness of the steel was low, which is not unusual for as-rolled A 515 plate. A rough estimate of the fracture toughness of the steel and of the stress intensity at a small flaw at the toe of the nozzle weld had very good agreement—within 10%. This agreement indicated that, under the conditions existing at the time of the hydrotest, the flaw could have initiated a brittle fracture of the vessel. Failure could have been prevented by eliminating the flaw, lowering the stresses, or increasing the toughness of the steel.

Summary

The fundamental basis for designing a structure or mechanical component to be fracture resistant is to study the interrelationship among the important elements in fracture. The most basic fracture prevention model contains three elements: applied stress, crack length, and fracture toughness. Stated another way, the basic fracture prevention model should account for the applied stress (analysis), crack length (detection), and fracture toughness (design criteria).

Currently, the crack-tip stress-intensity factor, K, is most commonly used as a basis for the design and analysis of structures or mechanical components that contain cracks. The stress-intensity factor is calculated and used in design much the same as an applied stress, except that the calculation now takes into consideration the effects of a crack.

For a more detailed discussion of fracture mechanics concepts, see the following appendix.

Appendix—Fracture Mechanics Concepts

The concepts of fracture mechanics were first introduced in the 1920s by A.A. Griffith when he developed a quantitative energy-based relationship for fracture in flawed brittle solids. Griffith said the driving force to extend a crack is the difference between the energy that could be released if the crack is extended and the energy needed to create new surfaces. This approach did not find much use until G.R. Irwin (often called the father of fracture mechanics) later developed the concept of the crack tip field parameters based on a crack-tip stress analysis. This introduced the concept of the crack-tip stress intensity factor, K, which became the basic parameter used in linear-elastic fracture mechanics. Irwin showed that K related to the Griffith energy release approach.

Linear-Elastic Fracture Mechanics

There are fundamentally two categories of fracture mechanics: linear-elastic fracture mechanics (LEFM) and elastic-plastic fracture mechanics (EPFM). LEFM can be used as long as the crack tip is sharp and only a limited amount of plastic deformation occurs at or near the crack tip. When plastic deformation is more excessive, the assumption of elastic behavior results in unacceptable errors in the calculations, and EPFM is used in the stress analysis.

The LEFM approach to fracture analysis includes three major assumptions:

- Cracks and similar discontinuities are inherently present in parts or specimens.
- The crack is a flat, internal free surface in a linear-elastic stress field (a purely elastic stress field in a featureless solid).
- The quantity of stored energy released from a cracked specimen or part during crack propagation is a basic material property, independent of specimen or part size.

The results of many failure analyses of parts and components verify the first assumption. The second assumption allows a mathematical description of the stress in the vicinity of the crack tip. A crack in a structural component or specimen generates its own stress field, and it is possible to calculate the stress field using the coordinate system shown in Fig. 8. Fracture mechanics is based on the theoretical principle that the stress field ahead of a sharp crack can be characterized by a single parameter called the crack-tip stress intensity, K, which is related to energy release rate. For mode I loading, the crack-tip stress field differs from another only by a stress-intensity factor, K. The stress-intensity factor represents how much the stress intensifies at the crack tip, as shown in Fig. 9, which allows describing external-loading and geometry factors that influence fracture on a uniform basis.

The third assumption states that rapid crack propagation is controlled solely by a material constant. This constant, called the critical stress-intensity factor (K_c), is the value of the stress-intensity factor (K) at which crack propagation becomes rapid. The critical stress-intensity factor, also called plane-strain fracture toughness, is generally expressed in units of MPa$\sqrt{m}$ (ksi$\sqrt{in.}$), and is directly related to the energy release rate for rapid crack propagation by the formula:

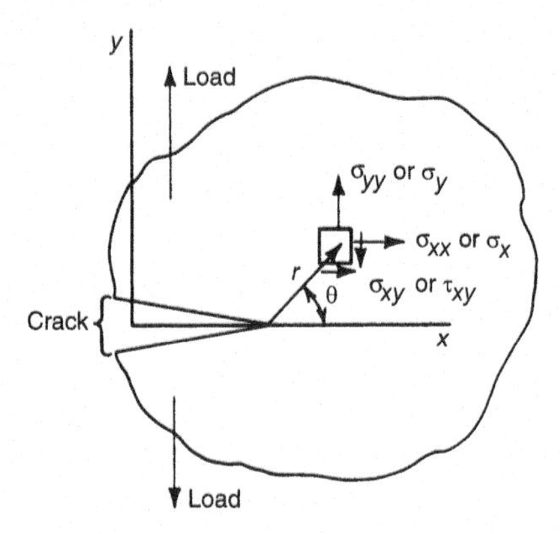

Fig. 8 Arbitrary body and coordinate system used to calculate a stress field at the tip of a crack under mode I loading.

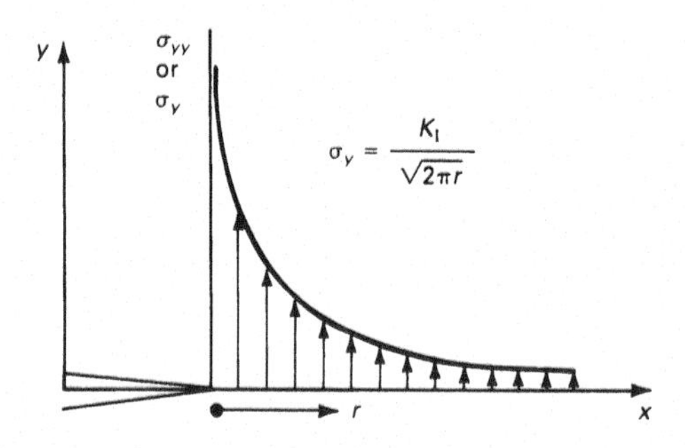

Fig. 9 Crack-tip stress distribution. The stress field ahead of a sharp crack is characterized by a single parameter, K (or K_I for mode I loading). Under linear-elastic conditions, the K_I stress field increases from the nominal stress, approaching infinite magnitude at the crack tip ($r = 0$).

$$K_c^2 = E'G_c \tag{Eq 1}$$

where E' is the elastic modulus (in MPa or ksi) and G_c is the critical strain-energy release rate for unstable crack propagation (in MJ/m^2, or in.-lb/in.^{2}). The existence of a relationship between G and K is very important in the fracture mechanics technology of today because it allows engineers to use K, which is mathematically analyzable, to perform structural sizing and failure analysis.

When the region of plastic deformation around the crack is small compared with the size of the crack (which often is the case with large structures and with high-strength materials), the magnitude of the stress field around the crack commonly is expressed as the stress-intensity factor:

$$K = K_{Ic} = f(g), \sigma\sqrt{\pi a} \tag{Eq 2}$$

where σ is remote applied stress, a is characteristic flaw-size dimension, and $f(g)$ is a function that accounts for crack geometry and structural configuration.

Plane Strain and Plane Stress

A two-dimensional state of stress is assumed when one of the dimensions of a body is small relative to the others. For a very thick part, lateral constraint is very high, and plane-strain conditions prevail. For a relatively thin component (or test specimen), there essentially is no lateral constraint (in the direction of the thickness), and plane-stress conditions dominate. Using either of these constraints in stress-field analysis allows two-dimensional solutions of three-dimensional problems.

The effects of plane strain and plane stress can be explained using a stress cube shown in Fig. 10 (where σ is stress and ε is strain). For a thick part, σ_x, σ_y, σ_z, ε_x, and ε_y have values, but ε_z is essentially zero. This is a plane-strain condition (ε_x and ε_y have a magnitude and are "in plane"). In the plane-strain state, a material is at its lowest point of resistance to unstable fracture. Conversely, for a thin specimen, σ_z is essentially zero, and σ_x and σ_y, and ε_x, and ε_z all have a magnitude, or value. This is a plane-stress condition (σ_x and σ_y have magnitude and are in plane).

Fracture Toughness Testing

For K_{Ic} measurements, ASTM E399 (Ref 2) describes procedures using test specimens like those shown in Fig. 11. The crack-tip plastic region is small compared with crack length (flaw size) and with the specimen dimension in the constraint direction. A compact-type C(T) specimen, shown in Fig. 11(d), often is used to experimentally determine fracture toughness and other fracture properties. From a record of load versus

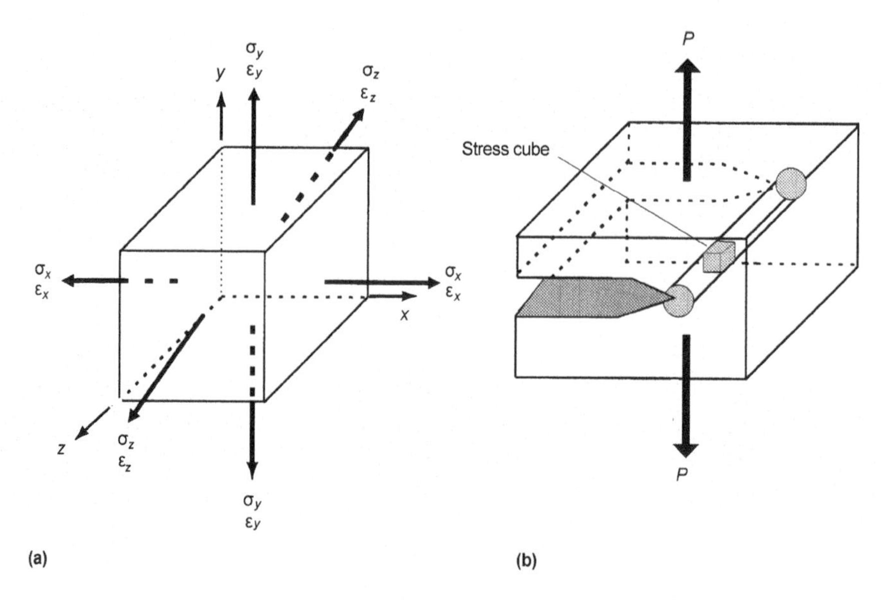

Fig. 10 Solid mechanics-type approach to describing stress and strain using a stress cube. (a) Expanded stress cube depicting normal (perpendicular to respective planes) stresses (σ) and strains (ε). (b) Schematic of a plate with a crack under mode I loading. The cylinder of material in front of the crack tip represents the plastic zone, which varies in size and shape with plate thickness (the plastic zone is not cylindrical in actual components). *P*, load

crack opening and from previously determined relations of crack configuration to stress intensity, plane-strain fracture toughness can be accurately measured if all the criteria for a valid test are met.

Today, laboratory testing for fracture toughness relies more on servohydraulic equipment, which consists of mechanical test apparatus with sophisticated computer data acquisition and controls. Compliance-based fracture testing uses a crack-mouth opening displacement (CMOD) gage. Direct current signals are amplified and conditioned to control and monitor the test, as shown schematically in Fig. 12. Generally, the load is monitored using a load cell mounted within the test frame in the load train.

Crack Growth: The Fracture Mechanics Approach to Fatigue

As previously mentioned, the three stages that occur during the fatigue life of a structure are crack initiation, crack propagation, and final rupture. The second stage, fatigue-crack growth, or propagation, is of primary importance.

Crack growth testing is performed on samples with established K_I versus crack length, a, characteristics. Under a specified controlled load using two dynamic variables, crack length is measured at successive intervals to determine the extension over the last increment of cycles. Crack length

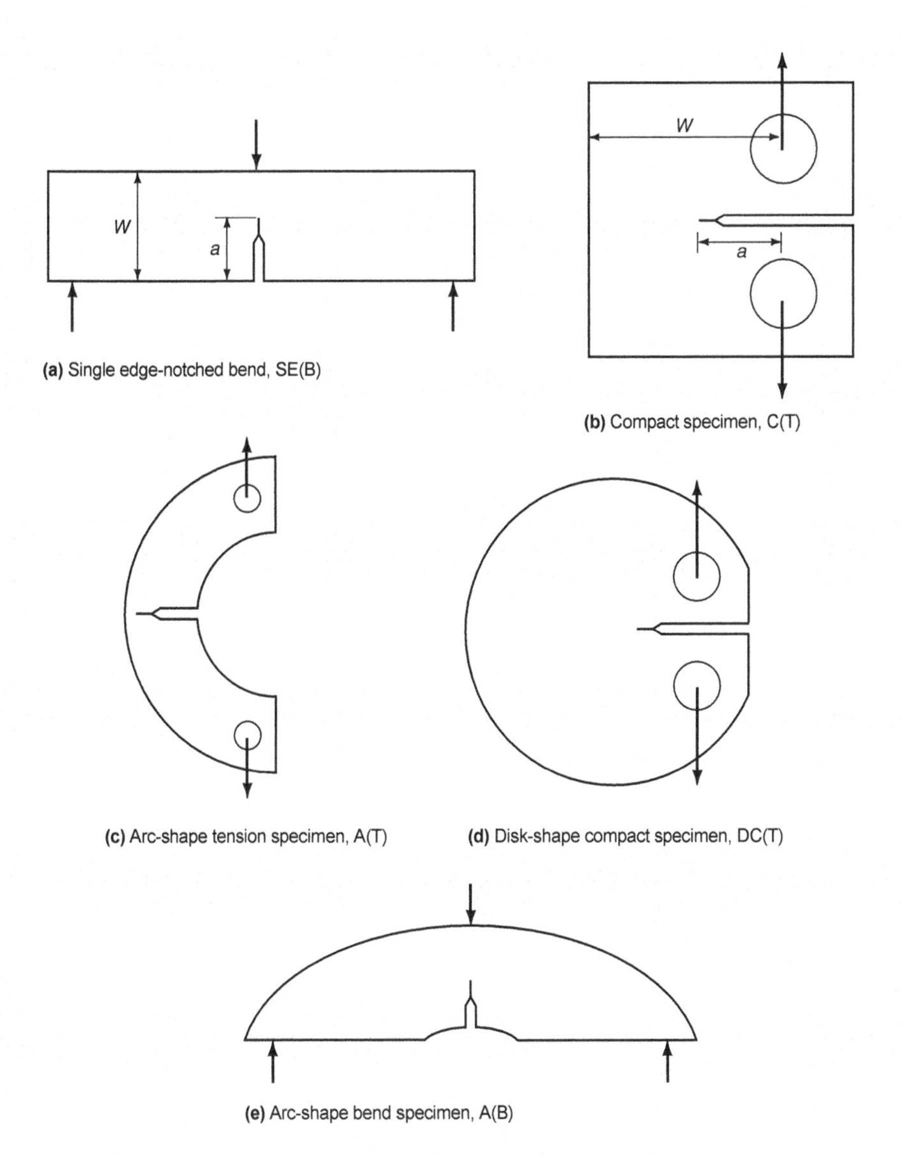

Fig. 11 Specimen types used in plane-strain fracture-toughness (K_{Ic}) testing (ASTMA 399). *a*, crack length; *W*, specimen width

measurement can be done visually, mechanically, or electronically. The phenomenon of fatigue-crack propagation in which the crack extends at each applied cycle is shown in Fig. 13. The amount of crack growth (Δa) per stress cycle is denoted as da/dN.

The generation of da/dN versus ΔK data is considerably more involved than either *S-N* or ε-*N* testing. Features at each end of the da/dN versus ΔK curve are shown in Fig. 14. At the upper limit of ΔK, it reaches the point of instability, and the crack growth rates become extremely large as fracture is approached. The lower end of the ΔK range where crack growth rates essentially decrease to zero is identified as the fatigue-crack growth

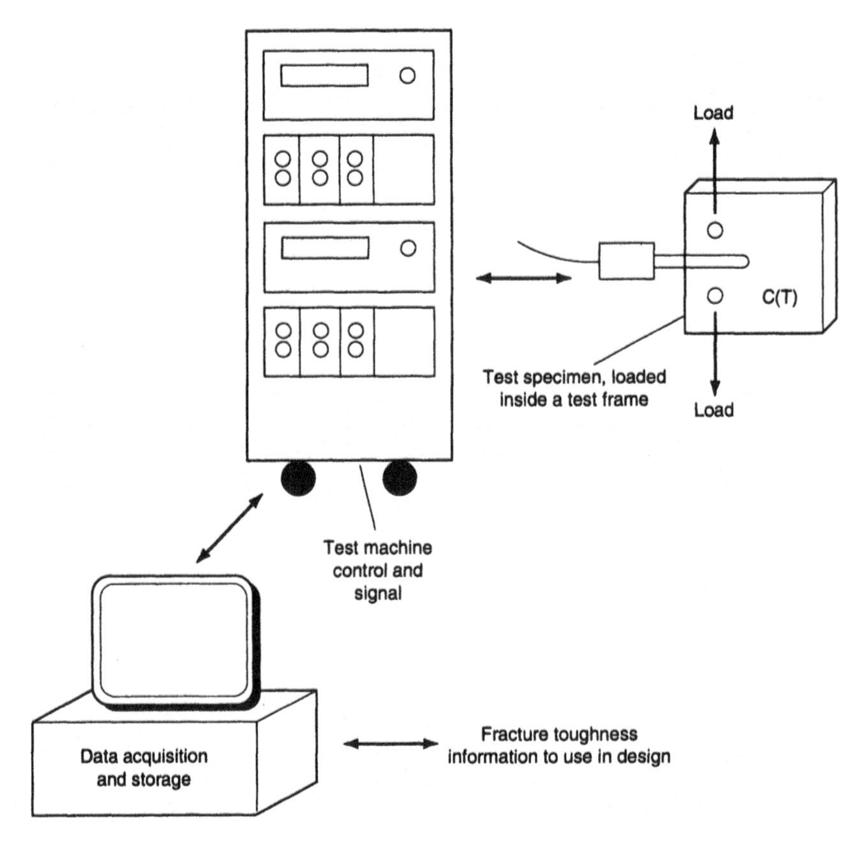

Fig. 12 Common fracture-toughness testing setup showing the interaction of the test specimen with the control and data acquisition instruments. A crack-mouth opening displacement gage is mounted in the compact-type (C(T)) specimen. Current systems generally use servohydraulic test systems

threshold, ΔK_{th}. The threshold behavior at low ΔK values is somewhat analogous to the fatigue limit of some ferrous materials in the *S-N* test.

Modeling of the central portion of the *da*/*dN* versus ΔK curve is frequently done using the Paris equation:

$$da/dN = C_0(\Delta K)^n \qquad \text{(Eq 3)}$$

where *da*/*dN* is the fatigue-crack growth rate; C_0 and n are constants that depend on material, relative average load, and frequency of loading; and ΔK is the range of the stress-intensity factor during one loading cycle.

Applications of Fracture Mechanics

It is important to know whether a part can operate at the intended stresses given a particular flaw size, or whether it is necessary to derate the

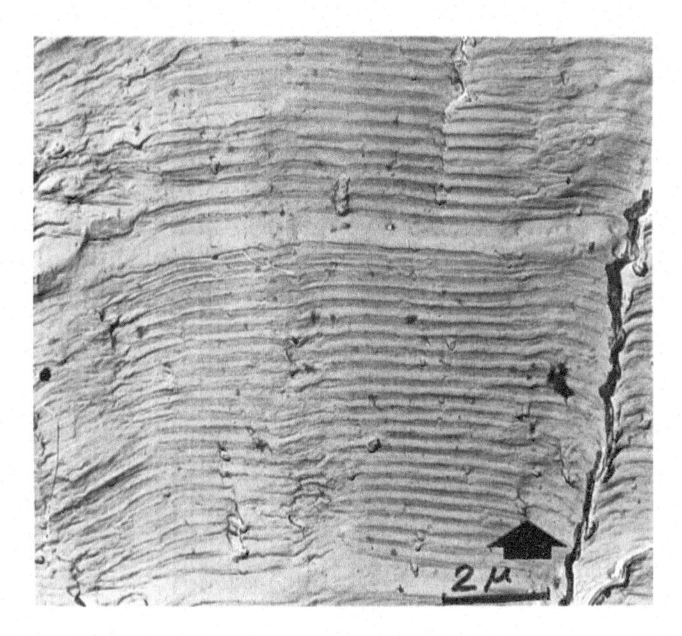

Fig. 13 Typical scanning electron microscope fractograph showing fatigue-crack propagation. Each striation, or ridge, on the fracture surface corresponds to one fatigue load cycle. The arrow indicates the crack propagation direction.

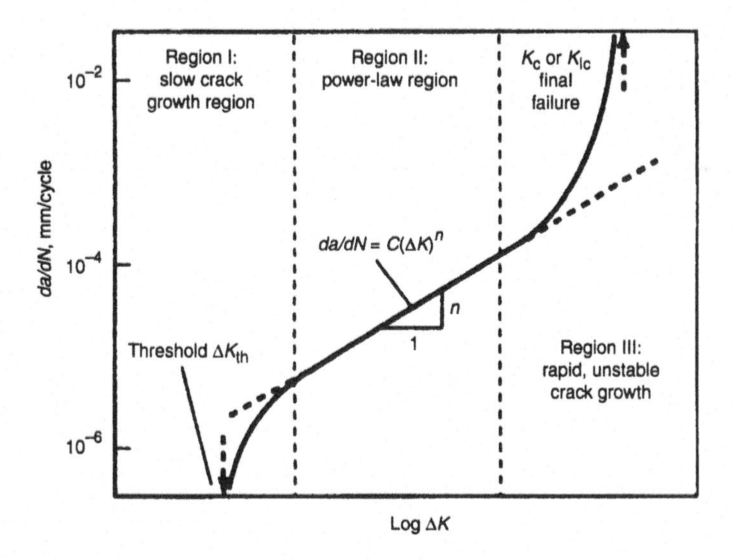

Fig. 14 A typical fatigue-crack growth-rate curve consists of three regions: a slow-growing region (threshold), a linear region (the middle section of the curve), and a terminal region toward the end of the curve where ΔK approaches K_c. The Paris power-law equation, $da/dN = C(\Delta K)^n$, describes fatigue-crack growth rate in the middle, or power-law, region. In other words, there is a linear relationship (in log-log scale) between da/dN and ΔK.

operating stress to a safe level. The following examples show how fracture mechanics can be used to solve design and operating problems.

Example 1: Calculation of the Maximum Safe Flaw Size. Maraging steel (350 grade) has a yield strength of approximately 2450 MPa (355 ksi) and a fracture toughness of 55 MPa$\sqrt{m}$ (50 ksi$\sqrt{in.}$). A landing gear is to be fabricated from this material and the design stresses are 70% of the yield strength (1715 MPa, or 248.6 ksi). Assuming that the flaws must be 2.54 mm (0.1 in.) to be detectable, can the part operate safely at this stress? Assume that edge cracks are present. The stress-intensity parameter for this crack geometry is:

$$K = 1.12\sigma\sqrt{\pi a} \qquad \text{(Eq 4)}$$

where σ is applied stress and a is the crack size.

Solution. The flaw size at which fracture occurs is calculated by rearranging the equation and noting that at fracture, $K = K_{IC}$:

$$a_f = 0.797/\pi\left(\frac{K_{Ic}}{\sigma_0}\right)^2 = \frac{0.797}{\pi}\left(\frac{55}{1715}\right)^2 = 0.26\text{ mm (0.01 in.)} \qquad \text{(Eq 5)}$$

The critical flaw size (the size that can lead to fracture) is smaller than the minimum detection limit. Therefore, even though the design tensile stresses for the part are below the yield strength, the stress is too high to ensure safe operation of the landing gear. Operating stress must be reduced to the point at which the critical flaw size is greater than the minimum detectable crack size of 2.54 mm (0.1 in.).

Example 2: Calculation of the Maximum Stress to Fracture. Suppose that the fracture toughness of a titanium alloy is 44 MPa$\sqrt{m}$, and a circular crack of diameter 16 mm (0.63 in.) is located in a thick plate that will be used in uniaxial tension (mode I loading). If plane-strain conditions are assumed and material yield stress is 900 MPa (130 ksi), then the maximum allowable stress, σ_f, without fracture is calculated as follows. The stress-intensity parameter for a circular crack is:

Solution.

$$K = 2\sigma\left(\frac{a}{\pi}\right)^{1/2} \qquad \text{(Eq 6)}$$

where a is the crack length and σ is the applied stress.

Solution. At fracture, $K = K_{Ic}$. Rearranging the equation and substituting appropriate values gives:

$$\sigma = K/2(\pi/a)^{1/2} = 44/2(\pi/0.008)^{1/2} \qquad \text{(Eq 7)}$$

$$\sigma_f = 436\text{ MPa (63 ksi)} \qquad \text{(Eq 7a)}$$

Therefore, fracture will occur well below the yield strength of the material. This calculation shows that there is no guarantee that fracture will not occur simply because the nominal applied stresses are below the yield strength.

Fracture Mechanics in Fatigue Loading. It is important to know the maximum load that can be applied without failure when assuming that there is a preexisting crack. However, a more typical situation is that there is a preexisting crack, and cyclically applied loads are present whose magnitude is below that which would cause immediate fracture. In this case, the repeated application of a load (such that $K < K_{Ic}$) causes the crack to grow, slowly at first but more rapidly as the crack increases in length. How many cycles can be applied before the crack becomes so long that complete separation occurs?

To determine the number of cycles, the crack growth rate as a function of the stress-intensity parameter is required. This is usually available for materials of engineering interest in the form:

$$\frac{da}{dN} = f(\Delta K) \qquad \text{(Eq 8)}$$

where N is the number of cycles and $\Delta K = K_{max} - K_{min}$. In this equation, ΔK is known as the stress-intensity parameter range, which characterizes the cyclic stresses and strains ahead of the crack tip. The Paris equation describes crack growth behavior over a fairly broad range of ΔK.

The cyclic life is computed by integration of the crack growth-rate equation or by numerical integration of crack growth-rate data. This is illustrated by integration of the Paris equation, as shown in Example 3.

Example 3: Estimation of Fatigue Life using Paris Equation. The crack growth rate of 7075-T6 aluminum is given by:

Solution.

$$\frac{da}{dN} = 5\times10^{-10}(\Delta K)^4 \qquad \text{(Eq 9)}$$

where ΔK is given in units of ksi · in. and da/dN is given in units of in./cycle. Assume that a part contains a center crack that is 5 mm (0.20 in.) long. The stresses vary from 0 to 207 MPa (0 to 30 ksi) and the fracture toughness is 27.5 MPa$\sqrt{m}$ (25 ksi$\sqrt{in.}$). For this crack geometry, the stress-intensity parameter is given by:

$$K = \sigma\sqrt{\pi a} \qquad \text{(Eq 10)}$$

$$\Delta K = \Delta\sigma\sqrt{\pi a} \qquad \text{(Eq 10a)}$$

Solution. Using the information given in the problem statement and the above expression, the crack growth rate is given by:

$$\frac{da}{dN}(\text{in./cycle}) = (5\times10^{-10})(30^4)(\pi^2)(a^2) = 4\times10^{-3}(a^2) \qquad \text{(Eq 11)}$$

This equation can be integrated from the initial condition of $N = 0$ and $a - a_0 = 2.54$ mm (0.10 in.) to the final condition of $N = N_f$ and $a = a_f$. The final crack length (a_f) is the crack length at which fracture occurs, and it corresponds to the condition $K_{max} = K_{Ic}$; K_{max} corresponds to the value of K at maximum stress. Integrating and rearranging terms gives:

$$N_f = 250\int_{a_0}^{a_f}\frac{da}{a^2} \qquad \text{(Eq 12)}$$

Integration of the right-hand side of the equation gives:

$$N_f = 250\left(\frac{1}{a_0} - \frac{1}{a_f}\right) \qquad \text{(Eq 13)}$$

The next step is to compute the crack length at failure. Final crack length (a_f) depends on the fracture toughness, and for the geometry being considered, this is given by the expression:

$$a_f = \frac{1}{\pi}\left(\frac{K_{Ic}}{\sigma}\right)^2 = \left(\frac{25}{30}\right)^2 = 0.22\text{ in. (5.6 mm)} \qquad \text{(Eq 14)}$$

Substituting a_f into the expression for life (N_f) gives:

$$N_f = 250\left(\frac{1}{0.1} - \frac{1}{0.22}\right) = 1363\text{ cycles} \qquad \text{(Eq 15)}$$

Therefore, based on the information given, the part would be expected to last over 1300 cycles.

REFERENCES

1. "Standard Terminology Relating to Fatigue and Fracture Testing", E 1823–11, *Annual Book of ASTM Standards* Vol. 03.01, ASTM, 2012
2. "Standard Test Method for Linear-Elastic, Plane-Strain Fracture Toughness K_{Ic} of Metallic Materials", E399-09, *Annual Book of ASTM Standards*, Vol. 03.01, ASTM, 2012.

SELECTED REFERENCES

The information in this chapter is largely taken from:

- T.L. Anderson, *Fracture Mechanics, Fundamentals and Applications*, 3rd ed., CRC Taylor & Francis, 2005

- K.A. Esaklul, *Handbook of Case Histories in Failure Analysis,* Vol 1, ASM International, 1992
- *Failure Analysis and Prevention,* Vol 10, *Metals Handbook,* American Society for Metals, 1975
- *Failure Analysis and Prevention,* Vol 11, *ASM Handbook,* ASM International, 1986
- *Fatigue and Fracture,* Vol 19, *ASM Handbook,* ASM International, 1996
- A.F. Liu, *Structural Life Assessment Methods,* ASM International, 1998
- *Materials Selection and Design,* Vol 20, *ASM Handbook,* ASM International, 1997
- *Mechanical Testing,* Vol 8, *ASM Handbook,* ASM International, 1985
- *Properties and Selection: Irons, Steels, and High-Performance Alloys,* Vol 1, *ASM Handbook,* ASM International, 1990
- "Test Method for Plane-Strain Fracture Toughness of Metallic Materials," E 399-90, *Annual Book of ASTM Standards,* ASTM, 1996

Glossary

A

addendum. That portion of a gear tooth between the pitch line and the tip of the tooth. Plural is "*addenda*."

ambient. Surrounding. Usually used in relation to temperature, as "ambient temperature" surrounding a certain part or assembly.

anode. The electrode at which oxidation or corrosion occurs. It is the opposite of *cathode*.

anodizing. Forming a surface coating for wear protection or aesthetic purposes on a metal surface. Usually applied to aluminum, in which an aluminum oxide coating is formed in an electrolytic bath.

applied stress. The stress applied to a part or assembly as a result of external forces or loads.

arc strike. The location where a welding electrode has contacted a metal surface, melting a small volume of metal.

asperity. A peak or projection from one surface. Used as a term in wear technology or tribology.

austenite. An elevated-temperature parent phase in ferrous metals from which all other low-temperature structures are derived. The normal condition of certain types of stainless steels.

axial. Longitudinal, or parallel to the axis or centerline of a part. Usually refers to axial compression or axial tension.

B

bainite. An intermediate transformation product from *austenite* in the heat treatment of steel. Bainite can somewhat resemble *pearlite* or *martensite*, depending on the transformation temperature.

beach marks. Macroscopic (visible) lines on a fatigue fracture that show the location of the tip of the fatigue crack at some point in time. Must not be confused with *striations*, which are extremely small and are formed in a different way.

body-centered cubic. See *cell.*

brittle. Permitting little or no plastic (permanent) deformation prior to fracture.

C

carbonitriding. An elevated-temperature process (similar to *carburizing*) by which a ferrous metal absorbs both carbon and nitrogen into the surface when exposed to an atmosphere high in carbon and nitrogen. The carbon and nitrogen atoms actually diffuse, or flow, into the metal to form a high-carbon, high-nitrogen zone near the surface.

carburizing. An elevated-temperature process by which a ferrous metal absorbs carbon into the surface when exposed to a high-carbon environment. Carbon atoms actually diffuse, or flow, into the metal to form a high-carbon zone near the surface.

case. In a ferrous metal, the outer portion that has been made harder than the interior, or *core*. The case is usually formed by diffusion of other atoms—particularly carbon and/or nitrogen—into the metal but may also be formed by localized heat treating of the surface, as by flame or *induction hardening*.

case crushing. See *subcase fatigue*.

case depth. The depth of the case, or hardened surface region, of a metal, usually steel. Since there are many ways of determining case depth, the method used should be stated.

cathode. The electrode at which reduction (and practically no corrosion) occurs. It is the opposite of *anode*.

cathodic protection. Reduction or elimination of corrosion by making the metal a cathode by means of an impressed direct current or attachment of a sacrificial anode (usually magnesium, aluminum, or zinc).

caustic embrittlement. Cracking as a result of the combined action of tensile stresses (applied or residual) and corrosion in alkaline solutions (as at riveted joints in boilers).

cavitation pitting fatigue. A type of pitting fatigue in which cavities, or regions of negative pressure, in a liquid implode, or collapse inward, against a metal surface to cause pits or cavities in the metal surface if repeated often enough at the same points on the surface.

cell. (1) A "building block" forming a grain or crystal. The cell (or "unit cell") is composed of a small number of atoms arranged in any of several different configurations, depending on the metal. The most common are cubic (with an atom at each corner); body-centered cubic (same as cubic, but also has an atom at the center of the cube); face-centered cubic (same as cubic, but also has an atom at the center of each face, or side); hexagonal; and tetragonal. (2) An electrical circuit consisting of an anode and a cathode in electrical contact with a solid or liquid electrolyte. Corrosion generally occurs only at anodic areas.

Charpy test. An impact test in which a V-notched, keyhole-notched, or U-notched specimen, supported at both ends, is struck behind the notch by a striker mounted at the lower end of a bar that can swing as a pendulum. See Chapter 15, Fig. 2, in this book. The energy that is absorbed in fracture is calculated from the height to which the striker would have risen had there been no specimen and the height to which it actually rises after fracture of the specimen.

chromizing. An elevated-temperature process by which a ferrous metal absorbs chromium into the surface when exposed to a high-chromium environment. Chromium atoms actually diffuse, or flow, into the metal to form a high-chromium surface layer.

circumferential. Around the circumference, or periphery, of a circle or a cylinder like a wheel or a shaft. Also called "tangential" or "hoop" when referring to stresses.

clad metal. A composite metal containing two or three layers that have been bonded together. The bonding may have been accomplished by rolling together, welding, casting, heavy chemical deposition, or heavy electroplating.

cleavage fracture. Splitting fracture of a metal along the edges of the cells but across the grains, or crystals. This is a brittle *transgranular fracture*, contrasted to a brittle *intergranular fracture*, in which the fracture is between the grains.

clevis joint. A U-shaped part with holes for a pin to hold another part between the sides of the U.

cohesive strength. The force that holds together the atoms in metal crystals. Analogous to *tensile strength*, but on a submicroscopic scale.

cold heading. Axial compression of the end of a metal cylinder to enlarge the cross section. Used to form the head of a nail or bolt.

cold shut. (1) A discontinuity that appears on the surface of cast metal as a result of two streams of liquid metal meeting but failing to unite. (2) A lap on the surface of a forging or billet that was closed without fusion during deformation. Same as forging lap.

cold work. To cause permanent deformation by application of an external force to a metal below its recrystallization temperature.

compressive. Pertaining to forces on a body or part of a body that tend to crush, or compress, the body.

compressive strength. In compression testing, the ratio of maximum load to the original cross-sectional area. Fracture may or may not occur, depending on the applied forces and the properties of the material.

concentration cell. A cell involving an electrolyte and two identical electrodes, with the electrical potential resulting from differences in the chemistry of the environments adjacent to the two electrodes.

conformal. Describing two surfaces that conform to each other; that is, they nest together, as does a convex surface that fits within a concave surface. An example is a bearing ball within an inner or outer *raceway*. Compare *counterformal*.

core. In a ferrous metal, the inner portion that is softer than the exterior, or *case*.

corrosion. Deterioration of a metal by chemical or electrochemical reaction with its environment.

corrosion fatigue. The combined action of corrosion and fatigue (cyclic stressing) in causing metal fracture.

counterformal. Describing two convex surfaces that are in contact but do not nest together. Examples are two gear teeth, or a roller bearing against an inner *raceway*. Compare *conformal*.

crack growth. Rate of propagation of a crack through a material due to a static or dynamic applied load.

crack opening displacement (COD). On a K_{Ic} specimen (see *stress-intensity factor*), the opening displacement of the notch surfaces at the notch and in the direction perpendicular to the plane of the notch and the crack. The displacement at the tip is called the crack tip opening displacement (CTOD); at the mouth, it is called the crack mouth opening displacement (CMOD).

crack size. A lineal measure of a principle planar dimension of a crack. This measure is commonly used in the calculation of quantities descriptive of the stress and displacement fields. In practice, the value of crack size is obtained from procedures for measurement of physical crack size, original crack size, or effective crack size, as appropriate to the situation under consideration.

crack-tip plane strain. A stress-strain field near a crack tip that approaches *plane strain* to the degree required by an empirical criterion.

creep. Time-dependent strain occurring under stress. Or, change of shape that occurs gradually under a steady load.

crevice corrosion. Localized corrosion resulting from the formation of a *concentration cell* in a crevice between a metal and a nonmetal or between two metals. See also *poultice corrosion*.

crystal. A three-dimensional array of atoms having a certain regularity in its internal arrangement. The crystal is composed of many *cells*, or lattices, in which the atoms are arranged in a given pattern, depending on the metal involved. Another name for crystal is *grain*, which is more commonly used in practical metallurgy.

crystallographic. Pertaining to the crystal structure of a metal.

cyclic load. Repetitive loading, as with regularly recurring stresses on a part, that sometimes leads to fatigue fracture.

cyclic stress. Same as *cyclic load.*

D

decarburization. Loss of carbon from the surface of a ferrous (iron-base) alloy as a result of heating in a medium that reacts with the carbon at the surface.

dedendum. That portion of a gear tooth between the pitch line and the root of the tooth. Plural is "dedenda."

deflection. Deformation within the elastic range caused by a load or force that does not exceed the *elastic limit* of the material. Temporary deformation, as that of a spring.

deoxidized metal. Metal that has been treated, when in the liquid state, with certain materials that tend to form oxides, thus removing the oxygen from the metal.

dimpled rupture fracture. A fractographic term describing ductile fracture that occurs by the formation and coalescence of microvoids along the fracture path. Seen at high magnification as tiny cups, or half-voids.

distortion. Change in the shape of a part due to the action of mechanical forces. Excludes removal of metal by wear or corrosion.

ductile. Permitting plastic (or permanent) deformation prior to eventual fracture.

ductility. The ability of a material to deform plastically (or permanently) prior to eventual fracture.

dynamic. Moving, or having high velocity. Frequently used with impact testing of metal specimens. Opposite of *static*, or essentially stationary, testing or service.

E

elastic. Able to return immediately to the original size and shape after being stretched or squeezed; springy.

elasticity. The property of being *elastic.*

elastic limit. The maximum stress to which a material may be subjected with no permanent deformation after release of the applied load.

elastic-plastic fracture mechanics. A design approach used for materials that fracture or behave in a "plastic" manner, such as lower-strength, high-toughness steels.

elastomer. A material with rubberlike properties—that is, quite elastic, returning to its original size and shape after being deformed.

electrochemical. Pertaining to combined electrical and chemical action. Deterioration (corrosion) of a metal occurs when an electrical current flows between cathodic and anodic areas on metal surfaces.

electrode. An electrical conductor, usually of metal or graphite, that leads current into or out of a solution (electrolyte).

electrolyte. A material, usually a liquid or paste, that will conduct an electric current.

endurance limit. See *fatigue limit.*

ε-N **curve.** Plot of strain versus number of load cycles indicating fatigue behavior of a metal test specimen, which takes into account both elastic and plastic responses to applied loadings.

eutectic alloy. An alloy having the composition indicated by the relatively low melting temperature on an equilibrium diagram of two metals.

F

face-centered cubic. See *cell.*

failure. Cessation of function or usefulness of a part or assembly. The major types of failure are *corrosion*, *distortion*, *fracture*, and *wear*.

false Brinelling. Fretting wear between a bearing ball and its *raceway*. Makes a dark depression in the race, similar to that made by an indentation from a Brinell hardness test. Properly called *fretting wear*.

fatigue. The phenomenon leading to fracture under repeated or fluctuating stresses having a maximum value less than the *tensile strength* of the material. Fatigue fractures are progressive, beginning as minute cracks that grow under the action of the fluctuating stresses.

fatigue life. The number of cycles of stress that can be sustained prior to failure for a stated condition.

fatigue limit. The maximum stress below which a material can presumably endure an infinite number of stress cycles. If the stress is not completely reversed, the value of the mean stress, the minimum stress, or the stress ratio should be stated.

fatigue strength. The maximum stress that can be sustained for a specified number of cycles without failure, the stress being completely reversed within each cycle unless otherwise stated.

ferrite. Essentially pure iron in the microstructure of an iron or steel specimen. It may have a small amount of carbon (less than 0.02 wt%). Also called "alpha iron."

ferrous. Describing a metal that is more than 50% iron, such as steel, stainless steel, cast iron, or ductile (nodular) cast iron.

fillet. A radius (curvature) imparted to inside meeting surfaces; a blended curve joining an internal corner to two lateral surfaces.

fractographic. Pertaining to photographic views of fracture surfaces, usually at high magnification.

fracture. A break, or separation, of a part into two or more pieces.

fracture mechanics. A quantitative analysis for evaluating structural behavior in terms of applied stress, crack length, and specimen or machine component geometry.

fracture toughness. A generic term for measures of resistance to extension of a crack. The term is sometimes restricted to results of *fracture* *mechanics* tests, which are directly applicable in fracture control. However, the term commonly includes results from simple tests of notched or precracked specimens not based on fracture mechanics analysis. Results from tests of the latter type are often useful for fracture control, based on either service experience or empirical correlations with fracture mechanics tests.

free-body diagram. A rectangle representing a theoretical point on the surface of a part under stress that shows, in a simplified way, the stresses and components of stresses acting on the part.

fretting wear. Surface damage to a metal part resulting from microwelding due to slight movement in a nearly stationary joint. Also called "fretting corrosion."

G

galvanic corrosion. Corrosion associated with the current of a galvanic cell consisting of two dissimilar conductors in an electrolyte or two similar conductors in dissimilar electrolytes. Where the two dissimilar metals are in contact, galvanic corrosion may occur.

galvanic series. A series of metals and alloys arranged according to their relative corrosive tendency in a given environment. The most common environment is seawater or other concentrations of salt in water.

gas porosity. A cavity caused by entrapped gas. Essentially a smooth-sided bubble within the metal, where the metal solidified before the gas could escape to the atmosphere. Also called "gas pocket."

gradient. A slope, such as a temperature gradient across a part in which one side is hotter than the other.

grain. The more common term for *crystal*, a three-dimensional array of atoms having a certain regularity in its internal arrangement. The grain is composed of many *cells*, or lattices, in which the atoms are arranged in a given pattern, depending on the metal involved.

grain boundary. The boundary between two grains.

graphitization. Formation of graphite in iron or steel, caused by precipitation of carbon from the iron-carbon alloy.

H

halides. A group of compounds containing one of the halogen elements—bromine, chlorine, fluorine, or iodine—that are sometimes damaging to metals. One of the most common halide compounds is sodium chloride, or ordinary salt.

hardness. Resistance of metal to *plastic deformation*, usually by indentation. However, the term may also refer to stiffness or temper or to resistance to scratching, abrasion, or cutting. Indentation hardness may be measured by various hardness tests, such as Brinell, Rockwell, Knoop, and Vickers. All indentation hardness tests employ arbitrary loads applied to arbitrarily shaped indenters, or penetrators.

HB. Abbreviation for "Hardness, Brinell," a hardness test. The number relates to the applied load and to the surface area of the permanent impression in a metal surface made by a hardened steel or carbide ball. Also known as "BHN" or Brinell hardness number.

heat treatment. Heating and cooling a metal or alloy in such a way as to obtain desired conditions or properties.

high-cycle fatigue. Fatigue that occurs at relatively large numbers of cycles, or stress applications. The numbers of cycles may be in the hundreds of thousands, millions, or even billions. There is no exact dividing line between low- and high-cycle fatigue, but for practical purposes, high-cycle fatigue is not accompanied by plastic, or permanent, deformation.

Hooke's law. Stress is proportional to strain. This law is valid only up to the proportional limit, or the end of the straight-line portion of the stress-strain curve.

hoop. See *circumferential*.

hot heading. Axial compression of the end of a metal cylinder at an elevated temperature to enlarge the cross section. Also called "upsetting."

HRB, HRC. Abbreviations for "Hardness, Rockwell B" and "Hardness, Rockwell C," respectively. The Rockwell B and C scales are two indentation hardness scales commonly used with metals. All Rockwell scales measure the depth of penetration of a diamond or hardened steel ball that is pressed into the surface of a metal under a standardized load.

hydrostatic. Describing three-dimensional compression similar to that imposed on a metal part immersed in a liquid under pressure.

hypoid. A type of bevel, or conical, gear in which the teeth are extremely curved within the conical shape. The teeth of a pinion, or driving gear, are more curved than a spiral bevel pinion, tending to wrap around the conical shape.

I

implode. Burst inward, such as in a collapsing cavity, or negative pressure region, such as during *cavitation pitting fatigue*.

inclusions. Nonmetallic particles, usually compounds, in a metal matrix. Usually considered undesirable, though in some cases, such as in free machining metals, inclusions may be deliberately introduced to improve machinability.

induction hardening. A method of locally heating the surface of a steel or cast iron part through the use of alternating electric current. It is usually necessary to rapidly cool, or quench, the heated volume to form *martensite*, the desired hard microstructure.

interface. The boundary between two contacting parts or regions of parts.

intergranular fracture. Brittle fracture of a metal in which the fracture is between the grains, or crystals, that form the metal. Contrasted to *transgranular fracture*.

intermetallic-phase precipitation. Formation of a very large number of particles of an intermediate phase in an alloy system.

K

keyway. A longitudinal groove, slot, or other cavity usually in a shaft, into which is placed a key to help hold a hub on the shaft. The key and keyway are used for alignment and/or mechanical locking.

L

lamellar. Platelike; made of a number of parallel plates or sheets. Usually applied to microstructures. The most common lamellar microstructure is *pearlite* in ferrous metals.

lateral. In a sideways direction.

lattice, lattice structure. Same as *cell*.

linear-elastic fracture mechanics. A method of fracture analysis that can determine the stress (or load) required to induce fracture instability in a structure containing a cracklike flaw of known size and shape.

longitudinal. Lengthwise, or in an *axial* direction.

low-cycle fatigue. Fatigue that occurs at relatively small numbers of cycles, or stress applications. The numbers of cycles may be in the tens, hundreds, or even thousands of cycles. There is no exact dividing line between low- and high-cycle fatigue, but for practical purposes, low-cycle fatigue may be accompanied by some plastic, or permanent, deformation.

M

macroscopic. Visible at magnifications up to about 25 to 50×.

martensite. The very hard structure in certain irons and steels that is usually formed by quenching (rapid cooling) from an elevated temperature. Martensite may or may not be tempered to reduce hardness and increase ductility and toughness.

martensitic transformation. Formation of *martensite*.

matrix. The principal phase of a metal in which another constituent is embedded. For example, in gray cast iron, the metal is the matrix in which the graphite flakes are embedded.

mechanical properties. The properties of a material that reveal its elastic and inelastic (plastic) behavior when force is applied, thereby indicating its suitability for mechanical (load-bearing) applications. Examples are elongation, *fatigue limit*, *hardness*, *modulus of elasticity*, *tensile strength*, and *yield strength*.

metal. An opaque, lustrous elemental chemical substance that is a good conductor of heat and electricity and, when polished, a good reflector of light. Most elemental metals are malleable and ductile and are, in general, heavier than the other elemental substances.

metallographic. Pertaining to examination of a metallic surface with the aid of a microscope. The surface is usually polished to make it flat and may be etched with various chemicals to reveal the microstructure.

microscopic. Visible only at magnifications greater than about 25 to 50×.

microstructure. The structure of polished and etched metals as revealed by a microscope at a magnification greater than 25 to 50×.

microvoid. A microscopic cavity that forms during fracture of a ductile metal. A very large number of microvoids form in the region with the highest stress; some of them join together to form the actual fracture surface, each side of which contains cuplike half-voids, usually called "dimples."

mode. One of the three classes of crack (surface) displacements adjacent to the crack tip. These displacement modes are associated with stress-strain fields around the crack tip.

modulus of elasticity. A measure of the stiffness of a metal in the elastic range—that is, the degree to which a metal will deflect when a given load is imposed on a given shape. Also called "Young's modulus."

monomolecular. Describing a film or surface layer one molecule thick.

monotonic. Pertaining to a single load application in a relatively short time, as in a monotonic tensile test. Same as *static*.

N

necking. The reduction in cross-sectional size that occurs when a part is stretched by a tensile stress.

nitriding. An elevated-temperature process (but lower than carburizing or carbonitriding) by which a ferrous metal absorbs nitrogen atoms into the surface when exposed to a high-nitrogen environment. Nitrogen atoms actually diffuse, or flow, into the metal to form a high-nitrogen surface layer.

nonferrous. Describing a metal that is less than 50% iron, such as aluminum, copper, magnesium, and zinc and their alloys.

normal stress. See *stress*.

notch. See *stress concentration*.

notched-bar impact test. A standardized mechanical test in which a metal test specimen with a specified notch is struck with a standardized swinging pendulum weight. The type of fracture and the energy absorbed by the fracturing process can be determined from the specimen.

notch toughness. An indication of the capacity of a metal to absorb energy when a notch, or *stress concentration*, is present.

P

pancake forging. Plastic deformation of a very ductile material under axial compressive forces between flat, parallel dies. The sides bulge outward, while the other surfaces become essentially flat and parallel.

Paris equation. A generalized fatigue-crack growth rate exponential-power law that shows the dependence of fatigue-crack growth rate on the *stress-intensity factor*, *K*, and has been verified by many investigations.

pearlite. A lamellar, or platelike, microstructure commonly found in steel and cast iron.

physical properties. The properties of a material that are relatively insensitive to structure and can be measured without the application of force. Examples are density, melting temperature, damping capacity, thermal conductivity, thermal expansion, magnetic properties, and electrical properties.

pitch line. The location on a gear tooth, approximately midway up the tooth, that crosses the pitch circle, or the equivalent-size disk that could geometrically replace the gear.

plastic deformation. Deformation that remains after removal of the load or force that caused the deformation, or change of shape. Same as permanent deformation.

polycrystalline. Pertaining to a solid metal composed of many crystals, such as an ordinary commercial metal.

polymeric. Pertaining to a polymer, or plastic.

poultice corrosion. Same as *crevice corrosion* but usually applies to a mass of particles or an absorptive material in contact with a metal surface that is wetted periodically or continuously. Corrosion occurs under the edges of the mass of particles or the absorptive material that retains moisture.

prestress. Stress on a part or assembly before any service or operating stress is imposed. Similar to internal or *residual stress*.

primary creep. The first, or initial, stage of *creep*, or time-dependent deformation.

proportional limit. The maximum stress at which strain remains directly proportional to stress; the upper end of the straight-line portion of the stress-strain or load elongation curve.

psi. Abbreviation for pounds per square inch, a unit of measurement for stress, strength, and modulus of elasticity.

Q

quasi-cleavage fracture. Term used to refer to a fracture mode that combines the characteristics of cleavage fracture and dimpled rupture fracture or tear ridges. The term is used to describe a microscale fracture appearance in steels that tends to result from (a) sudden or impact loading, (b) low temperature, (c) high levels of constraint (ambient temperature) or (d) in heavily cold worked parts (ambient temperature). The preferred term is "cleavage with ductile tear ridges."

R

raceway. The tracks or channels on which roll the balls or rollers in an antifriction rolling-element bearing. The inner race fits around a shaft, while the outer race fits within a hole in a larger part.

radial. In the direction of a radius between the center and the surface of a circle, cylinder, or sphere.

ratchet marks. Ridges on a fatigue fracture that indicate where two adjacent fatigue areas have grown together. Ratchet marks usually originate perpendicular to a surface and may be straight or curved, depending on the combination of stresses that is present.

reactive metals. Metals that tend to react with the environment, usually those near the anodic end of the galvanic series.

recrystallization. (1) The change from one crystal structure to another, such as occurs on heating or cooling through a critical temperature. (2) The formation of a new, strain-free grain structure from that existing in cold-worked metal, usually accomplished by heating.

residual stress. Internal stress; stress present in a body that is free from external forces or thermal gradients.

root (of a notch). The innermost part of a *stress concentration,* such as the bottom of a thread or groove.

rupture. Same as *fracture*.

S

service loads. Forces encountered by a part or assembly during operation in service.

shear. A type of force that causes or tends to cause two regions of the same part or assembly to slide relative to each other in a direction parallel to their plane of contact. May be considered on a microscale when planes of atoms slide across each other during permanent, or plastic, deformation. May also be considered on a macroscale when gross movement occurs along one or more planes, as when a metal is cut or "sheared" by another metal.

shear fracture. Fracture that occurs when shear stresses exceed shear strength before any other type of fracture can occur. Typical shear fractures are transverse fracture of a ductile metal under a torsional (twisting) stress, and fracture of a rivet cut by sliding movement of the joined parts in opposite directions, like the action of a pair of scissors (shears).

shear lip. A narrow, slanting ridge, nominally about 45° to the surface, along the edge of a fracture surface where the fracture emerged from the interior of the metal. In the fracture of a ductile tensile specimen, the shear lip forms the typical "cup-and-cone" fracture. Shear lips may be present on the edges of some predominantly brittle fractures to form a "picture frame" around the surface of a rectangular part.

shear stress. See *stress*.

shot peening. A carefully controlled process of blasting a large number of hardened spherical or nearly spherical particles (shot) against the softer surface of a part. Each impingement of a shot makes a small indentation in the surface of the part, thereby inducing compressive *residual stresses*, which are usually intended to resist fatigue fracture or *stress-corrosion cracking*.

shrinkage cavity. A void left in cast metals as a result of solidification shrinkage, because the volume of metal decreases during cooling. Shrinkage cavities usually occur in the last metal to solidify after casting.

sintered metal. Same as powdered metal. Type of metal part made from a mass of metal particles that are pressed together to form a compact and then sintered (or heated for a prolonged time below the melting point) to bond the particles together.

slant fracture. A type of fracture appearance, typical of ductile fractures of flat sections, in which the plane of metal separation is inclined at an angle (usually about 45°) to the axis of the applied stress.

***S-N* curve.** A plot of stress (S) against the number of cycles to failure (N). The stress can be the maximum stress (S_{max}) or the alternating stress amplitude (S_a). The stress values are usually nominal stress; that is, there is no adjustment for *stress concentration*. For S, a linear scale is used most often, but a log scale is sometimes used. Also called "*S-N* diagram."

spalling fatigue. See *subcase fatigue*.

spiral bevel gear. A type of bevel, or conical, gear in which the teeth are curved within the conical shape, rather than straight, as in a bevel gear. Compare *hypoid*.

spline. A shaft with a series of longitudinal, straight projections that fit into slots in a mating part to transfer rotation to or from the shaft.

static. Stationary, or very slow. Frequently used in connection with routine tensile testing of metal specimens. Same as *monotonic*. Opposite of *dynamic*, or impact, testing or service.

strain. A measure of relative change in the size or shape of a body. "Linear strain" is change (increase or decrease) in a linear dimension. Usually expressed in inches per inch (in./in.) or millimeters per millimeter (mm/mm).

strength gradient. Shape of the strength curve within a part. The strength gradient can be determined by hardness tests made on a cross section of a part; hardness values are then converted into strength values, usually in pounds per square inch (psi) or megapascals (MPa).

stress. Force per unit area, often thought of as a force acting through a small area within a plane. It can be divided into components, perpendicular and parallel to the plane, called "normal stress" and "shear stress," respectively. Usually expressed as pounds per square inch (psi) or megapascals (MPa).

stress concentration. Changes in contour, or discontinuities, that cause local increases in stress on a metal under load. Typical are sharp-cornered grooves, threads, fillets, holes, and the like. Effect is most critical when the stress concentration is perpendicular (normal) to the principal tensile stress. Same as notch or stress raiser.

stress corrosion. Preferential attack of areas under stress in a corrosive environment, where such an environment alone would not have caused corrosion.

stress-corrosion cracking. Failure by cracking under combined action of corrosion and a tensile stress, either external (applied) or internal (residual). Cracking may be either intergranular or transgranular, depending on the metal and the corrosive medium.

stress cube. A finite volume of material used to depict three-dimensional states of stress and strain (displacement) distributions at a crack tip.

stress field. Stress distribution generated ahead of a sharp crack present in a loaded part or specimen. The stress field is characterized by a single parameter called *stress-intensity factor*, *K*.

stress-field analysis. Mathematical analysis of an assumed two-dimensional state of stress (plane strain condition) at a crack tip in *linear-elastic fracture mechanics*.

stress gradient. Shape of a stress curve within a part when it is under load. In pure tension or compression, the stress gradient is uniform across the part, in the absence of stress concentrations. In pure torsion (twisting) or bending, the stress gradient is maximum at the surface and zero near the center, or neutral axis.

stress-intensity factor. A scaling factor, usually denoted by the symbol *K*, used in *linear-elastic fracture mechanics* to describe the intensification of applied stress at the tip of a crack of known size and shape. At the onset of rapid crack propagation in any structure containing a crack, the factor is called the "critical stress-intensity factor," or the *fracture toughness*. Various subscripts are used to denote different loading conditions or fracture toughness:

K_C. Plane-stress fracture toughness. The value of stress sections thinner than those in which plane-strain conditions prevail.

K_I. Stress-intensity factor for a loading condition that displaces the crack faces in a direction normal to the crack plane (also known as the opening mode of deformation).

K_{Ic}. Plane-strain fracture toughness. The minimum value of K_C for any given material and condition, which is attained when rapid crack propagation in the opening mode is governed by plane-strain conditions.

K_{Id}. Dynamic fracture toughness. The fracture toughness determined under dynamic loading conditions; it is used as an approximation of K_{Ic} for very tough materials.

K_{ISCC}. Threshold stress-intensity factor for stress-corrosion cracking. The critical plane-strain stress intensity at the onset of stress corrosion cracking under specified conditions.

K_Q. Provisional value for plane strain fracture toughness.

K_{th}. Threshold stress intensity for stress-corrosion cracking. The critical stress intensity at the onset of stress-corrosion cracking under specified conditions.

DK. The range of the stress-intensity factor during a fatigue cycle.

stress raiser. See *stress concentration*.

striations. Microscopic ridges or lines on a fatigue fracture that show the location of the tip of the fatigue crack at some point in time. They are locally perpendicular to the direction of growth of the fatigue crack. In ductile metals, the fatigue crack advances by one striation with each load application, assuming the load magnitude is great enough. Must not be confused with *beach marks*, which are much larger and are formed in a different way.

stringers. In metals that have been hot worked, elongated patterns of impurities, or *inclusions*, that are aligned longitudinally. Commonly the term is associated with elongated oxide or sulfide inclusions in steel.

subcase fatigue. A type of fatigue cracking that originates below a hardened *case* or in the *core*. Large pieces of metal may be removed from the surface because of very high compressive stresses, usually on gear teeth. Also called "spalling fatigue" and "case crushing."

T

tangential. See *circumferential*.

tensile. Pertaining to forces on a body that tend to stretch, or elongate, the body. A rope or wire under load is subject to tensile forces.

tensile strength. In tensile testing, the ratio of maximum load to the original cross-sectional area.

thermal cycles. Repetitive changes in temperature, that is, from a low temperature to a higher temperature, and back again.

through hardening. Hardening of a metal part, usually steel, in which the hardness across a section of the part is essentially uniform; that is, the center of the section is only slightly lower in hardness than the surface.

torque. A measure of the twisting moment applied to a part under a torsional stress. Usually expressed in terms of inch-pounds or foot-pounds, although the terms "pound-inches" and "pound-feet" are technically more accurate for torsional moments.

torsion. A twisting action applied to a generally shaftlike, cylindrical, or tubular member. The twisting may be either reversed (back and forth) or unidirectional (one way).

toughness. Ability of a material to absorb energy and deform plastically before fracturing. Toughness is proportional to the area under the stress-strain curve from the origin to the breaking point. In metals, toughness is usually measured by the energy absorbed in a notch impact test.

transgranular fracture. Through, or across, the crystals or grains of a metal. Same as transcrystalline and intracrystalline. Contrasted to *intergranular fracture*. The most common types of transgranular fracture are fatigue fractures, *cleavage fractures*, *dimpled rupture fractures*, and *shear fractures*.

transverse. Literally "across," usually signifying a direction or plane perpendicular to the axis of a part.

U

underbead crack. A subsurface crack in the base metal near a weld.

undercut. In welding, a groove melted into the base metal adjacent to the toe, or edge, of a weld and left unfilled.

W

wear. The undesired removal of material from contacting surfaces by mechanical action.

worm gear. A type of gear in which the gear teeth are wrapped around the shaftlike hub, somewhat as threads are wrapped around a bolt or screw.

Y

yield point. The first stress in a material, less than the maximum attainable stress, at which an increase in strain occurs without an increase in stress. Not a general term or property; only certain metals exhibit a yield point.

yield strength. The stress at which a material exhibits a specified deviation from proportionality of stress and strain. The specified deviation is usually 0.2% for most metals. A general term or property, preferred to *yield point*.

Young's modulus. Same as *modulus of elasticity*.

Index

A

B

C

D

E

F

G

H

I

N

O

P

T

U

V

W

X

Y

Z